Lecture Notes in Mathematics 1898

Horst Reinhard Beyer

Beyond Partial Differential Equations

On Linear and Quasi-Linear Abstract Hyperbolic Evolution Equations

Springer

Author

Horst Reinhard Beyer
Center for Computation & Technology
Louisiana State University
330 Johnston Hall
Baton Rouge
LA 70803
USA
e-mail: horst@cct.lsu.edu

Library of Congress Control Number: 2007921690

Mathematics Subject Classification (2000): 47J35, 47D06, 35L60, 35L45, 35L15

ISSN print edition: 0075-8434
ISSN electronic edition: 1617-9692
ISBN-10 3-540-71128-7 Springer Berlin Heidelberg New York
ISBN-13 978-3-540-71128-5 Springer Berlin Heidelberg New York

DOI 10.1007/978-3-540-71129-2

Springer is a part of Springer Science+Business Media
springer.com

Typesetting by the author and SPi using a Springer LaTeX macro package
Cover design: WMXDesign GmbH, Heidelberg

Printed on acid-free paper SPIN: 11962175 VA41/3100/SPi 5 4 3 2 1 0

Dedicated to God the Father

Preface

Semigroup Theory uses abstract methods of Operator Theory to treat initial boundary value problems for linear and nonlinear equations that describe the evolution of a system. Due to the generality of its methods, the class of systems that can be treated in this way exceeds by far those described by equations containing only local operators induced by partial derivatives, i.e., PDEs. In particular, that class includes the systems of Quantum Theory.

Another important application of semigroup methods is in field quantization. Simple examples are given by the cases of free fields in Minkowski spacetime like Klein-Gordon fields, the Dirac field and the Maxwell field, whose field equations are given by systems of linear PDEs. The second quantization of such a field replaces the field equation by a Schrödinger equation whose Hamilton operator is given by the second quantization of a non-local function of a self-adjoint linear operator. That operator generates the semigroup given by the time-development of the solutions of the field equation corresponding to arbitrary initial data as a function of time. More generally, in these cases the structures used in the formulation of a well-posed abstract initial value problem for the field equation also provide the mathematical framework for the quantization of the field. Quantum Theory is an abstract theory, therefore it should be expected that only an abstract approach to classical field equations using methods from Operator Theory is capable of providing the appropriate structures for quantization in the less simple cases of nonlinear fields, like the gravitational field described by Einstein's field equations.

A demonstration of the strength of semigroup methods can be seen in the first rigorous proof of well-posedness (local in time) of the initial value problem for quasi-linear symmetric hyperbolic systems by T. Kato in 1975 in [110]. This result is a particular application of a theorem on the well-posedness of the initial value problem for abstract quasi-linear equations[1], which has been successfully applied also to Einstein's equation's [102], the Navier-Stokes equations, the equations of Magnetohydrodynamics [109] and more. To my knowledge, there is no other approach to quasi-linear equations leading to a theorem of such generality.

---

[1] See Theorem 11.0.7 below.

The semigroup approach goes beyond that of a tool for deciding the well-posedness of initial boundary value problems. For a autonomous nonlinear equation the important question of the linearized stability of a particular solution leads on a spectral problem for the operator generating the semigroup given by the time-development of the solutions of the linearized equation around that solution corresponding to arbitrary initial data as a function of time.[2] These methods also provide, for autonomous linear equations, a representation of the solution of the initial value problem as an integral over the resolvent of the infinitesimal generator of the associated semigroup.[3] The resolvent operators can often be represented in the form of integral operators with kernels which are defined in terms of special functions. This is not only true in simple cases where the generator is a partial differential operator with constant coefficients, but also in a number of cases involving non-constant coefficients. In this way, an integral representation of the solution of the initial value problem is achieved for all such cases.[4] Finally, semigroup methods provide a framework which is general enough to include numerical forms of evolution equations, opening the possibility of computing true error estimates to the exact solution, rather than residual errors.

This should not give the impression that the semigroup approach could replace all 'hard' analysis facts. Instead, it reduces such application to a bare minimum, which gives the approach its efficiency.[5] For instance, results from harmonic analysis or the theory of singular integral operators have applications in so called 'commutator estimates' where the commutator of an intertwining operator with the principal part of a partial differential operator, usually two unbounded operators, has to be estimated. To achieve most general results, it is often necessary to choose intertwining operators as non-local operators.

These methods are especially attractive *if not inevitable* for theoretical physicists in view of their comprehension of classical physics and quantum physics. In addition, their efficiency is of advantage in view of time restrictions in the mathematics education of physicists. In spite of their power, efficiency and versatility, semigroup methods are surprisingly little used in theoretical physics.[6,7] This appears to be related to two misconceptions.

First, because of the requirements of Special Relativity and General Relativity, current problems in fundamental theoretical physics necessarily lead on hyperbolic

[2] For instance, see Chapter 5.4.

[3] Roughly speaking, for a precise statement see, e.g., Chapter 4.3.

[4] See, e.g., [25].

[5] But it is the belief of the author that the necessity of the use of 'hard' to achieve analytical facts in the solution of a problem indicates that its structure has not yet been fully understood.

[6] Paradoxically, in some sense, one could also hold the opposite view that these methods have been used for a long time in theoretical physics, mostly without realizing that they are rooted in in Spectral Theory.

[7] In addition, the author is not aware of a single introduction to PDE based solely on Semigroups/Operator Theory, although this would have been possible even before the appearance of classical introductions like [20] that use less general methods.

partial differential equation systems ('hyperbolic problems'). Standard texts on semigroups of linear operators mainly focus on applications to parabolic partial differential equation systems ('parabolic problems'). Differently to the hyperbolic case, this leads to the consideration of strongly continuous semigroups which are in addition analytic. Apparently, the consideration of parabolic problems originates from a focus on engineering applications. Engineering sciences predominantly apply classical Newtonian physics where signal propagation speeds are not limited by the speed of light in vacuum as this is the case in Special Relativity/General Relativity. For example, the evolution of a compactly supported temperature field in space at time $t = 0$ by the parabolic heat equation leads to a temperature field that has no compact support for every $t > 0$. As consequence, in the evolution signal propagation speeds occur that exceed the speed of light in vacuum, and hence the equation is incompatible with Special Relativity. The same is also true for Schrödinger equations.[8] The evolution of hyperbolic partial differential equations systems preserves the compactness of the support of the data under time evolution. To my experience, the focus on engineering applications has lead to the quite common misconception among physicists, and to some extent also among mathematicians working in the field of partial differential equations, that semigroup methods cannot be applied to hyperbolic problems.

Second, to my experience, another common misconception is that semigroups of linear operators can only be applied to systems of *linear* partial differential equations. This might be influenced by the fact that most standard texts on semigroups of linear operators, indeed, focus mainly on such applications.

As a consequence, the course should lead as rapidly as possible from autonomous linear equations to the nonlinear (quasi-linear) equations which are now seen in the hyperbolic problems emanating from current physics. Particular stress is on wave equations and Hermitian hyperbolic systems. The last cover the equations describing interacting fields in physics and therefore the major part of nonlinear equations occurring in fundamental physics. Throughout the course applications to problems from current relativistic ('hyperbolic') physics are provided, which display the potential of the methods in the solution of current problems in physics. These include problems from black hole physics, the formulation of outgoing boundary conditions for wave equations and the treatment of additional constraints. The last two are important current problems in the numerical evolution of Einstein's field equations for the gravitational field. Some of the examples contain new unpublished results of the author. To my knowledge, the major part of the material in the second part of the notes, including non-autonomous and quasi-linear Hermitian hyperbolic systems, has appeared only inside research papers.

The orientation of this course towards abstract quasi-linear evolution equations made it necessary to omit a number of topics that were not directly important for achieving its goals. On the other hand, texts on semigroups of linear operators that consider those topics are available [39,47,52,57,90,99,106,120,168,179,224]. Also,

---

[8] This was the reason for the development of Quantum Field Theory.

there is a theory of nonlinear semigroups that largely parallels that of semigroups of linear operators [15, 19, 88, 138, 146, 167].

This course assumes some basic knowledge of Functional Analysis which can be found, for instance, in the first volume of [179]. Some examples assume more specialized knowledge of properties of self-adjoint linear operators in Hilbert spaces which can be found, for instance, in the second volume of [179]. In addition, some applications assume basic knowledge of Sobolev spaces of the L^2-type as is provided, for instance, in [217]. Otherwise this course is self-contained. The material is presented in as compressed a form as possible and its results are formulated in view towards applications. In general, theorems contain their full set of assumptions, so that a study of their environment is not necessary for their understanding. For this reason and to limit the size of this text, shorter definitions appear as part of theorems. In addition, the abstract theory and its applications appear in separate chapters to allow the reader to estimate the necessary time to acquire the theory.

Chapters 2–5 constitute the notes of a course given at the Department of Mathematics of the Louisiana State University in Baton Rouge in Spring 2005. They provide a basic introduction into the properties of strongly continuous semigroups on Banach spaces and applications to autonomous linear hyperbolic systems of PDEs from General Relativity and Astrophysics. The theoretical part is kept to a minimum with a view to applications in the field of hyperbolic PDEs. It is formulated in a way which is expected to be natural to readers with a knowledge of the spectral theory of self-adjoint linear operators in Hilbert spaces. An exception, to this restriction to the minimum in these chapters, is the treatment of the integration of Banach space-valued maps which is more detailed than is usual in most other comparable texts. This is due to the fact that such integration is a basic tool which is usually not covered in standard Functional Analysis courses, at least not to an extent needed in the study of semigroups of operators. Otherwise, the theoretical material is standard and for this reason only few references to literature are given. For more comprehensive introductions into the theory of semigroups of linear operators that give a more exhaustive list of references, the reader is referred, for instance, to [57, 90, 99, 168].

Chapters 6–12 are the notes of a subsequent course at the same place in Spring 2006. They introduce into the field of abstract evolution equations with applications to non-autonomous linear and quasi-linear hyperbolic systems of PDEs. This second part of the notes follows closely the late Tosio Kato's 1993 paper *'Abstract evolution equations, linear and quasilinear, revisited'* in Proceedings of the International Conference in Memory of Professor Kosaku Yosida held at RIMS, Kyoto University, Japan, July 29-Aug. 2, 1991, [114]. Those results are more general than Kato's well-known older results[9] [107–109] in that they don't assume special properties of the underlying Banach spaces. Proofs of Kato's results are added along with detailed examples displaying their application to problems in current relativistic physics. In a few places, Lemmata were added to his outline that appeared necessary for those proofs.

---

[9] Those results assume the reflexivity and in some places also the separability of the underlying Banach spaces.

Acknowledgments

I am especially indebted to Olivier Sarbach, who diligently reviewed the major part of the notes and recommended valuable modifications and corrections. Also, I thank Frank Neubrander and Stephen Shipman for illuminating discussions in connection with the notes. Finally, I thank all who have worked on the book, especially the editorial and production staff of Springer-Verlag.

Baton Rouge, September 2006 *Horst Beyer*

Contents

1

Conventions

For every map f, the symbol $\mathrm{Ran} f$ denotes the set consisting of its assumed values. In particular, if f is a linear map between linear spaces, $\ker f$ denotes the subspace of the domain of f containing those elements that are mapped to the zero vector. For every non-empty set S, the symbol id_S denotes the identity map on S. We always assume the composition of maps (which includes addition, multiplication etc.) to be maximally defined. For instance, the addition of two maps is defined on the (possibly empty) intersection of their domains.

The symbols $\mathbb{N}$, $\mathbb{R}$, $\mathbb{C}$ denote the natural numbers (including zero), all real numbers and all complex numbers, respectively. The symbols $\mathbb{N}^*$, $\mathbb{R}^*$, $\mathbb{C}^*$ denote the corresponding sets without 0. We call $x \in \mathbb{R}$ positive (negative) if $x \geqslant 0$ ($x \leqslant 0$). We call $x \in \mathbb{R}$ strictly positive (strictly negative) if $x > 0$ ($x < 0$).

For every $n \in \mathbb{N}^*$, $e_1, \dots, e_n$ denotes the canonical basis of $\mathbb{R}^n$. For every $x \in \mathbb{R}^n$, $|x|$ denotes the euclidean norm of x. Further, in connection with matrices, the elements of $\mathbb{R}^n$ are considered as *column vectors*. Finally, for $\mathbb{K} \in \{\mathbb{R}, \mathbb{C}\}$, $M(n \times n, \mathbb{K})$ denotes the vector space of $n \times n$ matrices with entries from $\mathbb{K}$. Also, for every $A \in M(n \times n, \mathbb{K})$, $\det A$ denotes its determinant.

For each $k \in \mathbb{N}$, $n \in \mathbb{N}^*$, $\mathbb{K} \in \{\mathbb{R}, \mathbb{C}\}$ and each non-empty open subset M of $\mathbb{R}^n$, the symbol $C^k(M, \mathbb{K})$ denotes the linear space of continuous and k-times continuously differentiable $\mathbb{K}$-valued functions on M. Further, $C_0^k(M, \mathbb{K})$ denotes the subspace of $C^k(M, \mathbb{K})$ containing those elements that have a compact support in M. If M is *bounded*, $C^k(\bar{M}, \mathbb{K})$ is defined as the *subspace* of $C^k(M, \mathbb{K})$ consisting of those elements for which there is an extension to an element of $C^k(V, \mathbb{K})$ for some open subset V of $\mathbb{R}^n$ containing M. The superscript k is omitted if $k = 0$. We adopt the convention from General Relativity that for a map f defined on an open subset of $\mathbb{R}^n$

$$f_{,j} := \frac{\partial f}{\partial x_j}$$

for $j \in \{1, \dots, n\}$ or more generally

$$f_{,\alpha} := \frac{\partial^{|\alpha|} f}{\partial^\alpha x} := \frac{\partial^{|\alpha|} f}{\partial x_1^{\alpha_1} \dots \partial x_n^{\alpha_n}},$$

for $\alpha \in \mathbb{N}^n$ if existent.[1] For every map $f : U \to \mathbb{R}^n$ which is defined on some subset $U \subset \mathbb{R}^n$ as well as differentiable in $x \in U$, $f'(x) \in M(n \times n, \mathbb{R})$ denotes the derivative of f in x defined by

$$f'(x)_{ij} := \frac{\partial f_i}{\partial x_j}(x)$$

for all $i, j \in \{1, \ldots, n\}$. In addition, in the case $n = 1$, we define the gradient of f in x by

$$(\nabla f)(x) := \sum_{i=1}^{n} \frac{\partial f}{\partial x_i}(x)\, e_i \, .$$

Further, for every differentiable map γ from some non-trivial open interval I of $\mathbb{R}$ into $\mathbb{R}^n$, we define

$$\gamma'(\tau) := ((x_1 \circ \gamma)'(\tau), \ldots, (x_n \circ \gamma)'(\tau))$$

for every $\tau \in I$ where $x_1, \ldots, x_n$ denote the coordinate projections of $\mathbb{R}^n$. Further, $BC(\mathbb{R}^n, \mathbb{C})$ denotes the subspace of $C(\mathbb{R}^n, \mathbb{C})$ consisting of those functions which are bounded. $C_\infty(\mathbb{R}^n, \mathbb{C})$ denotes subspace of $C(\mathbb{R}^n, \mathbb{C})$ containing those functions f satisfying

$$\lim_{|x| \to \infty} f(x) = 0 \, .$$

Throughout the course, Lebesgue integration theory is used in the formulation of [182]. Compare also Chapter III in [101] and Appendix A in [216]. If not indicated otherwise, the terms *'almost everywhere'* (a.e.), *'measurable'*, *'summable'*, etc. refer to the Lebesgue measure v^n on $\mathbb{R}^n$, $n \in \mathbb{N}^*$. The appropriate n will be clear from the context. Nevertheless, we often mimic the notation of the Riemann-integral to improve readability. We follow common usage and don't differentiate between a function which is almost everywhere defined (with respect to the chosen measure) on some set and the associated equivalence class consisting of all functions which are almost everywhere defined on that set and differ from f only on a set of measure zero. In this sense, for $p > 0$ the symbol $L^p_{\mathbb{C}}(M, \rho)$, where ρ is some strictly positive real-valued continuous function on M, denotes the vector space of all complex-valued measurable functions f which are defined on M and such that $|f|^p$ is integrable with respect to the measure $\rho\, v^n$. For every such f, we define the L^p-norm $\|f\|_p$ corresponding to f by

$$\|f\|_p := \left(\int_M |f|^p \, dv^n \right)^{1/p} \, .$$

Equipped with $\| \, \|_p$, $L^p_{\mathbb{C}}(M, \rho)$ is a Banach space. In addition, we define in the special case $p = 2$ a scalar product $\langle \, | \, \rangle_2$ on $L^2_{\mathbb{C}}(M, \rho)$ by

$$\langle f | g \rangle_2 := \int_M \rho f^* g \, dv^n \, ,$$

[1] But, we don't use Einstein's summation convention.

for all $f, g \in L^2_{\mathbb{C}}(M,\rho)$. Here * denotes complex conjugation on $\mathbb{C}$. As a consequence, $\langle | \rangle_2$ is antilinear in the first argument and linear in its second. This convention is used for sesquilinear forms in general. It is a standard result of Functional Analysis that $C^k_0(M,\mathbb{C})$ is dense in $L^p_{\mathbb{C}}(M,\rho)$. If ρ is constant of value 1, we omit any reference to ρ in the previous symbols. $L^\infty_{\mathbb{C}}(M)$ denotes the vector space of complex-valued measurable bounded functions on M. For every $f \in L^\infty_{\mathbb{C}}(M)$ we define

$$\|f\|_\infty := \sup_{x\in M} |f(x)| .$$

Equipped with $\| \ \|_\infty$, $L^\infty_{\mathbb{C}}(M)$ is a Banach space.

Finally, standard results and nomenclature of Operator Theory are used. For this, compare textbooks on Functional Analysis, e.g., [179] Vol. I, [182,225]. In particular, for every non-trivial normed vector space $(X, \| \ \|_{\mathrm{X}})$ and any normed vector space $(Y, \| \ \|_{\mathrm{Y}})$ over the same field, we denote by $L(X,Y)$ the vector space of continuous linear maps from X to Y. Equipped with the operator norm $\| \ \|_{\mathrm{Op,X,Y}}$, defined by

$$\|A\|_{\mathrm{Op,X,Y}} := \sup_{\xi\in X, \|\xi\|_{\mathrm{X}}=1} \|A\xi\|_{\mathrm{Y}}$$

for all $A \in L(X,Y)$, $L(X,Y)$, is a normed vector space which is complete if $(Y, \| \ \|_{\mathrm{Y}})$ is complete. Frequently, the indices in $\| \ \|_{\mathrm{Op,X,Y}}$ are omitted if there is no confusion possible. Finally, for every non-void subset U of some normed vector space $(X, \| \ \|_{\mathrm{X}})$ and any normed vector space $(Y, \| \ \|_{\mathrm{Y}})$, the symbol $C(U,Y)$ denotes the vector space of continuous functions from U to Y, and the symbol $\mathrm{Lip}(U,Y)$ denotes its subspace consisting of all Lipschitz continuous elements, i.e., of those elements f for which there is a so called Lipschitz constant $C \in [0,\infty)$ such that

$$\|f(\xi) - f(\eta)\|_{\mathrm{Y}} \leqslant C\|\xi - \eta\|_{\mathrm{X}}$$

for all $\xi, \eta \in U$.

2

Mathematical Introduction

2.1 Quantum Theory

We give a very brief sketch of the mathematical structure of quantum theory. For more comprehensive introductions, see [134, 157, 175]. For mathematical introductions to quantum field theory, see [13, 38, 95, 191, 202] and the second volume of [179]. For such introductions to quantum field theory in curved space-times see [81] for the field theoretic aspects and [214] for the geometrical aspects.

The *states*[1] of a physical system are *rays*[2] $\mathbb{C}.\xi$, $\xi \in X$, in a *complex Hilbert space* $(X, \langle \,|\, \rangle)$ called a *representation space*. Such a space is unique up to Hilbert space isomorphisms, only. Hence its elements *are not observable*[3] and in this sense 'abstract'.

Observables of the system correspond to *densely-defined, linear and self-adjoint*[4] *operators* in X.

A '*quantization*' of a classical physical system is the association of such operators to its classical observables like positions, momenta, angular momenta of its constituents. In this canonically conjugate observables of the classical system are required to be mapped into observables satisfying Heisenberg's commutation relations or their Weylian form. In addition, any densely-defined, linear and self-adjoint operator in X is considered to be an observable of the quantum system.

The interpretation of *measurements* is probabilistic. The elementary events of a measurement of an observable A are the values of its spectrum $\sigma(A)$. The last consists of all those complex numbers a for which the corresponding operator $A - a$ is not bijective. Since A is self-adjoint, those values are necessarily *real*. The

[1] For a brief review of the axioms of Quantum Theory see, for e.g., [31].

[2] The use of rays in this definition becomes important in the description of particles with spin.

[3] In particular, 'wave functions' are *not* observable. On the other hand, wave functions become observable in the 'quasi-classical limit'.

[4] The assumption of symmetry is in general insufficient since such operators can have non-real spectral values.

non-normalized *probability* of finding the measured value to be part of a bounded subset $\Omega \subset \mathbb{R}$[5] is given by

$$\psi_\xi(\Omega) = \int_{\mathbb{R}} \chi_\Omega \, d\psi_\xi = \langle \xi | \chi_\Omega(A)\xi \rangle \ .$$

Here $\mathbb{C}.\xi$, $\xi \in X$, is the state of the system, ψ_ξ is the *spectral measure*[6] associated to A and ξ, χ_Ω denotes the characteristic function of Ω and $\chi_\Omega(A)$ is the orthogonal projection which is associated to $\chi_\Omega|_{\sigma(A)}$ by the functional calculus for A.[7] Non spectral values of A do not contribute to this probability since $\mathbb{R}\backslash\sigma(A)$ is a ψ_ξ-zero set.[8] In particular, note for the normalization that

$$\psi_\xi(\mathbb{R}) = \psi_\xi(\sigma(A)) = \int_{\mathbb{R}} d\psi_\xi = \langle \xi | \xi \rangle \ .$$

After a measurement of A finds its value to be part of Ω, the system is in the state

$$\mathbb{C}.\chi_\Omega(A)\xi \ .$$

Note that in the particular case that Ω is the disjoint union of two bounded subsets $\Omega_1, \Omega_2 \subset \mathbb{R}$[9] that

$$\mathbb{C}.\chi_\Omega(A)\xi = \mathbb{C}.\left(\chi_{\Omega_1}(A)\xi + \chi_{\Omega_2}(A)\xi\right) \ ,$$

i.e., that the system is in the state corresponding to the *superposition* of $\chi_{\Omega_1}(A)\xi$ and $\chi_{\Omega_2}(A)\xi$. If A is not found to be part of Ω, it is in the state

$$\mathbb{C}.\left(\mathrm{id}_X - \chi_\Omega(A)\right)\xi = \mathbb{C}.\chi_{\mathbb{R}\backslash\Omega}(A)\xi \ .$$

More generally, according to the usual rules of probability[10], for any bounded function $f : \sigma(A) \to \mathbb{R}$[11] being 'universally measurable'[12] the non-normalized *expectation value* for the measurement of the observable $f(A)$ is given by

$$\int_{\mathbb{R}} f \, d\psi_\xi = \langle \xi | f(A)\xi \rangle \ .$$

---

[5] In addition, Ω is assumed to be a countable union of bounded intervals of $\mathbb{R}$.

[6] ψ_ξ is an additive, monotone and regular interval function defined on the set of all bounded subintervals of $\mathbb{R}$.

[7] See, e.g., [179] Vol. I, Theorem VIII.5.

[8] This is the case for all $\xi \in X$.

[9] For instance this has application in the double-slit experiment.

[10] $\sigma(A)$ is the sample space and ψ_ξ is the probability distribution for the random variable $\mathrm{id}_{\sigma(A)}$.

[11] f is a random variable.

[12] In particular pointwise limits of sequences of continuous functions on $\sigma(A)$ are universally measurable.

The so called *Hamilton operator* is the associate of the Hamiltonian of the classical system. It is the generator of the time evolution of the states in the following sense[13,14] If the system is in the state $\mathbb{C}.\xi$, $\xi \in X$, at time $t_0 \in \mathbb{R}$, it will be/was in the state

$$\mathbb{C}.U(t - t_0)\xi$$

at time $t \in \mathbb{R}$. Here

$$U(t) := \exp\left(-i\,(t/\hbar).\mathrm{id}_{\sigma(H)}\right)(H)$$

is unitary for every $t \in \mathbb{R}$, and $\hbar$ is the reduced Planckian constant. Note that $U : \mathbb{R} \to L(X, X)$ is in particular strongly continuous. The *unitarity* of time evolution corresponds to *conservation of probability*.

If $\xi \in D(H)$, the unique solution $u : \mathbb{R} \to D(H)$ of the '*Schrödinger equation*'

$$i\hbar.u'(t) = Hu(t)$$

such that $u(t_0) = \xi$, where $'$ denotes the ordinary derivative of a X-valued path, is given by

$$u(t) := U(t - t_0)\xi$$

for all $t \in \mathbb{R}$.[15]

A simple example for a non-relativistic quantum system is given by a spinless point-particle of mass $m > 0$ solely interacting with an external potential $V \in L^\infty(\mathbb{R}^3)$. In the *position representation*

$$H = \left(W^2_{\mathbb{C}}(\mathbb{R}^3) \to L^2_{\mathbb{C}}(\mathbb{R}^3), f \mapsto -\frac{\hbar^2}{2m}\triangle f + Vf\right)$$

where $\triangle$ denotes the Laplacian in 3 dimensions. The operator corresponding to the measurement of the k-th, $k \in \{1, 2, 3\}$, component of position is given by the maximal multiplication operator T_{x_k} in $L^2_{\mathbb{C}}(\mathbb{R}^3)$ with the k-th coordinate projection $x_k : \mathbb{R}^3 \to \mathbb{R}$ defined by $x_k(\bar{x}) := \bar{x}_k$ for all $\bar{x} \in \mathbb{R}^3$. Its spectrum consists of all real numbers and is purely absolutely continuous. Further, for any $f \in L^2_{\mathbb{C}}(\mathbb{R}^3)$ the corresponding spectral measure ψ_f is given by

$$\psi_f(I) = \int_{x_k^{-1}(I)} |f|^2\, dv^3$$

for every bounded interval I of $\mathbb{R}$. The quantity $\psi_f(I)$ gives the non-normalized probability in a position measurement of finding the k-th coordinate to be in the range I if the particle is in the state $\mathbb{C}.f$.

---

13 This is true if the system is closed. Otherwise the Hamiltonian can depend on time.

14 Here we are using the '*Schrödinger picture*'. In the equivalent '*Heisenberg picture*' the observables undergo time-evolution, whereas the states of the system stay the same. Heisenberg's picture is generally used in Quantum Field Theory.

15 See Chapter VIII.4 in Vol. I of [179] on 'Stone's theorem'.

The operator corresponding to the measurement of the k-th, $k \in \{1,2,3\}$, component of the momentum is given by the closure of the densely-defined, linear, symmetric and essentially self-adjoint operator $p_k : C_0^\infty(\mathbb{R}^3, \mathbb{C}) \to L^2_{\mathbb{C}}(\mathbb{R}^3)$ given by

$$p_k f := \frac{\hbar}{i} \frac{\partial f}{\partial x_k}$$

for every $f \in C_0^\infty(\mathbb{R}^3, \mathbb{C})$. The Hilbert space isomorphism to the *momentum representation* is given by the unitary Fourier transformation $F_2 : L^2_{\mathbb{C}}(\mathbb{R}^3) \to L^2_{\mathbb{C}}(\mathbb{R}^3)$. The operator in that representation corresponding to the measurement of the k-th, $k \in \{1,2,3\}$, component of the momentum is given by

$$F_2 T_{\bar{p}_k} F_2^{-1} = T_{\hbar x_k} .$$

2.2 Wave Equations

Wave equations are an important prototype of hyperbolic equations frequently appearing in applications. In the following, we study simple examples of such equations whose corresponding initial value problems can be solved by applications of the spectral theorem for self-adjoint linear operators in Hilbert spaces. All examples describe *non-dissipative* systems. In all cases there is a conserved total energy. The theorems below are essentially known and proofs are left to the reader. For applications of Theorem 2.2.1 below to problems in General Relativity and Astrophysics, see [22–24, 69, 116, 187, 212, 213]. For the treatment of damped wave equations by semigroup methods, see Chapter 5.4 and the references given in that section.

Theorem 2.2.1. Let $(X, \langle|\rangle)$ be some non-trivial complex Hilbert space. Further, let $A : D(A) \to X$ be some densely-defined, linear, strictly positive[16] self-adjoint operator in X. Finally, let $\xi, \eta \in D(A)$.

(i) There is a uniquely determined twice continuously differentiable map $u : \mathbb{R} \to X$ assuming values in $D(A)$ and satisfying

$$u''(t) = -A\, u(t) \tag{2.2.1}$$

for all $t \in \mathbb{R}$ and

$$u(0) = \xi \ , \ u'(0) = \eta \ .$$

For this u, the corresponding function (describing the total energy of the field) $E_u : \mathbb{R} \to [0, \infty)$, defined by

$$E_u(t) := \frac{1}{2} \left(\langle u'(t)|u'(t)\rangle + \langle u(t)|Au(t)\rangle \right)$$

for all $t \in \mathbb{R}$, is constant.

---

[16] That is, the spectrum of A is a subset of the open interval $(0, \infty)$.

(ii) Let $B : D(B) \to X$ be some square root of A, i.e., some densely-defined, linear, self-adjoint operator commuting with A which satisfies[17]

$$B^2 = A \, ,$$

(for example, $B = A^{1/2}$) . Then u is given by

$$u(t) = \cos(tB)\xi + \frac{\sin(tB)}{B}\eta \tag{2.2.2}$$

for all $t \in \mathbb{R}$ where $\cos(tB), \sin(tB)/B$ denote the bounded linear operators that are associated by the functional calculus for B to the restrictions of cos, sin $/\mathrm{id}_{\mathbb{R}}$ to the spectrum of B.

Proof. The proof is left to the reader. For this, note that (2.2.2) is suggested by considering the one-parameter unitary groups generated by B and $-B$, respectively, and the observation that, for any $\xi, \eta \in D(A)$ the maps

$$\left(\mathbb{R} \to X, \, t \mapsto e^{itB}\xi\right) \quad \text{and} \quad \left(\mathbb{R} \to X, \, t \mapsto e^{-itB}\eta\right)$$

satisfy (2.2.1). The remaining part of the proof mainly consists in a simple application of the functional calculus for B and A. For the case $B = A^{1/2}$, it can be found, e.g., in [105]. See also [142] Chapter 24, §8. □

Theorem 2.2.1 has important applications in *Quantum Field Theory*.[18] There the (distributional) 'kernel' of the operator

$$\frac{\sin((t'-t)A^{1/2})}{A^{1/2}} \, ,$$

$t, t' \in \mathbb{R}$, is referred to as '*commutator function*' for the Klein-Gordon field. For such an application see, for example, [21]. For mathematical introductions to quantum field theory, see [13, 38, 95, 191, 202] and the second volume of [179]. For such introductions to quantum field theory in curved space-times see [81] for the field theoretic aspects and [214] for the geometrical aspects.

Theorem 2.2.1 has the corollary

Corollary 2.2.2. More generally, let $A : D(A) \to X$ be semibounded. Then for any ξ, η from $D(A)$ there is a uniquely determined twice continuously differentiable map $u : \mathbb{R} \to X$ assuming values in $D(A)$ and satisfying

$$u''(t) = -A\, u(t)$$

for all $t \in \mathbb{R}$ as well as

$$u(0) = \xi \, , \; u'(0) = \eta \, .$$

---

[17] For the interpretation of the following equation compare the conventions.

[18] For instance, see [32].

For this u, the corresponding function $E_u : \mathbb{R} \to \mathbb{R}$, defined by

$$E_u(t) := \frac{1}{2}\left(\langle u'(t)|u'(t)\rangle + \langle u(t)|Au(t)\rangle\right)$$

for all $t \in \mathbb{R}$, is constant. Moreover, this u is given by

$$u(t) = [\cos(t\sqrt{})]_e(A)\xi + [\sin(t\sqrt{})/\sqrt{}]_e(A)\eta$$

for all $t \in \mathbb{R}$ where $[\cos(t\sqrt{})]_e, [\sin(t\sqrt{})/\sqrt{}]_e$ denote the restrictions to the spectrum of A of the analytic extensions of $\cos(t\sqrt{})$ and $\sin(t\sqrt{})/\sqrt{}$, respectively.

Proof. Again, the proof is left to the reader. The proof of part (i) is straightforward. The proof of part (ii) proceeds by decomposition of X into closed invariant subspaces of A in which the densely-defined, linear and self-adjoint operators induced by A are positive and bounded, respectively. □

Simple examples of such A are, for instance,

$$A = \left(W^2_{\mathbb{C}}(\mathbb{R}^n) \to L^2_{\mathbb{C}}(\mathbb{R}^n), f \mapsto -\triangle f + Vf\right)$$

where $n \in \mathbb{N}^*$, $\triangle$ denotes the Laplacian in n dimensions and V is a real-valued element of $L^\infty_{\mathbb{C}}(\mathbb{R}^n)$.

For instance, for the Klein-Gordon field of mass $m > 0$ on the real line, $X = L^2_{\mathbb{C}}(\mathbb{R})$,

$$A = F_2^{-1}\,(k^0)^2\,F_2\ \left[= \left(W^2_{\mathbb{C}}(\mathbb{R}) \to X\ ,\ f \mapsto -f'' + m^2 f\right)\right]$$

where $(k^0)^2$ denotes the maximal multiplication operator in X by the function $(\mathbb{R} \to [0,\infty)\,,\, k \mapsto k^2 + m^2)$ and $F_2 : X \to X$ the unitary Fourier transformation. In particular, the uniquely determined positive self-adjoint square root $A^{1/2}$ of A is given by

$$A^{1/2} = F_2^{-1}\,k^0\,F_2$$

and for any bounded continuous complex-valued function on the *spectrum* $[m,\infty)$ *of* $A^{1/2}$:

$$f(A^{1/2}) = F_2^{-1}\,f(k^0)\,F_2\ . \tag{2.2.3}$$

Finally, it follows from (2.2.3) and some calculation that

$$\begin{aligned}[u(t)]\,(x) &= \left[\cos(tA^{1/2})f + \frac{\sin(tA^{1/2})}{A^{1/2}}g\right](x) \\ &= \frac{1}{2}\left[f(x+t) + f(x-t) - m\,t\int_{x-t}^{x+t}\frac{J_1(m\,[\,t^2-(x-x')^2\,]^{1/2})}{[\,t^2-(x-x')^2\,]^{1/2}}\,f(x')\,dx'\right. \\ &\quad\left. + \int_{x-t}^{x+t} J_0(m\,[\,t^2-(x-x')^2\,]^{1/2})\,g(x')\,dx'\right]\end{aligned}$$

for all $t \in \mathbb{R}$, $f, g \in X$ and almost all $x \in I$. Here J_0, J_1 denote Bessel functions of the first kind defined according to [1]. The verification of these results is left to the reader.

For the case of the Klein-Gordon field of mass $m > 0$ in 4-dimensional Minkowski space $X = L^2_{\mathbb{C}}(\mathbb{R}^3)$,

$$A = F_2^{-1}\,(k^0)^2\,F_2\ \left[= \left(W^2_{\mathbb{C}}(\mathbb{R}^3) \to X\ ,\ f \mapsto -\Delta f + m^2 f\right)\right]$$

where $(k^0)^2$ denotes the maximal multiplication operator in X by the function $(\mathbb{R}^3 \to [0,\infty),\ k \mapsto |k|^2 + m^2)$ and $F_2 : X \to X$ the unitary Fourier transformation. In particular, the uniquely determined positive self-adjoint square root $A^{1/2}$ of A is given by

$$A^{1/2} = F_2^{-1}\,k^0\,F_2$$

and for any bounded continuous complex-valued function on the *spectrum* $[m,\infty)$ *of* $A^{1/2}$:

$$f(A^{1/2}) = F_2^{-1}\,f(k^0)\,F_2\,. \tag{2.2.4}$$

Finally, it follows from (2.2.4) by some calculation that

$$\left[e^{-\varepsilon A^{1/2}}\,\frac{e^{itA^{1/2}}}{A^{1/2}}\,f\right](x) = \frac{m}{2\pi^2}\int_{\mathbb{R}^3}\frac{K_1(m\,[\,|x-x'|^2-(t+i\varepsilon)^2\,]^{1/2})}{[\,|x-x'|^2-(t+i\varepsilon)^2\,]^{1/2}}\,f(x')\,dx'$$

$$\left[e^{-\varepsilon A^{1/2}}\,e^{itA^{1/2}}\,f\right](x) = \frac{im^2(t+i\varepsilon)}{2\pi^2}\int_{\mathbb{R}^3}\frac{K_2(m\,[\,|x-x'|^2-(t+i\varepsilon)^2\,]^{1/2})}{|x-x'|^2-(t+i\varepsilon)^2}\,f(x')\,dx'$$

and that

$$\left[\cos(tA^{1/2})f\right](x) = -\frac{m^2}{2\pi^2}\lim_{\varepsilon\to 0+}\int_{\mathbb{R}^3}\operatorname{Im}\left((t+i\varepsilon)\,\frac{K_2(m\,[\,|x-x'|^2-(t+i\varepsilon)^2\,]^{1/2})}{|x-x'|^2-(t+i\varepsilon)^2}\right)f(x')\,dx'$$

$$\left[\frac{\sin(tA^{1/2})}{A^{1/2}}\,g\right](x) = \frac{m}{2\pi^2}\lim_{\varepsilon\to 0+}\int_{\mathbb{R}^3}\operatorname{Im}\left(\frac{K_1(m\,[\,|x-x'|^2-(t+i\varepsilon)^2\,]^{1/2})}{[\,|x-x'|^2-(t+i\varepsilon)^2\,]^{1/2}}\right)g(x')\,dx'$$

for all $t \in \mathbb{R}$, $f, g \in X$ and almost all $x \in \mathbb{R}^3$. Here the limits are to be taken in X, $[\ \ ]^{1/2}$ denotes the principal branch of the complex square root function and K_0, K_1 denote modified Bessel functions defined according to [1]. Again, the verification of these results is left to the reader. Note that the representations (2.2.2) involve only *continuous* linear operators which are functions of $A^{1/2}$ depending on time. Hence by the spectral theorem for self-adjoint operators in Hilbert space[19], this gives the continuous dependence of the solutions on the data also in time and *hence the well-posedness of the initial value problem for (2.2.1) in the sense of Hadamard.* In addition, it provides a much wider class of what may be considered as 'generalized' or 'weak' solutions' of (2.2.1).

Corollary 2.2.3. (Reformulation of (2.2.1) as a first order system in time) If A is in addition strictly positive (i.e, the spectrum of A is contained in $(0,\infty)$) and B is as in (ii), then

(i) $Y := (D(B) \times X, (\,|\,))$, where $(\,|\,) : Y^2 \to \mathbb{C}$ is defined by

$$(\xi|\eta) := \langle B\xi_1|B\eta_1\rangle + \langle \xi_2|\eta_2\rangle$$

for all $\xi = (\xi_1, \xi_2), \eta = (\eta_1, \eta_2) \in Y$, is a complex Hilbert space.

(ii) $A_B : D(A) \times D(B) \to Y$, defined by

$$A_B(\xi, \eta) := (-\eta, A\xi)$$

for all $\xi \in D(A)$ and $\eta \in D(B)$, is the generator of a one-parameter unitary group $U : \mathbb{R} \to L(Y, Y)$. U is given by

$$U(t)(\xi, \eta) = \left(\cos(tB)\xi + \frac{\sin(tB)}{B}\eta, -\sin(tB)B\xi + \cos(tB)\eta\right)$$

for all $(\xi, \eta) \in Y$ and $t \in \mathbb{R}$. Hence for any $\xi \in D(A), \eta \in D(B)$ there is a unique continuously differentiable map $(u, v) : \mathbb{R} \to Y$ assuming values in $D(A) \times D(B)$ and satisfying

$$u'(t) = v(t)\ ,\ v'(t) = -A\,u(t)$$

for all $t \in \mathbb{R}$ as well as

$$u(0) = \xi\ ,\ v(0) = \eta\ .$$

This map is given by

$$(u, v)(t) = U(t)(\xi, \eta)$$

for all $t \in \mathbb{R}$.

Proof. The proof is left to the reader. For the proof of the case $B = A^{1/2}$, see, e.g., [179]. □

[19] See the functional calculus form, Theorem VIII.5, in [179] Vol. I.

3

Prerequisites

3.1 Linear Operators in Banach Spaces

In the following, we introduce basics of the language of Operator Theory that can also be found in most textbooks on Functional Analysis [179, 182, 186, 216, 225]. For the convenience of the reader, we also include associated proofs.

Lemma 3.1.1. (Direct sums of Banach and Hilbert spaces)

(i) Let $(X, \| \|_X)$ and $(Y, \| \|_Y)$ be Banach spaces over $\mathbb{K} \in \{\mathbb{R}, \mathbb{C}\}$ and $\| \|_{X \times Y} : X \times Y \to \mathbb{R}$ be defined by

$$\|(\xi, \eta)\|_{X \times Y} := \sqrt{\|\xi\|_X^2 + \|\eta\|_Y^2}$$

for all $(\xi, \eta) \in X \times Y$. Then $(X \times Y, \| \|_{X \times Y})$ is a Banach space.

(ii) Let $(X, \langle | \rangle_X)$ and $(Y, \langle | \rangle_Y)$ be Hilbert spaces over $\mathbb{K} \in \{\mathbb{R}, \mathbb{C}\}$ and $\langle | \rangle_{X \times Y} : (X \times Y)^2 \to \mathbb{K}$ be defined by

$$\langle (\xi, \eta) | (\xi', \eta') \rangle_{X \times Y} := \langle \xi | \xi' \rangle_X + \langle \eta | \eta' \rangle_Y$$

for all $(\xi, \eta), (\xi', \eta') \in X \times Y$. Then $(X \times Y, \langle | \rangle_{X \times Y})$ is a Hilbert space.

Proof. '(i)': Obviously, $\| \|_{X \times Y}$ is positive definite and homogeneous. Further, it follows for $(\xi, \eta), (\xi', \eta') \in X \times Y$ by the Cauchy-Schwarz inequality for the Euclidean scalar product for $\mathbb{R}^2$ that

$$\begin{aligned}
&\|(\xi, \eta) + (\xi', \eta')\|_{X \times Y}^2 \\
&= \|\xi + \xi'\|_X^2 + \|\eta + \eta'\|_Y^2 \\
&\leqslant (\|\xi\|_X + \|\xi'\|_X)^2 + (\|\eta\|_Y + \|\eta'\|_Y)^2 = (a + a')^2 + (b + b')^2 \\
&= a^2 + b^2 + a'^2 + b'^2 + 2\,(a\,a' + b\,b') \\
&\leqslant a^2 + b^2 + a'^2 + b'^2 + 2\,\sqrt{a^2 + b^2} \cdot \sqrt{a'^2 + b'^2} \\
&= \left(\sqrt{a^2 + b^2} + \sqrt{a'^2 + b'^2}\right)^2 = (\|(\xi, \eta)\|_{X \times Y} + \|(\xi', \eta')\|_{X \times Y})^2,
\end{aligned}$$

where $a := \|\xi\|_X, a' := \|\xi'\|_X, b := \|\eta\|_Y, b' := \|\eta'\|_Y$, and hence that

$$\|(\xi,\eta) + (\xi',\eta')\|_{X\times Y} \leqslant \|(\xi,\eta)\|_{X\times Y} + \|(\xi',\eta')\|_{X\times Y}.$$

The completeness of $(X \times Y, \| \, \|_{X\times Y})$ is an obvious consequence of the completeness of X and Y.

'(ii)': Obviously, $\langle \, | \, \rangle_{X\times Y}$ is a positive definite symmetric bilinear, positive definite hermitian sesquilinear form, respectively. Further, the induced norm on $X \times Y$ coincides with the norm defined in (i). □

Definition 3.1.2. (Linear Operators) Let $(X, \| \, \|_X)$ and $(Y, \| \, \|_Y)$ be Banach spaces over $\mathbb{K} \in \{\mathbb{R}, \mathbb{C}\}$. Then we define

(i) A map A is called a *Y-valued linear operator in X* if its *domain* $D(A)$ is a subspace of X, $\operatorname{Ran} A \subset Y$ and A is linear. If $(Y, \| \, \|_Y) = (X, \| \, \|_X)$ such a map is also called a *linear operator in X*.
(ii) If in addition A is a Y-valued linear operator in X:
  a) The *graph* $G(A)$ of A by

  $$G(A) := \{(\xi, A\xi) \in X \times Y : \xi \in D(A)\}.$$

  Note that $G(A)$ is a subspace of $X \times Y$.
  b) A is *densely-defined* if $D(A)$ is in particular dense in X.
  c) A is *closed* if $G(A)$ is a closed subspace of $(X \times Y, \| \, \|_{X\times Y})$.
  d) A Y-valued linear operator B in X is said to be an *extension* of A, symbolically denoted by

  $$A \subset B \quad \text{or} \quad B \supset A \, ,$$

  if $G(A) \subset G(B)$.
  e) A is *closable* if there is a closed extension. In this case,

  $$\bigcap_{B\supset A, B \text{ closed}} G(B)$$

  is a closed subspace of $X \times Y$ which, obviously, is the graph of a unique Y-valued closed linear extension $\bar{A}$ of A, called the *closure* of A. By definition, every closed extension B of A satisfies $B \supset \bar{A}$.
  f) If A is closed, a *core* of A is a subspace D of its domain such that the closure of $A|_D$ coincides with A, i.e., if

  $$\overline{A|_D} = A \, .$$

Theorem 3.1.3. (Elementary properties of linear operators) Let $(X, \| \, \|_X)$, $(Y, \| \, \|_Y)$ be Banach spaces over $\mathbb{K} \in \{\mathbb{R}, \mathbb{C}\}$, A a *Y-valued linear operator in X* and $B \in L(X, Y)$.

(i) $(D(A), \| \|_A)$, where $\| \|_A : D(A) \to \mathbb{R}$ is defined by

$$\|\xi\|_A := \|(\xi, A\xi)\|_{X\times Y} = \sqrt{\|\xi\|_X^2 + \|A\xi\|_Y^2}$$

for every $\xi \in D(A)$, is a normed vector space. Further, the inclusion $\iota_A : (D(A), \| \|_A) \hookrightarrow X$ is continuous and $A \in L((D(A), \| \|_A), Y)$.

(ii) A is closed if and only if $(D(A), \| \|_A)$ is complete.

(iii) If A is closable, then $G(\bar{A}) = \overline{G(A)}$.

(iv) (*Inverse mapping theorem*) If A is closed and bijective, then $A^{-1} \in L(Y, X)$.

(v) (*Closed graph theorem*) In addition, let $D(A) = X$. Then A is bounded if and only if A is closed.

(vi) If A is closable, then $A + B$ is also closable and

$$\overline{A+B} = \bar{A} + B\,.$$

Proof. '(i)': Obviously, $(D(A), \| \|_A)$ is a normed vector space. Further, because of

$$\|\iota_A\,\xi\|_X = \|\xi\|_X \leqslant \sqrt{\|\xi\|_X^2 + \|A\xi\|_Y^2} = \|\xi\|_A$$

and

$$\|A\xi\|_Y \leqslant \sqrt{\|\xi\|_X^2 + \|A\xi\|_Y^2} = \|\xi\|_A$$

for every $\xi \in D(A)$, it follows that $\iota_A \in L((D(A), \| \|_A), X)$ and $A \in L((D(A), \| \|_A), Y)$.

'(ii)': Let A be closed and $\xi_0, \xi_1, \dots$ a Cauchy sequence in $(D(A), \| \|_A)$. Then $(\xi_0, A\xi_0), (\xi_1, A\xi_1), \dots$ is a Cauchy sequence in $G(A)$ and hence by Lemma 3.1.1 along with the closedness of $G(A)$ convergent to some $(\xi, A\xi) \in G(A)$. This implies that

$$\lim_{\nu\to\infty} \|\xi_\nu - \xi\|_A = 0$$

and the convergence of $\xi_0, \xi_1, \dots$ in $(D(A), \| \|_A)$. Let $(D(A), \| \|_A)$ be complete and $(\xi, \eta) \in \overline{G(A)}$. Then there is a sequence $(\xi_0, A\xi_0), (\xi_1, A\xi_1), \dots$ in $G(A)$ which is convergent to (ξ, η). Hence $(\xi_0, A\xi_0), (\xi_1, A\xi_1), \dots$ is a Cauchy sequence in $X \times Y$. As a consequence, $\xi_0, \xi_1, \dots$ is a Cauchy sequence in $(D(A), \| \|_A)$ and therefore convergent to some $\xi' \in D(A)$. In particular,

$$\lim_{\nu\to\infty} \|(\xi_\nu, A\xi_\nu) - (\xi', A\xi')\|_{X\times Y} = 0$$

and hence $(\xi, \eta) = (\xi', A\xi') \in G(A)$.

'(iii)': Let A be closable. Then the closed graph of every closed extension of A contains $G(A)$ and hence also $\overline{G(A)}$. Therefore $G(\bar{A}) \supset \overline{G(A)}$. This implies in particular that $\overline{G(A)}$ is the graph of a map $\tilde{A}$. Further, $D(\tilde{A}) = \mathrm{pr}_1\overline{G(A)}$, where $\mathrm{pr}_1 := (X \times Y \to X, (\xi, \eta) \mapsto \xi)$, is a subspace of X and $\tilde{A}$ is in particular a linear closed extension of A. Hence $\tilde{A} \supset \bar{A}$ and $\overline{G(A)} = G(\tilde{A}) \supset G(\bar{A})$.

'(iv)': Let A be closed and bijective. Then it follows by (ii) that $(D(A), \| \|_A)$ is a Banach space and that $A \in L((D(A), \| \|_A), Y)$. Hence it follows by the 'inverse

mapping theorem', for e.g. see Theorem III.11 in Vol. I of [179], that $A^{-1} \in L(Y,(D(A),\| \|_{A}))$ and by the continuity of ι_{A} that $A^{-1} \in L(Y,X)$.
'(v)': Let $D(A) = X$. If A is bounded and $\xi_0, \xi_1, \dots$ is some Cauchy sequence in $(X, \| \|_{A})$, it follows by the continuity of ι_{A} that $\xi_0, \xi_1, \dots$ is a Cauchy sequence in X and hence convergent to some $\xi \in X$. Since A is continuous, it follows the convergence of $A\xi_0, A\xi_1, \dots$ to $A\xi$ and therefore also the convergence of $\xi_0, \xi_1, \dots$ in $(X, \| \|_{A})$ to ξ. Hence $(X, \| \|_{A})$ is complete and A is closed by (ii). If A is closed, it follows by (ii) that $(X, \| \|_{A})$ is a Banach space and that the bijective X-valued linear operator ι_{A} is continuous. Hence ι_{A} is closed by the previous part of the proof. Therefore, the inverse of ι_{A} is continuous by (iv) and hence A is bounded.
'(vi)': Let A be closable. In a first step, we prove that $\bar{A} + B$ is closed. For this, let $(\xi, \eta) \in \overline{G(\bar{A} + B)}$. Then there is a sequence $\xi_0, \xi_1, \dots$ in $D(\bar{A})$ which is convergent to ξ and such that $(\bar{A}+B)\xi_0, (\bar{A}+B)\xi_1, \dots$ is convergent to η. Since B is continuous, it follows that $\bar{A}\xi_0, \bar{A}\xi_1, \dots$ is convergent to $\eta - B\xi$. Since $\bar{A}$ is closed, it follows that $\xi \in D(\bar{A})$ as well as $\bar{A}\xi = \eta - B\xi$ and hence that $\xi \in D(\bar{A} + B)$ as well as $(\bar{A} + B)\xi = \eta$. Hence $\bar{A} + B$ is closed, and therefore $A + B$ is closable such that $\bar{A} + B \supset \overline{A + B}$. Further, it follows by the previous part of the proof that $\overline{A + B} - B$ is a closed extension of A. Hence $\overline{A + B} - B \supset \bar{A}$ and therefore also $\overline{A + B} \supset \bar{A} + B$. Finally, it follows that $\overline{A + B} = \bar{A} + B$. □

Theorem 3.1.4. (Existence of a discontinuous linear functional on every infinite dimensional normed vector space) Let $(X, \| \|)$ be some infinite dimensional normed vector space over $\mathbb{K} \in \{\mathbb{R}, \mathbb{C}\}$. Then there is a discontinuous linear functional $\omega : X \to \mathbb{K}$.

Proof. For this, let B be a Hamel basis of X, i.e., a maximal linearly independent set, whose existence follows by an application of Zorn's lemma or the equivalent axiom of choice. Without restriction, it can be assumed that B contains only elements of norm 1. Since $(X, \| \|)$ is infinite dimensional, B contains infinitely many elements $e_0, e_1, e_2, \dots$. We define $\omega : B \to \mathbb{N}$ by $\omega(e_n) := n$ for all $n \in \mathbb{N}$ and $\omega(e) := 0$ for all other $e \in B$. Then there is a unique extension of ω to a linear functional on X. This functional is unbounded and hence discontinuous. □

Example 3.1.5. (Example for a non-closable linear operator) Let $a, b \in \mathbb{R}$ be such that $a < b$ and I the open interval between a and b. Define $\omega : C(\bar{I}, \mathbb{C}) \to \mathbb{C}$ by

$$\omega(f) := \lim_{x \to a} f(x)$$

for all $f \in C(\bar{I}, \mathbb{C})$. Obviously, ω is a densely-defined $\mathbb{C}$-valued linear operator in $L^2_{\mathbb{C}}(I)$. Further, let $g \in C(\bar{I}, \mathbb{C})$ be such that $g(a) \neq 0$. Then the sequence $g_0, g_1, \dots$ in $C(\bar{I}, \mathbb{C})$, defined by $g_\nu := g$ for all $\nu \in \mathbb{N}$, is converging in $L^2_{\mathbb{C}}(I)$ to g and

$$\lim_{\nu \to \infty} \omega(g_\nu) = g(a) \neq 0 .$$

Since $C_0(I, \mathbb{C})$ is dense in $L^2_{\mathbb{C}}(I)$, there is a sequence $h_0, h_1, \dots$ in $C_0(I, \mathbb{C})$ such that

$$\lim_{\nu \to \infty} \|h_\nu - g\|_2 = 0 .$$

For such a sequence

$$\lim_{\nu \to \infty} \omega(h_\nu) = 0 .$$

Hence $\overline{G(\omega)}$ contains the differing elements $(g, 0), (g, g(a))$ and therefore ω is not closable. Note that the non-closability of ω is caused by its discontinuity in g.

Theorem 3.1.6. (An application of the closed graph theorem) Let $(X, \| \|_{\mathrm{X}})$, $(Y, \| \|_{\mathrm{Y}})$, $(Z, \| \|_{\mathrm{Z}})$ be Banach spaces over $\mathbb{K} \in \{\mathbb{R}, \mathbb{C}\}$, A a closed bijective Y-valued linear operator in X and B a *closable* Z-valued linear operator in X such that $D(B) \supset D(A)$. Then there is $C \in [0, \infty)$ such that

$$\|B\xi\|_{\mathrm{Z}} \leqslant C \, \|A\xi\|_{\mathrm{Y}}$$

for all $\xi \in D(A)$ and hence in particular $B|_{D(A)} \in L((D(A), \| \|_{\mathrm{A}}), Z)$.

Proof. First, it follows by Theorem 3.1.3 (iv) that $A^{-1} \in L(Y, X)$. Further, $B \circ A^{-1}$ is a Z-valued linear operator on Y since A^{-1} maps into the domain of B. Let $(\eta, \zeta) \in \overline{G(\, B \circ A^{-1}\,)}$. Then there is a sequence $(\eta_0, B(A^{-1}\eta_0)), (\eta_1, B(A^{-1}\eta_1)), \ldots$ in $G(B \circ A^{-1})$ converging to (η, ζ). In particular,

$$\lim_{\nu \to \infty} \eta_\nu = \eta$$

and therefore also

$$\lim_{\nu \to \infty} A^{-1}\eta_\nu = A^{-1}\eta .$$

Since B is closable, it follows that $(A^{-1}\eta, \zeta) \in G(\bar{B})$ and hence because of $A^{-1}\eta \in D(A) \subset D(B)$ that $(A^{-1}\eta, \zeta) \in G(B)$. Therefore also $BA^{-1}\eta = \zeta$ and $(\eta, \zeta) \in G(B \circ A^{-1})$. Hence $B \circ A^{-1}$ is in addition closed and therefore by Theorem 3.1.3 (v) bounded. As a consequence, it follows

$$\|B\xi\|_{\mathrm{Z}} = \|B \circ A^{-1} A\xi\|_{\mathrm{Z}} \leqslant C \, \|A\xi\|_{\mathrm{Y}}$$

for every $\xi \in D(A)$ where $C \in [0, \infty)$ is some bound for $B \circ A^{-1}$. □

Theorem 3.1.7. (Definition and elementary properties of the adjoint) Let $(X, \langle | \rangle_{\mathrm{X}})$ and $(Y, \langle | \rangle_{\mathrm{Y}})$ be Hilbert spaces over $\mathbb{K} \in \{\mathbb{R}, \mathbb{C}\}$, A a *densely-defined* Y-valued linear operator in X and $U : X \times Y \to Y \times X$ the Hilbert space isomorphism defined by $U(\xi, \eta) := (-\,\eta, \xi)$ for all $(\xi, \eta) \in X \times Y$.

(i) Then the closed subspace

$$[U(G(A))]^{\perp}$$

of $Y \times X$ is the graph of an uniquely determined X-valued linear operator A^* in Y which is in particular closed and called the *adjoint* of A. *If in addition* $(X, \langle | \rangle_{\mathrm{X}}) = (Y, \langle | \rangle_{\mathrm{Y}})$, we call A *symmetric* if $A^* \supset A$ and *self-adjoint* if $A^* = A$.

(ii) If B is a Y-valued linear operator B in X such that $B \supset A$, then

$$B^* \subset A^* .$$

(iii) If A^* is densely-defined, then $A \subset A^{**} := (A^*)^*$ and hence A is in particular closable.
(iv) If A is closed, then A^* is densely-defined and $A^{**} = A$.
(v) If A is closable, then $\bar{A} = A^{**}$.

If in addition $(X, \langle\,|\,\rangle_X) = (Y, \langle\,|\,\rangle_Y)$:

(vi) (*Maximality of self-adjoint operators*) If A is self-adjoint and $B \supset A$ is symmetric, then $B = A$.
(vii) If A is symmetric, then $\bar{A}$ is symmetric, too. Therefore, we call a symmetric A *essentially self-adjoint* if $\bar{A}$ is self-adjoint.
(viii) (*Hellinger-Toeplitz*) If $D(A) = X$ and A is self-adjoint, then $A \in L(X,X)$.

Proof. '(i)': First, it follows that

$$[U(G(A))]^{\perp} = \{(\eta,\xi) \in Y \times X : \langle(\eta,\xi)|U(\xi',A\xi')\rangle_{Y\times X} = 0 \text{ for all } \xi' \in D(A)\}$$

and hence that

$$[U(G(A))]^{\perp} = \{(\eta,\xi) \in Y \times X : \langle\eta|A\xi'\rangle_Y = \langle\xi|\xi'\rangle_X \text{ for all } \xi' \in D(A)\} .$$

In particular, it follows for $(\eta,\xi_1), (\eta,\xi_2) \in [U(G(A))]^{\perp}$ that

$$\langle\xi_1 - \xi_2|\xi'\rangle_X = 0$$

for all $\xi' \in D(A)$ and hence that $\xi_1 = \xi_2$ since $D(A)$ is dense in X. As a consequence, by

$$A^* : \mathrm{pr}_1[U(G(A))]^{\perp} \to X ,$$

where $\mathrm{pr}_1 := (Y \times X \to Y, (\eta,\xi) \mapsto \eta)$, defined by

$$A^*\eta := \xi ,$$

for all $\eta \in \mathrm{pr}_1[U(G(A))]^{\perp}$, where $\xi \in X$ is the unique element such that $(\eta,\xi) \in [U(G(A))]^{\perp}$, there is defined a map such that

$$G(A^*) = [U(G(A))]^{\perp} .$$

Note that the domain of A^* is a subspace of Y. In particular, it follows for all $\eta,\eta' \in D(A^*)$ and $\lambda \in \mathbb{K}$

$$\begin{aligned}\langle\eta+\eta'|A\xi'\rangle_Y &= \langle\eta|A\xi'\rangle_Y + \langle\eta'|A\xi'\rangle_Y = \langle A^*\eta|\xi'\rangle_X + \langle A^*\eta'|\xi'\rangle_X \\ &= \langle A^*\eta + A^*\eta'|\xi'\rangle_X \\ \langle\lambda.\eta|A\xi'\rangle_Y &= \lambda^{(*)}\cdot\langle\eta|A\xi'\rangle_Y = \lambda^{(*)}\cdot\langle A^*\eta|\xi'\rangle_X = \langle\lambda.A^*\eta|\xi'\rangle_X\end{aligned}$$

for all $\xi' \in D(A)$ and hence also the linearity of A^*.
'(ii)': Obvious.
'(iii)': For this, let A^* be densely-defined. Then, it follows

$$(Y \times X \to X \times Y, (\eta, \xi) \mapsto (-\xi, \eta)) = -U^{-1}$$

and hence

$$\begin{aligned} G(A^{**}) &= \left[-U^{-1}(G(A^*))\right]^{\perp} = \left[U^{-1}(G(A^*))\right]^{\perp} = \left[U^{-1}[U(G(A))]^{\perp}\right]^{\perp} \\ &= \left[[U^{-1}U(G(A))]^{\perp}\right]^{\perp} = G(A)^{\perp\perp} = \overline{G(A)} \supset G(A)\,. \end{aligned} \tag{3.1.1}$$

'(iv)': For this, let A be closed. Then, it follows for $\eta \in [D(A^*)]^{\perp}$

$$\begin{aligned} (0, \eta) &\in \left[U^{-1}(G(A^*))\right]^{\perp} = \left[U^{-1}[U(G(A))]^{\perp}\right]^{\perp} = \left[[U^{-1}U(G(A))]^{\perp}\right]^{\perp} \\ &= G(A)^{\perp\perp} = \overline{G(A)} = G(A) \end{aligned}$$

and hence $\eta = 0$. Hence $D(A^*)$ is dense in X, and it follows by (3.1.1) that $G(A^{**}) = \overline{G(A)} = G(A)$.
'(v)': For this, let A be closable. Since $\bar{A}$ is densely defined and closed, it follows by (iv) that $\bar{A}^*$ is densely-defined. Because of $A \subset \bar{A}$, this implies that $A^* \supset \bar{A}^*$ and hence that A^* is densely-defined, too. Therefore, it follows by (iii) that $A \subset A^{**}$ and by (3.1.1) that $G(A^{**}) = \overline{G(A)} = G(\bar{A})$ and hence, finally, that $A^{**} = \bar{A}$. In the following, it is assumed that $(X, \langle|\rangle_X) = (Y, \langle|\rangle_Y)$.
'(vi)': For this, let A be self-adjoint and B a symmetric extension of A. Then, it follows by using $G(B) \supset G(A)$ that

$$G(B) \subset G(B^*) = [U(G(B))]^{\perp} \subset [U(G(A))]^{\perp} = G(A^*) = G(A)$$

and hence $B \subset A \subset B$ and therefore, finally, that $B = A$.
'(vii)': For this, let A be symmetric. Then $A^* \supset A$ and hence also $A^* \supset \bar{A}$.

$$\begin{aligned} G(\bar{A}^*) &= [U(G(\bar{A}))]^{\perp} = [U\,\overline{G(A)}\,]^{\perp} = [\,\overline{U(G(A))}\,]^{\perp} = [[U(G(A))]^{\perp\perp}]^{\perp} \\ &= \overline{G(A^*)} = G(A^*) \supset G(\bar{A})\,. \end{aligned}$$

'(viii)': For this, let A be self-adjoint and $D(A) = X$. Then, $A = A^*$ is in particular closed and hence by Theorem 3.1.3 (v) bounded. □

Theorem 3.1.8. Let $(X, \langle|\rangle_X)$ be a Hilbert space over $\mathbb{K} \in \{\mathbb{R}, \mathbb{C}\}$ and Y a dense subspace of X. Further, let $\langle|\rangle_Y : Y^2 \to \mathbb{K}$ be a scalar product for Y such that $(Y, \langle|\rangle_Y)$ is a Hilbert space over $\mathbb{K}$ and such that there is $\kappa > 0$ such that

$$\|\xi\|_Y^2 \geqslant \kappa\, \|\xi\|^2$$

for all $\xi \in Y$ where $\|\;\|_Y : Y \to \mathbb{R}$ is the norm induced on Y by $\langle|\rangle_Y$. Then, there is a uniquely determined densely-defined, linear and self-adjoint operator in X such that $D(A)$ is a dense subspace of $(Y, \langle|\rangle_Y)$ and

$$\langle\xi|A\xi\rangle = \langle\xi|\xi\rangle_Y$$

for all $\xi \in D(A)$. This A is given by

$$D(A) = \{\xi \in Y : \langle\xi|\cdot\rangle_Y \in L((Y, \| \|), \mathbb{K})\}$$

and for every $\xi \in D(A)$

$$A\xi = \hat{\xi}$$

where $\hat{\xi} \in X$ is the, by the denseness of Y in X, the linear extension theorem and Riesz' lemma, uniquely determined element such that

$$\langle\xi|\cdot\rangle_Y = \langle\hat{\xi}|\cdot\rangle\big|_Y . \tag{3.1.2}$$

In particular, A is semibounded from below with bound κ, i.e.,

$$\langle\xi|A\xi\rangle \geqslant \kappa\,\langle\xi|\xi\rangle$$

for all $\xi \in D(A)$.

Proof. For $\xi \in X$, it follows

$$|\langle\xi|\eta\rangle| \leqslant \|\xi\| \cdot \|\eta\| \leqslant \kappa^{-1/2}\,\|\xi\| \cdot \|\eta\|_Y$$

for every $\eta \in Y$ and hence by Riesz' lemma the existence of a uniquely determined $\hat{\xi} \in Y$ such that

$$\langle\xi|\eta\rangle = \langle\hat{\xi}|\eta\rangle_Y$$

for all $\eta \in Y$. Hence by $B\xi := \hat{\xi}$ for every $\xi \in X$ there is given map $B : X \to X$ such $\text{Ran}B \subset Y$. Further, B is obviously linear and because of

$$\langle B\xi|\eta\rangle = \langle\eta|B\xi\rangle^{(*)} = \langle B\eta|B\xi\rangle_Y^{(*)} = \langle B\xi|B\eta\rangle_Y = \langle\xi|B\eta\rangle$$

for all $\xi, \eta \in X$ also symmetric. Further, for $\xi \in \ker B$, it follows that vanishing of the restriction of $\langle\xi|\cdot\rangle$ to Y and hence by the denseness of Y in X and Riesz' lemma that $\xi = 0$. Hence B is injective. In addition, $\text{Ran}B$ is dense in $(Y, \| \|_Y)$ since for $\xi \in \text{Ran}B^{\perp_Y}$

$$0 = \langle\xi|B\eta\rangle_Y = \langle B\eta|\xi\rangle_Y^{(*)} = \langle\eta|\xi\rangle^{(*)} = \langle\xi|\eta\rangle$$

for every $\eta \in X$ and hence $\xi = 0$. As a consequence,

$$\overline{\text{Ran B}} = \text{Ran B}^{\perp_Y \perp_Y} = \{0\}^{\perp_Y} = Y$$

where the closure is performed in $(Y, \| \|_Y)$. Therefore, since the inclusion $\iota_{Y \hookrightarrow X}$ of $(Y, \| \|_Y)$ into X is continuous and Y is dense in X, it follows also that $\text{Ran}B$ is dense in X. In the following, we define A to be the inverse of the restriction of B in image to its range. Then A is a densely-defined, linear and symmetric operator in X with range X. In particular, it follows for $(\xi_1, \xi_2) \in G(A^*)$ and $\xi \in D(A)$

$$\begin{aligned}\langle\xi_1|A\xi\rangle &= \langle\xi_2|\xi\rangle = \langle B\xi_2|\xi\rangle_Y = \langle B\xi_2|BA\xi\rangle_Y = \langle BA\xi|B\xi_2\rangle_Y^{(*)} = \langle A\xi|B\xi_2\rangle^{(*)} \\ &= \langle B\xi_2|A\xi\rangle\end{aligned}$$

and hence that $\xi_1 - B\xi_2$ is orthogonal to $\text{Ran}A = X$. As a consequence, it follows that $\xi_1 = B\xi_2$ and therefore that $\xi_1 \in D(A)$ and $A\xi_1 = \xi_2$. Hence $A^* \subset A$. Finally, since the symmetry of A implies that $A^* \supset A$, it follows that A is self-adjoint. In particular, it follows for $\xi \in D(A)$

$$\langle\xi|A\xi\rangle = \langle A\xi|\xi\rangle^{(*)} = \langle BA\xi|\xi\rangle_Y^{(*)} = \langle\xi|\xi\rangle_Y^{(*)} = \langle\xi|\xi\rangle_Y \geqslant \kappa\,\|\xi\|^2$$

and hence that A is in particular semibounded from below with bound κ. Further, if $\xi \in D(A)$, then

$$\langle\xi|\eta\rangle_Y = \langle BA\xi|\eta\rangle_Y = \langle A\xi|\eta\rangle$$

for all $\eta \in Y$ and hence $\langle\xi|\cdot\rangle_Y \in L((Y, \|\ \|), \mathbb{K})$. On the other hand, if $\xi \in Y$ is such that $\langle\xi|\cdot\rangle_Y \in L((Y, \|\ \|), \mathbb{K})$ and $\hat{\xi} \in X$ such that (3.1.2) is true, it follows that $B\hat{\xi} = \xi$ and hence $\xi \in D(A)$ and $A\xi = \hat{\xi}$. Finally, if A' is a densely-defined, linear and self-adjoint operator in X such that $D(A')$ is a dense subspace of $(Y, \langle|\rangle_Y)$ and

$$\langle\xi|A'\xi\rangle = \langle\xi|\xi\rangle_Y$$

for all $\xi \in D(A')$, then it follows by polarization that

$$\langle\xi|\eta\rangle_Y = \langle\xi|A'\eta\rangle = \langle A'\xi|\eta\rangle$$

for all $\xi, \eta \in D(A')$. Hence it follows for $\xi \in D(A)$ by the denseness of $D(A)$ in $(Y, \|\ \|_Y)$ and the continuity of $\iota_{Y \hookrightarrow X}$ that

$$\langle\xi|\eta\rangle_Y = \langle A'\xi|\eta\rangle$$

for all $\eta \in Y$. Hence $\langle\xi|\cdot\rangle_Y \in L((Y, \|\ \|), \mathbb{K})$ and by the foregoing $\xi \in D(A)$ and $A\xi = A'\xi$. As a consequence, it follows $A \supset A'$ and since A, A' are both self-adjoint that $A' = A$. □

Theorem 3.1.9. Let $(X, \langle|\rangle_X)$ and $(Y, \langle|\rangle_Y)$ be Hilbert spaces over $\mathbb{K} \in \{\mathbb{R}, \mathbb{C}\}$, A a densely-defined Y-valued, linear and closed operator in X. Then A^*A (as usual maximally defined) is a densely-defined linear, self-adjoint and positive operator in X. In particular, $D(A^*A)$ is a core for A and $\ker A = \ker(A^*A)$.

Proof. Since A is closed, it follows that $(D(A), \langle\ \rangle_A)$, where $\langle\ \rangle_A : (D(A))^2 \to \mathbb{K}$ is defined by

$$\langle\xi|\eta\rangle_A := \langle\xi|\eta\rangle_X + \langle A\xi|A\eta\rangle_Y$$

for all $\xi, \eta \in D(A)$, is a Hilbert space. In particular, it follows

$$\|\xi\|_A = \sqrt{\|\xi\|_X^2 + \|A\xi\|_Y^2} \geqslant \|\xi\|_X$$

for all $\xi \in D(A)$ where $\|\ \|_A$ denotes the norm induced on $D(A)$ by $\langle|\rangle_A$. Hence by the previous theorem, $B : D(B) \to X$, defined by

$$D(B) = \{\xi \in D(A) : \langle A\xi|A\cdot\rangle_Y \in L((D(A), \|\ \|_X), \mathbb{K})\}$$

and for every $\xi \in D(B)$ by

$$B\xi = \hat{\xi} \,,$$

where $\hat{\xi} \in X$ is the uniquely determined element such that

$$\langle A\xi | A\cdot \rangle_{\mathrm{Y}} = \langle \hat{\xi} - \xi | \cdot \rangle_{\mathrm{X}} \big|_{D(A)} \,,$$

is densely-defined, linear, self-adjoint and semibounded from below with bound 1 and $D(B)$ is a dense subspace of $(D(A), \| \, \|_{\mathrm{A}})$. Hence it follows by the definition of A^*A that

$$A^*A = B - \mathrm{id}_{\mathrm{X}} \,.$$

As a consequence, A^*A is a densely-defined, linear, self-adjoint and positive operator in X. Further, since $D(B)$ is dense in $(D(A), \| \, \|_{\mathrm{A}})$, for $\xi \in D(A)$ there is a sequence $\xi_0, \xi_1, \dots$ in $D(B)$ such that

$$\lim_{\nu \to \infty} \| \xi_\nu - \xi \|_{\mathrm{A}} = 0 \,.$$

Hence it follows also that

$$\lim_{\nu \to \infty} \| \xi_\nu - \xi \|_{\mathrm{X}} = 0 \,, \ \lim_{\nu \to \infty} \| A\xi_\nu - A\xi \|_{\mathrm{Y}} = 0$$

and therefore that $D(A^*A)$ is a core for A. Further, for $\xi \in \ker(A^*A)$ it follows

$$\langle A\xi | A\xi \rangle_{\mathrm{Y}} = \langle A^*A\xi | \xi \rangle_{\mathrm{X}} = 0$$

and hence $\xi \in \ker A$. Since $\ker A \subset \ker(A^*A)$, it follows finally that $\ker A = \ker(A^*A)$. □

Theorem 3.1.10. Let $(X, \langle | \rangle_{\mathrm{X}})$ and $(Y, \langle | \rangle_{\mathrm{Y}})$ be complex Hilbert spaces, A a densely-defined Y-valued, linear and closed operator in X. Then $\ker A$ is a closed subspace of X and the orthogonal projection $P_0 \in L(X, X)$ onto $\ker A$ is given by

$$P_0 = \mathrm{s}-\lim_{\nu \to \infty} (1 + \nu A^*A)^{-1}$$

where 's$-$lim' denotes the strong limit. Note in particular that, because of

$$(1 + \nu A^*A)\xi = \xi$$

for all $\xi \in \ker A$, the elements of $\ker A$ are *fixed points* of $(1 + \nu A^*A)^{-1}$ for every $\nu \in \mathbb{N}$.

Proof. From the linearity of A, it follows that $\ker A$ is a subspace of X. Further, for $\xi \in \overline{\ker A}$ there is a sequence $\xi_0, \xi_1, \dots$ in $\ker A$ converging to ξ. The corresponding sequence $A\xi_0, A\xi_1, \dots$ is converging to 0. Hence it follows by the closedness of A that $\xi \in \ker A$ and therefore that $\ker A = \overline{\ker A}$ is a closed subspace of X. According to the previous theorem, $\ker A$ equals the kernel of the densely-defined, linear, self-adjoint and positive operator $B := A^*A$ in X. In the next step, we prove for $\nu \in \mathbb{N}$ by using the functional calculus for B that

$$(1+\nu B)^{-1} = \frac{1}{1+\nu\,\mathrm{id}_{\sigma(B)}}\,(B)$$

where $\sigma(B)$ denotes the spectrum of B. For this, we first note that $1+\nu B$ is a densely-defined, linear and self-adjoint operator in X which is semibounded from below with bound 1. Hence the spectrum of $1+\nu B$ is contained in $[1,\infty)$ and therefore $1+\nu B$ is in particular bijective. Further, $1/(1+\nu\,\mathrm{id}_{\sigma(B)})$ is a bounded real-valued function on $\sigma(B)$ which is measurable with respect to every additive, monotone and regular interval function on $\mathbb{R}$. By the functional calculus for B, the Cayley transform $U_B = (B-i)(B+i)^{-1} = 1-2i(B+i)^{-1}$ of B is given by

$$U_B = \frac{\mathrm{id}_{\sigma(B)}-i}{\mathrm{id}_{\sigma(B)}+i}\,(B) = 1-2i\,\frac{1}{\mathrm{id}_{\sigma(B)}+i}\,(B)\,.$$

Hence it follows that

$$(B+i)^{-1} = \frac{1}{\mathrm{id}_{\sigma(B)}+i}\,(B)\,.$$

Further,

$$\begin{aligned}1+\nu B &= 1-\nu i+\nu(B+i) = \left[(1-\nu i)(B+i)^{-1}+\nu\right](B+i)\\ &= \left[\frac{1-\nu i}{\mathrm{id}_{\sigma(B)}+i}+\nu\right](B)\,(B+i) = \frac{1+\nu\,\mathrm{id}_{\sigma(B)}}{\mathrm{id}_{\sigma(B)}+i}\,(B)\,(B+i)\end{aligned}$$

and hence

$$(1+\nu B)^{-1} = \frac{1}{\mathrm{id}_{\sigma(B)}+i}\,(B)\,\frac{\mathrm{id}_{\sigma(B)}+i}{1+\nu\,\mathrm{id}_{\sigma(B)}}\,(B) = \frac{1}{1+\nu\,\mathrm{id}_{\sigma(B)}}\,(B)\,.$$

In a further step, we prove that the orthogonal projection $P_0 \in L(X,X)$ onto $\ker B$ is given by

$$\left(\chi_{\{0\}}|_{\sigma(B)}\right)(B)$$

where $\chi_{\{0\}}$ denotes the characteristic function of $\{0\}$. First, it follows by the functional calculus for B that $\left(\chi_{\{0\}}|_{\sigma(B)}\right)(B)$ is an idempotent, self-adjoint, bounded and linear on X and hence an orthogonal projection. For $\xi \in X$, it follows that

$$\begin{aligned}(B+i)^{-1}\left(\chi_{\{0\}}|_{\sigma(B)}\right)(B)\xi &= \frac{1}{\mathrm{id}_{\sigma(B)}+i}\,(B)\left(\chi_{\{0\}}|_{\sigma(B)}\right)(B)\xi\\ &= \frac{1}{i}\left(\chi_{\{0\}}|_{\sigma(B)}\right)(B)\xi\end{aligned}$$

and hence that

$$\left(\chi_{\{0\}}|_{\sigma(B)}\right)(B)\xi \in \ker B\,.$$

In particular, it follows in the case that 0 is no eigenvalue of B that $\left(\chi_{\{0\}}|_{\sigma(B)}\right)(B)$ is the zero operator which projects onto $\ker B = \{0\}$. Further, if 0 is an eigenvalue of B, it follows for $\xi \in \ker B$ by the functional calculus for B that

$$\left(\chi_{\{0\}}|_{\sigma(B)}\right)(B)\xi = \left(\chi_{\{0\}}|_{\sigma(B)}\right)(0)\xi = \xi$$

and hence also in this case that $\operatorname{Ran}\left(\chi_{\{0\}}|_{\sigma(B)}\right)(B) = \ker B$. Finally,

$$\left(\frac{1}{1+\nu\,\mathrm{id}_{\sigma(B)}}\right)_{\nu\in\mathbb{N}}$$

is a sequence which is uniformly bounded by 1 and everywhere on $\sigma(B)$ pointwise convergent to $\chi_{\{0\}}|_{\sigma(B)}$. Hence it follows by the functional calculus for B that

$$\text{s}-\lim_{\nu\to\infty}\frac{1}{1+\nu\,\mathrm{id}_{\sigma(B)}}(B) = \left(\chi_{\{0\}}|_{\sigma(B)}\right)(B)\,.$$

□

Theorem 3.1.11. (Elementary properties of the resolvent) Let $(X, \|\ \|_X)$ be a non-trivial Banach space over $\mathbb{K} \in \{\mathbb{R}, \mathbb{C}\}$ and A a densely-defined closed linear operator in X.

(i) We define the *resolvent set* $\rho(A) \subset \mathbb{K}$ of A by

$$\rho(A) := \{\lambda \in \mathbb{K} : A - \lambda \text{ is bijective}\}\,.$$

Then $\rho(A)$ is an *open* subset of $\mathbb{K}$. Therefore, its complement $\sigma(A) := \mathbb{K}\backslash\rho(A)$, which is called the *spectrum* of A, is a *closed* subset of $\mathbb{K}$.

(ii) We define the *resolvent* $R_A : \rho(A) \to L(X,X)$ of A by

$$R_A(\lambda) := (A-\lambda)^{-1}$$

for every $\lambda \in \rho(A)$. Then R_A satisfies the *first resolvent formula*

$$R_A(\mu) - R_A(\lambda) = (\mu - \lambda)\, R_A(\mu) R_A(\lambda) \tag{3.1.3}$$

for every $\lambda, \mu \in \rho(A)$ and the *second resolvent formula*

$$R_A(\lambda) - R_B(\lambda) = R_A(\lambda)(B-A)R_B(\lambda) \tag{3.1.4}$$

for every $\lambda \in \rho(A) \cap \rho(B)$ where B is some closed linear operator in X having the *same domain* as A, i.e., $D(B) = D(A)$.

(iii) For every $\xi \in X$, $\omega \in L(X,\mathbb{K})$ is the corresponding function

$$\omega \circ R_A \xi$$

real-analytic/holomorphic. Here $R_A\xi : \rho(A) \to X$ is defined by $(R_A\xi)(\lambda) := R_A(\lambda)\xi$.

Proof. '(i), (iii)': For this, let $\lambda_0 \in \rho(A)$. Then $A - \lambda_0$ is a closed densely-defined bijective linear operator in X and hence $R_A(\lambda_0) \in L(X,X)\backslash\{0\}$. Then it follows for every $\lambda \in U_{1/\|R_A(\lambda_0)\|}(\lambda_0)$

$$A - \lambda = \left[\mathrm{id}_X - (\lambda - \lambda_0)\,.\,R_A(\lambda_0)\right](A - \lambda_0)$$

and therefore, since

$$\mathrm{id}_X - (\lambda - \lambda_0).R_A(\lambda_0)$$

is bijective as a consequence of

$$\|(\lambda - \lambda_0).R_A(\lambda_0)\| < 1 \; ,$$

see e.g. [128] Chapter IV, §2, Theorem 2, that $A - \lambda$ is bijective as a composition of bijective maps and

$$R_A(\lambda) = \sum_{k=0}^{\infty} (\lambda - \lambda_0)^k \, [R_A(\lambda_0)]^{k+1} \; . \tag{3.1.5}$$

Hence $\lambda \in \rho(A)$ and in particular for every $\xi \in X$, $\omega \in L(X, \mathbb{K})$

$$(\omega \circ R_A \xi)(\lambda) = \sum_{k=0}^{\infty} \omega \left([R_A(\lambda_0)]^{k+1} \xi \right) (\lambda - \lambda_0)^k \; . \tag{3.1.6}$$

'(ii)': For $\lambda, \mu \in \rho(A)$ and every $\xi \in D(A)$, it follows

$$(A - \mu)\xi = (A - \lambda)\xi + (\lambda - \mu)\xi$$

and hence for every $\eta \in X$

$$(A - \mu)R_A(\lambda)\eta = \eta + (\lambda - \mu)R_A(\lambda)\eta \; .$$

The last implies

$$R_A(\lambda) = R_A(\mu) + (\lambda - \mu)R_A(\mu)R_A(\lambda)$$

and hence (3.1.3). Finally, let $B : D(A) \to X$ be some closed linear operator in X. Then it follows for every $\mu \in \rho(A)$, $\lambda \in \rho(B)$ and every $\xi \in D(A)$

$$(A - \mu)\xi = (A - B)\xi + (B - \lambda)\xi + (\lambda - \mu)\xi$$

and hence for every $\eta \in X$

$$(A - \mu)R_B(\lambda)\eta = (A - B)R_B(\lambda)\eta + \eta + (\lambda - \mu)R_B(\lambda)\eta \; .$$

The last implies

$$R_B(\lambda) = R_A(\mu)(A - B)R_B(\lambda) + R_A(\mu) + (\lambda - \mu)R_A(\mu)R_B(\lambda)$$

and hence (3.1.4). □

3.2 Weak Integration of Banach Space-Valued Maps

The integration of Banach space-valued maps [49, 52, 225] is an essential tool in the study of semigroups of operators. Most authors use for this the Bochner integral.

Instead, the so called weak (or Pettis) integral is developed in the following up to the level needed for the remainder of the course. The use of the more general weak integral is mainly due to the validity of Theorem 3.2.2 below which seems to favour the approach via weak integration in the important special case of Hilbert space-valued maps. On the other hand, in the following only integrals of maps are needed which are a.e. defined and a.e. continuous on open subsets of $\mathbb{R}^n$ for some $n \in \mathbb{N}^*$. For this class of functions, it can easily be seen that the weak integral and the Bochner integral coincide if existent.

Definition 3.2.1. (Weak Integral/Pettis' integral) Let $n \in \mathbb{N}^*$ and $(X, \| \|)$ be a Banach space over $\mathbb{K} \in \{\mathbb{R}, \mathbb{C}\}$. We define for every X-valued map f which is a.e. defined on $\mathbb{R}^n$:

(i) f is *weakly measurable* if $\omega \circ f$ is measurable for all $\omega \in L(X, \mathbb{K})$,
(ii) f is *weakly summable* if $\omega \circ f$ is summable for every $\omega \in L(X, \mathbb{K})$ and if there is $\xi \in X$ such that

$$\omega(\xi) = \int_{\mathbb{R}^n} \omega \circ f \, dv^n$$

for every $\omega \in L(X, \mathbb{K})$. Such ξ, if existent, is unique since $L(X, \mathbb{K})$ separates points on X.[1] For this reason, we define in that case the *weak* (or *Pettis*) *integral* of f by

$$\int_{\mathbb{R}^n} f \, dv^n := \xi \, .$$

Theorem 3.2.2. (Existence of the weak integral for reflexive Banach spaces) Let $n \in \mathbb{N}^*$, $(X, \| \|)$ be a reflexive Banach space over $\mathbb{K} \in \{\mathbb{R}, \mathbb{C}\}$ and f a X-valued map which is a.e. defined on $\mathbb{R}^n$. Then f is weakly summable if and only if $\omega \circ f$ is summable for every $\omega \in L(X, \mathbb{K})$.

Proof. If f is weakly summable, by definition, $\omega \circ f$ is summable for every $\omega \in L(X, \mathbb{K})$. If, on the other hand, $\omega \circ f$ is summable for every $\omega \in L(X, \mathbb{K})$, we define $A : L(X, \mathbb{K}) \to L^1_{\mathbb{K}}(\mathbb{R}^n)$ by

$$A\omega := \omega \circ f$$

for every $\omega \in L(X, \mathbb{K})$. Obviously, A is linear. A is in addition closed. For this, let $\omega \in L(X, \mathbb{K})$, $\omega_1, \omega_2, \dots$ be a sequence in $L(X, \mathbb{K})$ such that $\omega_1 \circ f, \omega_2 \circ f, \dots$ is convergent to some $g \in L(X, \mathbb{K})$. Then a subsequence of $\omega_1 \circ f, \omega_2 \circ f, \dots$ is converging a.e. pointwise on $\mathbb{R}^n$ to g. Hence $\omega \circ f$ is a.e. equal to g on $\mathbb{R}^n$ and therefore $A\omega = g$. Hence $A \in L(L(X, \mathbb{K}), L^1_{\mathbb{K}}(\mathbb{R}^n))$ by the closed graph theorem, Theorem 3.1.3 (v). As a consequence, $I_A : L(X, \mathbb{K}) \to \mathbb{K}$, defined by

$$I_A(\omega) := \int_{\mathbb{R}^n} \omega \circ f \, dv^n \, ,$$

is an element of $L(L(X, \mathbb{K}), \mathbb{K})$. Since X is reflexive, it follows the existence of $\xi \in X$ such that $I_A(\omega) = \omega(\xi)$ for all $\omega \in L(X, \mathbb{K})$ and therefore, finally, the weak summability of f. □

[1] See, e.g., [186] Theorem 3.4.

Remark 3.2.3. For an example of an actual calculation of a weak integral, compare the proof of Lemma 10.2.1 (v).

Theorem 3.2.4. (Elementary properties of the weak integral) Let $n \in \mathbb{N}^*$, $\mathbb{K} \in \{\mathbb{R}, \mathbb{C}\}$, $(X, \| \|_X)$, $(Y, \| \|_Y)$ be Banach spaces over $\mathbb{K}$, f, g be X-valued maps which are a.e. defined on $\mathbb{R}^n$ and weakly summable, $\lambda \in \mathbb{K}$ and $T \in L(X, Y)$.

(i) If f is weakly integrable and g is a.e. equal to f, then g is weakly integrable and

$$\int_{\mathbb{R}^n} g \, dv^n = \int_{\mathbb{R}^n} f \, dv^n .$$

(ii) Then $f + g$, λf and $T \circ f$ are weakly integrable and

$$\int_{\mathbb{R}^n} f + g \, dv^n = \int_{\mathbb{R}^n} f \, dv^n + \int_{\mathbb{R}^n} g \, dv^n \ , \quad \int_{\mathbb{R}^n} \lambda f \, dv^n = \lambda \int_{\mathbb{R}^n} f \, dv^n \ ,$$

$$\int_{\mathbb{R}^n} T \circ f \, dv^n = T \int_{\mathbb{R}^n} f \, dv^n .$$

(iii) For every $f \in L^1_{\mathbb{K}}(\mathbb{R}^n)$ and every $\xi \in X$:

$$\int_{\mathbb{R}^n} f.\xi \, dv^n = \left(\int_{\mathbb{R}^n} f \, dv^n \right) .\xi$$

where $f.\xi$ is defined by $(f.\xi)(x) := f(x).\xi$ for all x in the domain of f.

Proof. '(i)': Obvious.
'(ii)': For every $\omega \in L(X, \mathbb{K})$, $\omega \circ (f + g) = \omega \circ f + \omega \circ g$, $\omega \circ (\lambda f) = \lambda \omega \circ f$ is summable and

$$\int_{\mathbb{R}^n} \omega \circ (f + g) \, dv^n = \omega \left(\int_{\mathbb{R}^n} f \, dv^n + \int_{\mathbb{R}^n} g \, dv^n \right) \ ,$$

$$\int_{\mathbb{R}^n} \omega \circ (\lambda f) \, dv^n = \omega \left(\lambda \int_{\mathbb{R}^n} f \, dv^n \right) .$$

Further, it follows for every $\omega \in L(Y, \mathbb{K})$ that $\omega \circ T \in L(X, \mathbb{K})$ and hence the summability of $\omega \circ (T \circ f) = (\omega \circ T) \circ f$ and

$$\int_{\mathbb{R}^n} \omega \circ (T \circ f) \, dv^n = (\omega \circ T) \left(\int_{\mathbb{R}^n} f \, dv^n \right) = \omega \left(T \int_{\mathbb{R}^n} f \, dv^n \right) .$$

'(iii)': For this, let $f \in L^1_{\mathbb{K}}(\mathbb{R}^n)$ and $\xi \in X$. Then it follows for every $\omega \in L(X, \mathbb{K})$ that $\omega \circ (f.\xi) = \omega(\xi) . f$ is summable and that

$$\int_{\mathbb{R}^n} \omega \circ (f.\xi) \, dv^n = \omega \left(\left(\int_{\mathbb{R}^n} f \, dv^n \right) .\xi \right) .$$

□

Theorem 3.2.5. (Existence of the weak integral) Let $\mathbb{K} \in \{\mathbb{R}, \mathbb{C}\}$, $(X, \| \|)$ be a $\mathbb{K}$-Banach space, $n \in \mathbb{N}^*$, Ω a non-empty open subset of $\mathbb{R}^n$ and $f : \Omega \to X$ almost everywhere continuous.

(i) There is a sequence $(s_\nu)_{\nu \in \mathbb{N}}$ of step functions such that $\text{supp}(s_\nu) \subset \Omega$, $\text{Ran}(s_\nu) \subset \text{Ran}(f) \cup \{0_X\}$ for all $\nu \in \mathbb{N}$ and for almost all $x \in \mathbb{R}^n$

$$\lim_{\nu \to \infty} s_\nu(x) = \hat{f}(x)$$

where $\hat{f} : \mathbb{R}^n \to X$ is defined by $\hat{f}(x) := f(x)$ for all $x \in \Omega$ and $\hat{f}(x) := 0_X$ for all $x \in \mathbb{R}^n \backslash \Omega$. (As a consequence, $\hat{f}$ is '*strongly measurable*'.)

(ii) $\hat{f}$ is *essentially separably-valued*, i.e., there is a zero set $M \subset \mathbb{R}^n$ along with an at most countable subset D of X such that $\hat{f}(\mathbb{R}^n \backslash M) \subset \overline{D}$.

(iii) If $\|\hat{f}(x)\| \leqslant h(x)$ for almost all $x \in \mathbb{R}^n$ and some a.e. on $\mathbb{R}^n$ defined summable function h, then $\hat{f}$ is weakly-summable, $\|\hat{f}\|$ is summable and

$$\left\| \int_{\mathbb{R}^n} \hat{f}\, dv^n \right\| \leqslant \int_{\mathbb{R}^n} \|\hat{f}\|\, dv^n \,. \tag{3.2.1}$$

Proof. '(i)': For this, we define for every $\nu \in \mathbb{N}^*$, $k \in \mathbb{Z}^n$ the interval I^ν_k of side length $1/\nu$ by

$$I^\nu_k := \left[\frac{k_1}{\nu}, \frac{k_1+1}{\nu}\right) \times \cdots \times \left[\frac{k_n}{\nu}, \frac{k_n+1}{\nu}\right) \,.$$

The family $\left(I^\nu_k\right)_{k \in \mathbb{Z}^n}$ gives a decomposition of $\mathbb{R}^n$ into pairwise disjoint bounded intervals of length $1/\nu$. We define for every $\nu \in \mathbb{N}^*$ a corresponding step function $s_\nu : \mathbb{R}^n \to X$ by

$$s_\nu(x) := f(x^\nu_k) \;,\; x \in I^\nu_k$$

for all $I^\nu_k \subset U_\nu(0) \cap \Omega$ where x^ν_k is some chosen element of I^ν_k. For all other $x \in \mathbb{R}^n$, we define $s_\nu(x) := 0_X$. Note that $\text{Ran}(s_\nu) \subset \text{Ran}(f) \cup \{0_X\}$. Then it follows for every point $x \in \Omega$ of continuity of f that $\lim_{\nu \to \infty} s_\nu(x) = f(x)$: Since f is continuous in x and Ω is open, for given $\varepsilon > 0$, there is $\delta > 0$ such that $U_\delta(x) \subset \Omega$ and at the same time such that $f(y) \in U_\varepsilon(f(x))$ for all $y \in U_\delta(x)$. Hence for $\nu > \max\{|x| + \delta, \sqrt{n}/\delta\}$ it follows that $x \in U_\nu(0) \cap \Omega$,

$$x \in I^\nu_{([\nu x_1],\dots,[\nu x_n])} \subset B_{\sqrt{n}/\nu}(x) \subset U_\delta(x) \subset U_\nu(0) \cap \Omega$$

where $[\,] : \mathbb{R} \to \mathbb{Z}$ is the floor function defined by $[y] := \max\{k \in \mathbb{Z} : k \leqslant y\}$, and hence $\|s_\nu(x) - f(x)\| = \|f(x^\nu_k) - f(x)\| < \varepsilon$ where $k := ([\nu x_1], \dots, [\nu x_n])$. Finally, for $x \notin \Omega$, it follows that $\lim_{\nu \to \infty} s_\nu(x) = 0_X$ because $s_\nu(x) = 0_X$ for all $\nu \in \mathbb{N}^*$.
'(ii)': Let M consist of those $x \in \mathbb{R}^n$ for which $(s_\nu(x))_{\nu \in \mathbb{N}^*}$ fails to converge to $\hat{f}(x)$. By (i) M is a zero set. In addition, let D be the union of the ranges of all s_ν, $\nu \in \mathbb{N}^*$. Then D is at most countable, and $f(\mathbb{R}^n \backslash M)$ is contained in the closure of D.
'(iii)': For this, let h be as described in (iii) and $(s_\nu)_{\nu \in \mathbb{N}^*}$ be as defined defined in the proof of (i). Then it follows that $\|\hat{f}\|$ is measurable since a.e. on $\mathbb{R}^n$ pointwise limit of a sequence of measurable functions and hence also summable since a.e. on $\mathbb{R}^n$

majorized by the summable function h. In the following, let $\varepsilon > 0$. Then we define for every $\nu \in \mathbb{N}^*$ the step function

$$t_\nu(x) := \begin{cases} s_\nu(x) & \text{if } \|s_\nu(x)\| \leqslant (1+\varepsilon)\,\|\hat{f}(x)\| \\ 0 & \text{if } \|s_\nu(x)\| > (1+\varepsilon)\,\|\hat{f}(x)\| \end{cases}$$

for every $x \in \mathbb{R}^n$. Then also

$$\lim_{\nu\to\infty} t_\nu(x) = \hat{f}(x)\,,$$

for almost all $x \in \mathbb{R}^n$. Further, $\|t_\nu - \hat{f}\|$ is Lebesgue summable for every $\nu \in \mathbb{N}^*$. To prove this, we notice that for any $\mu \in \mathbb{N}^*$ the corresponding function $\|t_\nu - t_\mu\|$ is a step function, and that $(\|t_\nu - t_\mu\|)_{\mu\in\mathbb{N}^*}$ converges almost everywhere on $\mathbb{R}^n$ pointwise to $\|t_\nu - \hat{f}\|$. Hence $\|t_\nu - \hat{f}\|$ is measurable. In addition, $(2+\varepsilon)h$ is a summable majorant for $\|t_\nu - \hat{f}\|$ and hence $\|t_\nu - \hat{f}\|$ is also summable. Further, $(\|t_\nu - \hat{f}\|)_{\nu\in\mathbb{N}^*}$ is almost everywhere on $\mathbb{R}^n$ convergent to 0 and is majorized by the summable function $(2+\varepsilon)h$. Hence it follows by Lebesgue's dominated convergence theorem that

$$\lim_{\nu\to\infty} \int_{\mathbb{R}^n} \|t_\nu - \hat{f}\|\, dv^n = 0\,. \tag{3.2.2}$$

In addition, it follows for $\mu, \nu \in \mathbb{N}^*$ that

$$\left\| \int_{\mathbb{R}^n} t_\mu\, dv^n - \int_{\mathbb{R}^n} t_\nu\, dv^n \right\| = \left\| \int_{\mathbb{R}^n} (t_\mu - t_\nu)\, dv^n \right\| \leqslant \int_{\mathbb{R}^n} \|t_\mu - t_\nu\|\, dv^n$$
$$\leqslant \int_{\mathbb{R}^n} \|t_\mu - \hat{f}\|\, dv^n + \int_{\mathbb{R}^n} \|t_\nu - \hat{f}\|\, dv^n$$

and hence by (3.2.2) and the completeness of X that

$$\lim_{\nu\to\infty} \int_{\mathbb{R}^n} t_\nu\, dv^n = \xi$$

for some $\xi \in X$. Note in particular that

$$\left\| \int_{\mathbb{R}^n} t_\nu\, dv^n \right\| \leqslant \int_{\mathbb{R}^n} \|t_\nu\|\, dv^n \leqslant (1+\varepsilon) \int_{\mathbb{R}^n} \|\hat{f}\|\, dv^n$$

and hence that

$$\|\xi\| \leqslant (1+\varepsilon) \int_{\mathbb{R}^n} \|\hat{f}\|\, dv^n\,. \tag{3.2.3}$$

Further, it follows by Lebesgue's dominated convergence theorem for every $\omega \in L(X, \mathbb{C})$

$$\int_{\mathbb{R}^n} \omega \circ \hat{f}\, dv^n = \lim_{\nu\to\infty} \int_{\mathbb{R}^n} \omega \circ t_\nu\, dv^n = \omega\left(\lim_{\nu\to\infty} \int_{\mathbb{R}^n} t_\nu\, dv^n \right) = \omega(\xi)\,.$$

Hence $\hat{f}$ is weakly-summable and

$$\int_{\mathbb{R}^n} \hat{f}\, dv^n = \xi\,.$$

Finally, (3.2.1) follows by (3.2.3). □

Remark 3.2.6. It is not difficult to see that a function f satisfying the assumptions of Theorem 3.2.5 and the additional assumption of Theorem 3.2.5 (iii) is Bochner integrable and that its Bochner integral and its weak integral coincide.

Corollary 3.2.7. (Fubini's theorem for a class of weakly integrable functions) Let $\mathbb{K} \in \{\mathbb{R}, \mathbb{C}\}$, $(X, \| \ \|)$ a $\mathbb{K}$-Banach space, $m, n \in \mathbb{N}^*$, Ω be a non-empty open subset of $\mathbb{R}^{m+n}$, $f : \Omega \to X$ be almost everywhere continuous and such that $\|\hat{f}\|$ is a.e. on $\mathbb{R}^{m+n}$ majorized by a summable function h where $\hat{f} : \mathbb{R}^{m+n} \to X$ is defined by $\hat{f}(x) := f(x)$ for all $x \in \Omega$ and $\hat{f}(x) := 0_X$ for all $x \in \mathbb{R}^{m+n} \backslash \Omega$. Then there is a zero set $N_1 \subset \mathbb{R}^m$ such that

(i) $\hat{f}(x, \cdot)$ is weakly summable for all $x \in \mathbb{R}^m \backslash N_1$.
(ii)

$$\left(\mathbb{R}^m \backslash N_1 \to X, x \mapsto \int_{\mathbb{R}^n} \hat{f}(x, \cdot) \, dv^n \right)$$

is weakly summable and

$$\int_{\mathbb{R}^{m+n}} \hat{f} \, dv^{m+n} = \int_{\mathbb{R}^m} \left(\mathbb{R}^m \backslash N_1 \to X, x \mapsto \int_{\mathbb{R}^n} \hat{f}(x, \cdot) \, dv^n \right) dv^m .$$

Proof. '(i)': First, we note that by integration theory for any zero set $N \subset \mathbb{R}^{m+n}$, there is a zero set $N_1 \subset \mathbb{R}^m$ such that

$$N_x := \{y \in \mathbb{R}^n : (x, y) \in N\}$$

is a zero set for all $x \in \mathbb{R}^m \backslash N_1$. Further, by Theorem 3.2.5 it follows the weak summability of $\hat{f}$ and the summability of $\|\hat{f}\|$. Also, according to the proof of Theorem 3.2.5 (iii), there is a sequence $(s_\nu)_{\nu \in \mathbb{N}}$ of step functions on $\mathbb{R}^{m+n}$ such that $\operatorname{supp}(s_\nu) \subset \Omega$, $\operatorname{Ran}(s_\nu) \subset \operatorname{Ran}(f) \cup \{0_X\}$ for all $\nu \in \mathbb{N}$,

$$\lim_{\nu \to \infty} s_\nu(x) = \hat{f}(x)$$

for almost all $x \in \mathbb{R}^{m+n}$,

$$\|s_\nu(x)\| \leqslant 2 \, \|\hat{f}(x)\|$$

for all $\nu \in \mathbb{N}$, $x \in \mathbb{R}^{m+n}$ and

$$\lim_{\nu \to \infty} \int_{\mathbb{R}^{m+n}} s_\nu \, dv^{m+n} = \int_{\mathbb{R}^{m+n}} \hat{f} \, dv^{m+n} .$$

Hence there is a zero set $N_1 \subset \mathbb{R}^m$ such that for all $x \in \mathbb{R}^m \backslash N_1$ the corresponding sequence of step functions $(s_\nu(x, \cdot))_{\nu \in \mathbb{N}}$ satisfies

$$\lim_{\nu \to \infty} s_\nu(x, \cdot) = \hat{f}(x, \cdot)$$

almost everywhere on $\mathbb{R}^n$ and at the same time such that $\|\hat{f}(x, \cdot)\|$ is summable. In particular, it follows for such x that $\|s_\nu(x, \cdot) - \hat{f}(x, \cdot)\|$ is Lebesgue summable for every $\nu \in \mathbb{N}^*$. To prove this, we notice that for any $\mu \in \mathbb{N}^*$ the corresponding function $\|s_\nu(x,\cdot) - s_\mu(x,\cdot)\|$ is a step function and that $(\|s_\nu(x,\cdot) - s_\mu(x,\cdot)\|)_{\mu\in\mathbb{N}^*}$ converges almost everywhere on $\mathbb{R}^n$ pointwise to $\|s_\nu(x,\cdot) - \hat{f}(x,\cdot)\|$. Hence $\|s_\nu(x,\cdot) - \hat{f}(x,\cdot)\|$ is measurable. In addition, $2\|\hat{f}(x,\cdot)\|$ is a summable majorant for $\|s_\nu(x,\cdot) - \hat{f}(x,\cdot)\|$ and hence $\|s_\nu(x,\cdot) - \hat{f}(x,\cdot)\|$ is also summable. Further, $(\|s_\nu(x,\cdot) - \hat{f}(x,\cdot)\|)_{\nu\in\mathbb{N}^*}$ is almost everywhere on $\mathbb{R}^n$ convergent to 0 and is majorized by the summable function $2\|\hat{f}(x,\cdot)\|$. Hence it follows by Lebesgue's dominated convergence theorem that

$$\lim_{\nu\to\infty} \int_{\mathbb{R}^n} \|s_\nu(x,\cdot) - \hat{f}(x,\cdot)\| \, dv^n = 0\,.$$

Further,

$$\begin{aligned}\left\| \int_{\mathbb{R}^n} s_\mu(x,\cdot)\, dv^n - \int_{\mathbb{R}^n} s_\nu(x,\cdot)\, dv^n \right\| &\leqslant \int_{\mathbb{R}^n} \|s_\mu(x,\cdot) - s_\nu(x,\cdot)\| \, dv^n \\ &\leqslant \int_{\mathbb{R}^n} \|s_\mu(x,\cdot) - \hat{f}(x,\cdot)\| \, dv^n + \int_{\mathbb{R}^n} \|s_\nu(x,\cdot) - \hat{f}(x,\cdot)\| \, dv^n\end{aligned}$$

for all $\mu, \nu \in \mathbb{N}$, and hence it follows by the completeness of $(X, \| \ \|)$ the existence of $\xi_x \in X$ such that

$$\lim_{\nu\to\infty} \int_{\mathbb{R}^n} s_\nu(x,\cdot)\, dv^n = \xi_x\,.$$

In particular

$$\|\xi_x\| \leqslant \int_{\mathbb{R}^n} \|\hat{f}(x,\cdot)\| \, dv^n$$

since

$$\left\| \int_{\mathbb{R}^n} s_\nu(x,\cdot)\, dv^n \right\| \leqslant \int_{\mathbb{R}^n} \|s_\nu(x,\cdot)\| \, dv^n \leqslant \int_{\mathbb{R}^n} \|\hat{f}(x,\cdot)\| \, dv^n$$

for every $\nu \in \mathbb{N}$. Since

$$\omega(\xi_x) = \lim_{\nu\to\infty} \int_{\mathbb{R}^n} \omega \circ s_\nu(x,\cdot)\, dv^n = \int_{\mathbb{R}^n} \omega \circ \hat{f}(x,\cdot)\, dv^n$$

for all $\omega \in L(X, \mathbb{K})$, it follows the weak integrability of $\hat{f}(x,\cdot)$ and

$$\lim_{\nu\to\infty} \int_{\mathbb{R}^n} s_\nu(x,\cdot)\, dv^n = \int_{\mathbb{R}^n} \hat{f}(x,\cdot)\, dv^n\,.$$

'(ii)': Further, we define for every $\nu \in \mathbb{N}$ the corresponding step function t_ν on $\mathbb{R}^m$ by

$$t_\nu(x) := \int_{\mathbb{R}^n} s_\nu(x,\cdot)\, dv^n$$

for all $x \in \mathbb{R}^m$ and $F : \mathbb{R}^m \backslash N_1 \to X$ by

$$F(x) := \int_{\mathbb{R}^n} \hat{f}(x,\cdot)\, dv^n$$

for all $x \in \mathbb{R}^m \backslash N_1$. Then

$$\lim_{\nu \to \infty} t_\nu(x) = F(x)$$

and

$$\|t_\nu(x)\| \leqslant \int_{\mathbb{R}^n} \|\hat{f}(x,\cdot)\| \, dv^n \ , \ \|F(x)\| \leqslant \int_{\mathbb{R}^n} \|\hat{f}(x,\cdot)\| \, dv^n$$

for all $x \in \mathbb{R}^m \backslash N_1$. Note that

$$\left(\mathbb{R}^m \backslash N_1 \to \mathbb{R}, x \mapsto \int_{\mathbb{R}^n} \|\hat{f}(x,\cdot)\| \, dv^n \right)$$

is summable by Fubini's theorem. Also, it follows by Fubini's theorem that

$$\int_{\mathbb{R}^m} t_\nu \, dv^m = \int_{\mathbb{R}^{m+n}} s_\nu \, dv^{m+n}$$

for every $\nu \in \mathbb{N}$ and hence that

$$\lim_{\nu \to \infty} \int_{\mathbb{R}^m} t_\nu \, dv^m = \int_{\mathbb{R}^{m+n}} \hat{f} \, dv^{m+n} \, .$$

In particular, it follows that $\|t_\nu - F\|$ is Lebesgue summable for every $\nu \in \mathbb{N}^*$. To prove this, we notice that for any $\mu \in \mathbb{N}^*$ the corresponding function $\|t_\nu - t_\mu\|$ is a step function and that $(\|t_\nu - t_\mu\|)_{\mu \in \mathbb{N}^*}$ converges almost everywhere on $\mathbb{R}^n$ pointwise to $\|t_\nu - F\|$. Hence $\|t_\nu - F\|$ is measurable. In addition,

$$\left(\mathbb{R}^m \backslash N_1 \to \mathbb{R}, x \mapsto \int_{\mathbb{R}^n} 2 \, \|\hat{f}(x,\cdot)\| \, dv^n \right) \tag{3.2.4}$$

is a summable majorant for $\|t_\nu - F\|$ and hence $\|t_\nu - F\|$ is also summable. Further, $(\|t_\nu - F\|)_{\nu \in \mathbb{N}^*}$ is almost everywhere on $\mathbb{R}^m$ convergent to 0 and is majorized by the summable function (3.2.4). Hence it follows by Lebesgue's dominated convergence theorem that

$$\lim_{\nu \to \infty} \int_{\mathbb{R}^m} \|t_\nu - F\| \, dv^m = 0 \, .$$

As a consequence,

$$\omega \left(\int_{\mathbb{R}^{m+n}} \hat{f} \, dv^{m+n} \right) = \lim_{\nu \to \infty} \int_{\mathbb{R}^m} \omega \circ t_\nu \, dv^m = \int_{\mathbb{R}^m} \omega \circ F \, dv^m$$

for all $\omega \in L(X, \mathbb{K})$. Finally, this implies the weak integrability of F and that

$$\int_{\mathbb{R}^m} F \, dv^m = \int_{\mathbb{R}^{m+n}} \hat{f} \, dv^{m+n} \, .$$

□

Theorem 3.2.8. Let $\mathbb{K} \in \{\mathbb{R}, \mathbb{C}\}$, $(X, \| \|)$ a $\mathbb{K}$-Banach space and $f : [a, b] \to X$ be bounded and almost everywhere continuous. Then $F : [a, b] \to X$ defined by

$$F(x) := \int_a^x f(t)\, dt$$

for every $x \in [a, b]$ is continuous. Furthermore, if f is continuous in $x \in (a, b)$, then F is differentiable in x and

$$F'(x) = f(x) .$$

Proof. Obviously, by Theorem 3.2.5, it follows the weak integrability of $\chi_{[a,x]} . \hat{f}$ for all $x \in [a, b]$. Further, it follows for $x, y \in [a, b]$ that

$$\|F(y) - F(x)\| = \left\| \int_x^y f(t)\, dt \right\| \leqslant \int_x^y \|f(t)\|\, dt \leqslant M \cdot |y - x|$$

if $y \geqslant x$ as well as

$$\|F(y) - F(x)\| = \left\| \int_y^x f(t)\, dt \right\| \leqslant \int_y^x \|f(t)\|\, dt \leqslant M \cdot |y - x|$$

if $y < x$, where $M \geqslant 0$ is such that $\|f(t)\| \leqslant M$ for all $t \in [a, b]$, and hence the continuity of F. Further, let f be continuous in $x \in (a, b)$. Hence for given $\varepsilon > 0$, there is $\delta > 0$ such that

$$\|f(t) - f(x)\| < \varepsilon$$

for all $t \in [a, b]$ satisfying $|t - x| < \delta$. Now let $h \in \mathbb{R}^*$ be such that $|h| < \delta$ and small enough such that $x + h \in (a, b)$. We consider the cases $h > 0$ and $h < 0$. In the first case, it follows that

$$\left\| \frac{1}{h} . (F(x+h) - F(x)) - f(x) \right\| = \left\| \frac{1}{h} \left[\int_a^{x+h} f(t)\, dt - \int_a^x f(t)\, dt \right] - f(x) \right\|$$
$$= \left\| \frac{1}{h} \int_x^{x+h} [f(t) - f(x)]\, dt \right\| \leqslant \frac{1}{h} \int_x^{x+h} \|f(t) - f(x)\|\, dt \leqslant \varepsilon .$$

Analogously, in the second case,

$$\left\| \frac{1}{h} . (F(x+h) - F(x)) - f(x) \right\| = \left\| \frac{1}{h} \left[\int_a^{x+h} f(t)\, dt - \int_a^x f(t)\, dt \right] - f(x) \right\|$$
$$= \left\| -\frac{1}{h} \int_{x+h}^x [f(t) - f(x)]\, dt \right\| \leqslant \frac{1}{|h|} \int_{x+h}^x \|f(t) - f(x)\|\, dt \leqslant \varepsilon .$$

Hence it follows

$$\lim_{h \to 0, h \neq 0} \frac{1}{h} . (F(x+h) - F(x)) = f(x) .$$

□

Theorem 3.2.9. Let $\mathbb{K} \in \{\mathbb{R}, \mathbb{C}\}$, $(X, \| \, \|)$ be a $\mathbb{K}$-Banach space and $f : [a,b] \to X$ continuous where a and b are some elements of $\mathbb{R}$ such that $a < b$. Further, let $F : [a,b] \to X$ be continuous and differentiable on (a,b) such that $F'(x) = f(x)$ for all $x \in (a,b)$. Then

$$\int_a^b f(x)\,dx = F(b) - F(a)\ . \tag{3.2.5}$$

Proof. For this, let $\omega \in L(X, \mathbb{K})$. Then $\omega \circ f$, $\omega \circ F$ are continuous, and $\omega \circ F$ is differentiable on (a,b) with derivative $\omega \circ f|_{(a,b)}$. Hence it follows by the fundamental theorem of calculus that

$$\omega\left(\int_a^b f(x)\,dx\right) = \int_a^b (\omega \circ f)(x)\,dx = (\omega \circ F)(b) - (\omega \circ F)(a) = \omega(F(b) - F(a))$$

and hence (3.2.5) since $L(X, \mathbb{K})$ separates points on X. □

Theorem 3.2.10. (Substitution rule for weak integrals) Let $\mathbb{K} \in \{\mathbb{R}, \mathbb{C}\}$, $(X, \| \, \|)$ a $\mathbb{K}$-Banach space, $n \in \mathbb{N}^*$, Ω_1, Ω_2 non-empty open subsets of $\mathbb{R}^n$, $f : \Omega_2 \to X$ almost everywhere continuous and such that $\|f\|$ is summable. Finally, let $h : \Omega_1 \to \Omega_2$ be continuously differentiable such that $h'(x) \neq 0$ for all $x \in \Omega_1$ and bijective. Then $|\det(h')|.(f \circ h)$ is weakly summable and

$$\int_{\Omega_2} f\,dv^n = \int_{\Omega_1} |\det(h')|.(f \circ h)\,dv^n\ . \tag{3.2.6}$$

Proof. First, it follows by the inverse mapping theorem that $h^{-1} : \Omega_2 \to \Omega_1$ is continuously differentiable. Hence it follows by the substitution rule for Lebesgue integrals that $h^{-1}(N_f) \subset \Omega_1$ is a zero set where $N_f \subset \Omega_2$ denotes the set of discontinuities of f. In particular, $|\det(h')|.(f \circ h)$ is a.e. continuous and

$$\| |\det(h')|.(f \circ h) \| \leqslant |\det(h')| \cdot (\|f\| \circ h)\ .$$

Since $|\det(h')| \cdot (\|f\| \circ h)$ is summable, it follows that $|\det(h')|.(f \circ h)$ is weakly summable. Further, it follows by the substitution rule for Lebesgue integrals that

$$\begin{aligned}\omega\left(\int_{\Omega_2} f\,dv^n\right) &= \int_{\Omega_2} \omega \circ f\,dv^n = \int_{\Omega_1} |\det(h')|.[(\omega \circ f) \circ h]\,dv^n \\ &= \omega\left(\int_{\Omega_1} |\det(h')|.(f \circ h)\,dv^n\right)\end{aligned}$$

for every $\omega \in L(X, \mathbb{K})$ and hence (3.2.6). □

Theorem 3.2.11. (Integration of strongly continuous maps) Let $\mathbb{K} \in \{\mathbb{R}, \mathbb{C}\}$, $(X, \| \, \|_X)$, $(Y, \| \, \|_Y)$ be $\mathbb{K}$-Banach spaces, $n \in \mathbb{N}^*$ and Ω a non-empty open subset of $\mathbb{R}^n$. Further, let $f : \Omega \to L(X,Y)$ be such that for every $\xi \in X$ the corresponding map $f\xi := (\Omega \to Y, x \mapsto f(x)\xi)$ is almost everywhere continuous and for which there is some a.e. on $\mathbb{R}^n$ defined summable function h such that

$$\|\hat{f}(x)\| \leqslant h(x)$$

for almost all $x \in \mathbb{R}^n$. Then by

$$\int_{\mathbb{R}^n} \hat{f}\, dv^n := \left(X \to Y, \xi \mapsto \int_{\mathbb{R}^n} \hat{f}\xi\, dv^n \right), \tag{3.2.7}$$

there is defined a bounded linear operator on X satisfying

$$\left\| \int_{\mathbb{R}^n} \hat{f}\, dv^n \right\| \leqslant \|h\|_1 . \tag{3.2.8}$$

Proof. For this, let $\xi \in X$. Then $\hat{f}\xi$ is almost everywhere continuous and

$$\|\hat{f}\xi\|_{\mathrm{Y}} \leqslant \|\xi\|_{\mathrm{X}} . h .$$

Hence it follows by Theorem 3.2.5 that $\hat{f}\xi$ is weakly integrable, that $\|\hat{f}\xi\|_{\mathrm{Y}}$ is integrable as well as

$$\left\| \int_{\mathbb{R}^n} \hat{f}\xi\, dv^n \right\|_{\mathrm{Y}} \leqslant \int_{\mathbb{R}^n} \|\hat{f}\xi\|_{\mathrm{Y}}\, dv^n \leqslant \|h\|_1 \cdot \|\xi\|_{\mathrm{X}} . \tag{3.2.9}$$

Hence it follows that by (3.2.7) it is defined a map from X to Y which is linear by the linearity of the weak integral. Finally, the boundedness of that operator and (3.2.8) follows from (3.2.9). □

3.3 Exponentials of Bounded Linear Operators

This section defines the exponential function exp on $L(X, X)$ where X is a Banach space. The Theorems 3.3.1 and 4.1.1 at the beginning of the next section give a complete characterization of all semigroups which are continuous in the topology induced on $L(X, X)$ by the operator norm. For every such semigroup $T : [0, \infty) \to L(X, X)$, there is a uniquely determined $A \in L(X, X)$ such that $T(t) = \exp(tA)$ for every $t \in [0, \infty)$. Hence there is a unique extension of T to a homomorphism of $(\mathbb{R}, +)$ into $(L(X, X), \circ)$ given by $(\mathbb{R} \to L(X, X), t \mapsto \exp(tA))$. As a consequence of Theorem 3.3.1 (i), for every $\xi \in X$ the corresponding $u := (\mathbb{R} \to X, t \mapsto \exp(tA)\xi)$ satisfies $u(0) = \xi$ and

$$u'(t) = -Au(t) \tag{3.3.1}$$

for every $t \in \mathbb{R}$. Here $'$ denotes the ordinary derivative of functions with values in X. Applications of (3.3.1) with $A \in L(X, X)$ are usually restricted to finite dimensional X, i.e., to systems of ordinary differential equations of the first order. An exception to this is given in Chapter 5.2. Equations of the type (3.3.1) in infinite dimensions usually involve partial differential operators. In general, such operators induce unbounded linear operators in Banach spaces.

Theorem 3.3.1. (Definition and properties of the exponential function) Let $\mathbb{K} \in \{\mathbb{R}, \mathbb{C}\}$ and $(X, \| \|)$ a $\mathbb{K}$-Banach space. Then we define the exponential function $\exp : L(X,X) \to L(X,X)$ by

$$\exp(A) := \sum_{k=0}^{\infty} \frac{1}{k!} . A^k$$

where $A^0 := \mathrm{id}_X$ and $A^{k+1} := A \circ A^k$ for all $k \in \mathbb{N}$. Note that this series is absolutely convergent since $\|A^k\| \leqslant \|A\|^k$ for all $k \in \mathbb{N}$.

(i) The map $u_A : \mathbb{K} \to L(X,X)$, defined by

$$u_A(t) := \exp(t.A)$$

for every $t \in \mathbb{K}$, is differentiable with derivative

$$u_A'(t) = A \circ u_A(t)$$

for all $t \in \mathbb{K}$.

(ii) For all $A, B \in L(X,X)$ satisfying $A \circ B = B \circ A$

$$\exp(A + B) = \exp(A) \circ \exp(B) \ . \tag{3.3.2}$$

(iii) For all $A \in L(X,X)$ satisfying $\|A\| \leqslant 1$, $n \in \mathbb{N}$ and $\xi \in X$,

$$\left\| \exp\left(n.(A - \mathrm{id}_X)\right)\xi - A^n \xi \right\| \leqslant \sqrt{n} \cdot \|(A - \mathrm{id}_X)\xi\| \ . \tag{3.3.3}$$

Proof. '(i)': For this, let $A \in L(X,X)$. Then it follows for $t \in \mathbb{K}$, $h \in \mathbb{K}^*$, by using the bilinearity and continuity of the composition map on $((L(X,X))^2$, that

$$\begin{aligned}
&\left\| \frac{1}{h} . \left[\exp((t+h).A) - \exp(t.A) \right] - A \circ \exp(t.A) \right\| \\
&= \left\| \sum_{k=2}^{\infty} \frac{1}{k!} \left[\frac{(t+h)^k - t^k}{h} - k t^{k-1} \right] . A^k \right\| \\
&= \lim_{n \to \infty} \left\| \sum_{k=2}^{n} \frac{1}{k!} \left[\frac{(t+h)^k - t^k}{h} - k t^{k-1} \right] . A^k \right\| \ .
\end{aligned} \tag{3.3.4}$$

Further, for any $n \in \mathbb{N}, n \geqslant 2$:

$$\left\| \sum_{k=2}^{n} \frac{1}{k!} \left[\frac{(t+h)^k - t^k}{h} - k t^{k-1} \right] . A^k \right\| \leqslant \sum_{k=2}^{n} \frac{1}{k!} \left| \frac{(t+h)^k - t^k}{h} - k t^{k-1} \right| \|A\|^k \ , \tag{3.3.5}$$

and for any $k \in \mathbb{N}, k \geqslant 2$:

$$\begin{aligned}
\left| \frac{(t+h)^k - t^k}{h} - k t^{k-1} \right| &= \left| \frac{t+h-t}{h} \cdot \left[\sum_{l=0}^{k-1} (t+h)^l \cdot t^{k-(l+1)} \right] - k t^{k-1} \right| \\
&= \left| \sum_{l=1}^{k-1} \left[(t+h)^l \cdot t^{k-(l+1)} - t^{k-1} \right] \right| = \left| \sum_{l=1}^{k-1} t^{k-(l+1)} \left[(t+h)^l - t^l \right] \right|
\end{aligned}$$

$$= \left| \sum_{l=1}^{k-1} \sum_{m=0}^{l-1} (t+h)^m \cdot t^{k-(m+2)} \right| \cdot |h| \leqslant |h| \cdot \sum_{l=1}^{k-1} \sum_{m=0}^{l-1} (\,|t| + |h|\,)^{k-2}$$

$$= \frac{|h|}{2} \cdot k(k-1) \cdot (\,|t| + |h|\,)^{k-2} \, .$$

Inserting the last into (3.3.5) gives

$$\left\| \sum_{k=2}^{n} \frac{1}{k!} \left[\frac{(t+h)^k - t^k}{h} - kt^{k-1} \right] . A^k \right\| \leqslant \frac{|h|}{2} \sum_{k=2}^{n} \frac{1}{(k-2)!} \cdot (\,|t| + |h|\,)^{k-2} \, \|A\|^k$$

$$\leqslant \frac{|h| \cdot \|A\|^2}{2} \exp\left((|t| + |h|) \cdot \|A\| \right) .$$

Finally, inserting the last into (3.3.4) gives

$$\left\| \frac{1}{h} \cdot [\exp((t+h).A) - \exp(t.A)] - A \circ \exp(t.A) \right\|$$

$$\leqslant \frac{|h| \cdot \|A\|^2}{2} \exp\left((|t| + |h|) \cdot \|A\| \right)$$

and hence

$$\lim_{h \to 0, h \neq 0} \left\| \frac{1}{h} . [\exp((t+h).A) - \exp(t.A)] - A \circ \exp(t.A) \right\| = 0 \, .$$

'(ii)': For this, let $A, B \in L(X, X)$ be such that $A \circ B = B \circ A$ and $t \in \mathbb{K}, h \in \mathbb{K}^*$. Then

$$\Big\| \frac{1}{h} . (u_A(t+h) \circ u_B(t+h)$$

$$-u_A(t) \circ u_B(t)) - \left(u_A'(t) \circ u_B(t) + u_A(t) \circ u_B'(t) \right) \Big\|$$

$$= \Big\| \left[\frac{1}{h} . (u_A(t+h) - u_A(t)) - u_A'(t) \right] \circ u_B(t)$$

$$+ u_A(t) \circ \left[\frac{1}{h} . (u_B(t+h) - u_B(t)) - u_B'(t) \right]$$

$$+ \frac{1}{h} . (u_A(t+h) - u_A(t)) \circ (u_B(t+h) - u_B(t)) \Big\|$$

$$\leqslant \left\| \left[\frac{1}{h} . (u_A(t+h) - u_A(t)) - u_A'(t) \right] \right\| \cdot \|u_B(t)\|$$

$$+ \|u_A(t)\| \cdot \left\| \left[\frac{1}{h} . (u_B(t+h) - u_B(t)) - u_B'(t) \right] \right\|$$

$$+ \left\| \frac{1}{h} . (u_A(t+h) - u_A(t)) \right\| \cdot \left\| (u_B(t+h) - u_B(t)) \right\| .$$

Hence it follows by (i) the differentiability of $g_{A,B} : \mathbb{K} \to L(X,X)$ defined by $h_{A,B}(t) := u_{A+B}(t) - u_A(t) \circ u_B(t)$ for every $t \in \mathbb{K}$ and

$$\begin{aligned} h'_{A,B}(t) &= (A+B) \circ u_{A+B}(t) - A \circ u_A(t) \circ u_B(t) - u_A(t) \circ B \circ u_B(t) \\ &= (A+B) \circ u_{A+B}(t) - A \circ u_A(t) \circ u_B(t) - B \circ u_A(t) \circ u_B(t) \\ &= (A+B) \circ h_{A,B}(t) \end{aligned}$$

for all $t \in \mathbb{K}$ where the bilinearity and continuity of the composition map on $(L(X,X))^2$ has been used as well as that $A \circ B = B \circ A$ by assumption. Hence it follows by $h_{A,B}(0) = u_{A+B}(0) - u_A(0) \circ u_B(0) = 0$ along with Theorem 3.2.9, Theorem 3.2.5 that

$$\|h_{A,B}(t)\| \leqslant \|A+B\| \cdot \int_0^t \|h_{A,B}(s)\| \, ds$$

for all $t \in [0,\infty)$. As a consequence, it follows for $\varepsilon > 0$ that

$$\|h_{A,B}(t)\| < \varepsilon \, e^{t\|A+B\|} \tag{3.3.6}$$

for all $t \in [0,\infty)$. Because otherwise there is $t_0 \in (0,\infty)$ such that

$$\|h_{A,B}(t_0)\| \geqslant \varepsilon \, e^{t_0\|A+B\|}$$

and such that (3.3.6) is valid for all $t \in [0,t_0)$. Then

$$\begin{aligned} \|h_{A,B}(t_0)\| &\leqslant \|A+B\| \cdot \int_0^{t_0} \|h_{A,B}(s)\| \, ds \leqslant \|A+B\| \cdot \int_0^{t_0} \varepsilon \, e^{t\|A+B\|} \, ds \\ &= \varepsilon \cdot \left(e^{t_0\|A+B\|} - 1\right) < \varepsilon \cdot e^{t_0\|A+B\|} \,. \lightning \end{aligned}$$

From (3.3.6) it follows that $h_{A,B}(t) = 0$ for all $t \geqslant 0$ and hence (3.3.2).
'(iii)': For this, let $A \in L(X,X)$ be such that $\|A\| \leqslant 1$, $n \in \mathbb{N}$ and $\xi \in X$. Then

$$\begin{aligned} \left\| \exp\left(n.(A - \mathrm{id}_X)\right)\xi - A^n \xi \right\| &= e^{-n} \cdot \| \exp(n.A)\xi - e^n . A^n \xi \| \\ &= e^{-n} \cdot \lim_{m\to\infty} \left\| \sum_{k=0}^{m} \frac{n^k}{k!} (A^k - A^n)\xi \right\| \,. \end{aligned} \tag{3.3.7}$$

Further, it follows for $m \in \mathbb{N}$ by using the Cauchy-Schwarz inequality for the Euclidean scalar product on $\mathbb{R}^{m+1}$:

$$\begin{aligned} \left\| \sum_{k=0}^{m} \frac{n^k}{k!} (A^k - A^n)\xi \right\| &\leqslant \sum_{k=0}^{m} \frac{n^k}{k!} \|(A^k - A^n)\xi\| \leqslant \sum_{k=0}^{m} \frac{n^k}{k!} \left\|(A^{|k-n|} - \mathrm{id}_X)\xi\right\| \\ &= \sum_{k=0}^{m} \frac{n^k}{k!} \left\| \sum_{l=0}^{|k-n|-1} A^l \circ (A - \mathrm{id}_X)\xi \right\| \leqslant \|(A - \mathrm{id}_X)\xi\| \cdot \sum_{k=0}^{m} |k-n| \, \frac{n^k}{k!} \end{aligned}$$

$$
\begin{aligned}
&\leqslant \|(A-\mathrm{id}_X)\xi\| \cdot \left(\sum_{k=0}^{m}(k-n)^2\,\frac{n^k}{k!}\right)^{1/2} \cdot \left(\sum_{k=0}^{m}\frac{n^k}{k!}\right)^{1/2} \qquad (3.3.8)\\
&\leqslant \|(A-\mathrm{id}_X)\xi\| \cdot e^{n/2} \cdot \left(\sum_{k=0}^{\infty}(k-n)^2\,\frac{n^k}{k!}\right)^{1/2}\\
&= \|(A-\mathrm{id}_X)\xi\| \cdot e^{n/2} \cdot \left(\sum_{k=0}^{\infty}\left[k(k-1)-(2n-1)k+n^2\right]\frac{n^k}{k!}\right)^{1/2}\\
&= \|(A-\mathrm{id}_X)\xi\| \cdot e^{n/2}\left(\left[n^2-(2n-1)n+n^2\right]e^n\right)^{1/2} = \sqrt{n}\,e^n\,\|(A-\mathrm{id}_X)\xi\| \,.
\end{aligned}
$$

Finally, (3.3.3) follows from (3.3.7) and (3.3.8). □

4

Strongly Continuous Semigroups

In this chapter, we study strongly continuous semigroups of linear operators on Banach spaces. Important motivation for this comes from applications. A major goal of physics is the prediction of the future development of a system from given data in the past, for instance, at time $t = 0$. Denoting the data by the symbol ξ, for all $t \geqslant 0$ there is to be found the state $T(t)\xi$ of the system where $T(t)$ is a map on the space of the data X. If the system is autonomous, $T(t)$ should map X into X and, clearly,

$$T(0) = \mathrm{id}_X \ , \ T(t+s) = T(t) \circ T(s)$$

for all $t, s \in [0, \infty)$, i.e., the map T of $([0, \infty), +)$ into the set of mappings of X into X equipped with the operation of composition is a homomorphism. Such T is called a semigroup of operators. In many important cases, physical systems can be described to a good approximation by assuming the superposition principle. In such cases X carries the structure of a vector space and $T(t)$ is a linear operator for all $t \in [0, \infty)$. For instance, all systems described by Quantum Theory necessarily have to satisfy the superposition principle according to the rules of Quantum Theory.[1] Such T is called a semigroup of linear operators if in addition $T(t)$ is bounded for every $t \in [0, \infty)$. Semigroups of this type will be studied in the following. Remarkably, there is a theory of nonlinear semigroups which largely parallels that of semigroups of linear operators. For this, compare the literature [15, 19, 88, 138, 146, 167].

Usually, in physical applications the determination of the state $u(t) := T(t)\xi$ at time $t \in [0, \infty)$ corresponding to initial data $u(0) = \xi$ is achieved by finding the solution of an initial boundary value problem for a differential equation of the form

$$u'(t) = -Au(t) \tag{4.0.1}$$

for every $t \in \mathbb{R}$ describing the evolution of the system. Here $'$ denotes the ordinary derivative of functions with values in X. In this, A is often a partial differential

[1] See Chapter 2.1. Currently, one could even go so far to say that nature is linear on a fundamental level because all physical systems necessarily have to be described by Quantum Theory. The only nonlinear classical field still resisting a quantization is the gravitational field.

operator. In general, such operators induce unbounded linear operators in Banach spaces. Hence our experience from Chapter 3.3 suggests that in these cases it cannot be expected that T is continuous in the topology induced on $L(X, X)$ by the operator norm. In this, we assume that $[0, \infty)$ is equipped with the usual topology. Indeed, it turns out that in the majority of such cases T is continuous if $L(X, X)$ is equipped with the strong topology.[2]

4.1 Elementary Properties

Theorem 4.1.1. (Strongly continuous semigroups) Let $\mathbb{K} \in \{\mathbb{R}, \mathbb{C}\}$, $(X, \| \, \|)$ a $\mathbb{K}$-Banach space and $T : [0, \infty) \to L(X, X)$ a *strongly continuous semigroup*, i.e., such that

$$T(0) = \mathrm{id}_X \, , \ T(t + s) = T(t)T(s)$$

for all $t, s \in [0, \infty)$ and

$$T\xi := \big([0, \infty) \to X, \xi \mapsto T(t)\xi\big) \in C\big([0, \infty), X\big)$$

for all $\xi \in X$. We define the *(infinitesimal) generator* $A : D(A) \to X$ of T by

$$D(A) := \left\{ \xi \in X : \lim_{t \to 0, t > 0} \frac{1}{t} . [T(t) - \mathrm{id}_X]\xi \text{ exists} \right\}$$

and for every $\xi \in D(A)$:

$$A\xi := - \lim_{t \to 0, t > 0} \frac{1}{t} . [T(t) - \mathrm{id}_X]\xi \, .$$

(i) There are $\mu \in \mathbb{R}$ and $c \in [1, \infty)$ such that for every $t \in [0, \infty)$:

$$\|T(t)\| \leqslant c \, e^{\mu t} \, .$$

(ii) A is a densely-defined, linear and closed operator in X. Furthermore,

$$A \, T(t) \supset T(t)A$$

for all $t \in [0, \infty)$.

(iii) A is bounded if and only if T is continuous.

(iv) If $\mu \in \mathbb{R}$ and $c \in [1, \infty)$ are such that $\|T(t)\| \leqslant c \, e^{\mu t}$ for all $t \in [0, \infty)$, then any $\lambda \in \mathbb{K}$ such that $\mathrm{Re}(\lambda) < -\mu$ is contained in $\rho(A)$ and in particular

$$[R_A(\lambda)]^n = \frac{1}{(n-1)!} \int_0^\infty t^{n-1} e^{\lambda t} \, T(t) \, dt$$

as well as

$$\| \, [R_A(\lambda)]^n \, \| \leqslant \frac{c}{|\mathrm{Re}(\lambda) + \mu|^n}$$

for every $n \in \mathbb{N}^*$. Here we define $\mathrm{Re} := \mathrm{id}_{\mathbb{R}}$ in the case $\mathbb{K} = \mathbb{R}$.

---

[2] But, see [7,60,121,155] for initial value problems for autonomous linear partial differential equations of second order not belonging to this category and approaches [6,8,48,129,170, 205] to treat such cases.

(v) Let $t_0 \geqslant 0$, $t_1 > t_0$ and $\xi \in D(A)$. Then $u : (t_0, t_1) \to X$, defined by

$$u(t) := T(t - t_0)\xi$$

for all $t \in (t_0, t_1)$, is the uniquely determined differentiable map such that $\operatorname{Ran} u \subset D(A)$, $\lim_{t \to t_0} u(t) = \xi$ and

$$u'(t) = -A\,u(t)$$

for all $t \in (t_0, t_1)$.

(vi) (*Different semigroups have different generators*) If $S : [0, \infty) \to X$ is a strongly continuous semigroup different from T and B its generator, then $B \neq A$.

(vii) (*Infinitesimal generators are maximal*) If B is a closable linear extension of A such that $\bar{B}$ is the infinitesimal generator of a strongly continuous semigroup on X, then $B = A$.

Proof. '(i)': Since $T|_{[0,1]}$ is strongly continuous and $[0, 1]$ is compact, it follows the boundedness of

$$\{T(t)\xi : t \in [0, 1]\}$$

for every $\xi \in X$ and hence by the principle of uniform boundedness the existence of a $c \in [1, \infty)$ such that

$$\|T(t)\| \leqslant c$$

for all $t \in [0, 1]$. Further, it follows for $n \in \mathbb{N}$ and $t \in [n, n+1)$

$$\begin{aligned}
&\| T(t)\| = \| T\,((n+1)\cdot t/(n+1))\| \leqslant \| T\,(t/(n+1))\|^{n+1} \leqslant c^{n+1} \\
&= c\,e^{n\ln(c)} \leqslant c\,e^{\ln(c)\cdot t}
\end{aligned}$$

and hence

$$\| T(t)\| \leqslant c\,e^{\ln(c)\cdot t}$$

for all $t \in [0, \infty)$.

'(ii)': Obviously, $D(A)$ is a subspace of X, and A is a linear operator in X as a consequence of the definitions. In a first step, we conclude that for every $\xi \in X$

$$\int_0^{\tau} T(s)\xi\,ds \in D(A) \tag{4.1.1}$$

for all $\tau \in [0, \infty)$. For this, let $\xi \in X$ and $\tau \in [0, \infty)$. Then it follows for every $t \in (0, \infty)$

$$\begin{aligned}
&\frac{1}{t}\cdot\left(T(t)\int_0^{\tau} T(s)\xi\,ds - \int_0^{\tau} T(s)\xi\,ds\right) \\
&= \frac{1}{t}\cdot\left(\int_t^{\tau+t} T(s)\xi\,ds - \int_0^{\tau} T(s)\xi\,ds\right) \\
&= \frac{1}{t}\cdot\left(\int_{\tau}^{\tau+t} T(s)\xi\,ds - \int_0^{t} T(s)\xi\,ds\right)
\end{aligned}$$

$$= \frac{1}{t} \cdot \left(T(\tau) \int_0^t T(s)\xi \, ds - \int_0^t T(s)\xi \, ds \right)$$

$$= (T(\tau) - \mathrm{id}_X) \, \frac{1}{t} \cdot \int_0^t T(s)\xi \, ds \,. \tag{4.1.2}$$

Further,

$$\left\| \frac{1}{t} \cdot \int_0^t T(s)\xi \, ds - \xi \right\| = \frac{1}{t} \left\| \int_0^t (T(s)\xi - \xi) \, ds \right\| \leqslant \frac{1}{t} \int_0^t \|T(s)\xi - \xi\| \, ds$$
$$= \int_0^1 \|T(ts)\xi - \xi\| \, ds \,.$$

Obviously, it follows by Lebesgue's dominated convergence theorem that the last integral approaches zero for $t \to 0$. Hence it follows

$$\lim_{t \to 0+} \frac{1}{t} \cdot \int_0^t T(s)\xi \, ds = \xi \tag{4.1.3}$$

and therefore by (4.1.2) that

$$\lim_{t \to 0+} \frac{1}{t} \cdot \left(T(t) \int_0^\tau T(s)\xi \, ds - \int_0^\tau T(s)\xi \, ds \right) = (T(\tau) - \mathrm{id}_X)\, \xi$$

which implies (4.1.1). Finally, by (4.1.3) it also follows that $D(A)$ is dense in X. Further, we conclude for $\xi \in D(A)$, $s \in (0, \infty)$, $h > 0$

$$\left\| \frac{1}{h} \cdot (T(s+h)\xi - T(s)\xi) + T(s)A\xi \right\| = \left\| T(s) \left(\frac{1}{h} \cdot (T(h)\xi - \xi) + A\xi \right) \right\|$$

and for $h < 0$ such that $|h| \leqslant s/2$

$$\left\| \frac{1}{h} \cdot (T(s+h)\xi - T(s)\xi) + T(s)A\xi \right\|$$
$$= \left\| \frac{1}{|h|} \cdot (T(s)\xi - T(s-|h|)\xi) + T(s)A\xi \right\|$$
$$= \left\| T(s-|h|) \left(\frac{1}{|h|} \cdot (T(|h|)\xi - \xi) + A\xi + T(|h|)A\xi - A\xi \right) \right\|$$
$$\leqslant C_s \left[\left\| \frac{1}{|h|} \cdot (T(|h|)\xi - \xi) + A\xi \right\| + \|T(|h|)A\xi - A\xi\| \right]$$

where $C_s \in [0, \infty)$ is such that

$$\|T(t)\| \leqslant C_s$$

for all $t \in [s/2, s]$. Hence it follows the differentiability of $T\xi$ on $(0, \infty)$ and

$$(T\xi)'(s) = -(TA\xi)(s) \tag{4.1.4}$$

for all $s \in (0, \infty)$. In addition, since for every $s \in (0, \infty)$, $h > 0$

$$\left\| \frac{1}{h} . (T(s+h)\xi - T(s)\xi) + T(s)A\xi \right\| = \left\| -\frac{1}{h} . (T(h) - \mathrm{id}_X)\, T(s)\xi - T(s)A\xi \right\| ,$$

it also follows that

$$T(s)\xi \in D(A) \;\text{ and }\; AT(s)\xi = T(s)A\xi \tag{4.1.5}$$

for every $s \in [0, \infty)$. Since $T\xi$ and $TA\xi$ are in particular continuous, we conclude by integrating (4.1.4) that

$$T(t)\xi = \xi - \int_0^t T(s)A\xi \, ds \tag{4.1.6}$$

for all $t \in [0, \infty)$. In the final step, we conclude that A is also closed. For this, let $(\xi, \eta) \in \overline{G(A)}$ and $\xi_0, \xi_1, \ldots$ a sequence in $D(A)$ such that

$$\lim_{\nu \to \infty} \xi_\nu = \xi \;\text{ and }\; \lim_{\nu \to \infty} A\xi_\nu = \eta \, .$$

Then by (4.1.6) for $t \in (0, 1]$

$$\frac{1}{t} . (T(t)\xi_\nu - \xi_\nu) = -\frac{1}{t} . \int_0^t T(s)A\xi_\nu \, ds \, . \tag{4.1.7}$$

In addition,

$$\left\| \int_0^t T(s)A\xi_\nu \, ds - \int_0^t T(s)\eta \, ds \right\| \leqslant \int_0^t \| T(s)(A\xi_\nu - \eta) \| \, ds \leqslant Ct \, \| A\xi_\nu - \eta \|$$

where $C \in [0, \infty)$ is such that $\|T(s)\| \leqslant C$ for all $s \in [0, 1]$. Hence it follows from (4.1.7) that

$$\frac{1}{t} . (T(t)\xi - \xi) = -\frac{1}{t} . \int_0^t T(s)\eta \, ds \, .$$

Because of (4.1.3), this implies $\xi \in D(A)$ and $A\xi = \eta$.
'(iii)': First, we consider the case that A bounded. Then it follows by the denseness of $D(A)$ in X as well as the closedness of A that $D(A) = X$ and hence that $A \in L(X, X)$. Further, let $t \in [0, \infty)$, $h \in \mathbb{R}$ such that $|h| \leqslant 1$ and $C' \in [0, \infty)$ such that $\|T(s)\| \leqslant C'$ for all $s \in [t-1, t+1] \cap [0, \infty)$. Then it follows by (4.1.6) that

$$\| (T(t+h) - T(t))\xi \| \leqslant C' \, \|A\| \, |h| \, \|\xi\|$$

for every $\xi \in X$ and hence that

$$\| T(t+h) - T(t) \| \leqslant C' \, \|A\| \, |h| \, .$$

The last inequality implies the continuity of T in t. Finally, if T is continuous, we conclude as follows. Since $(L(X,X) \to X, B \mapsto B\xi) \in L(L(X,X),X)$ for every $\xi \in X$, it follows from (4.1.6) that

$$-\frac{1}{t}\,(T(t) - \mathrm{id}_X)\,\xi = \left(\frac{1}{t}\int_0^t T(s)\,ds\right) A\xi \tag{4.1.8}$$

for all $t > 0$ and $\xi \in D(A)$. Further,

$$\begin{aligned}\left\| \frac{1}{t}.\int_0^t T(s)\,ds - \mathrm{id}_X \right\| &= \frac{1}{t}\left\| \int_0^t (T(s) - \mathrm{id}_X)\,ds \right\| \leqslant \frac{1}{t}\int_0^t \|T(s) - \mathrm{id}_X\|\;ds \\ &= \int_0^1 \|T(ts) - \mathrm{id}_X\|\;ds\;.\end{aligned}$$

for every $t > 0$. Hence it follows by Lebesgue's dominated convergence theorem that

$$\lim_{t \to 0+} \frac{1}{t}.\int_0^t T(s)\,ds = \mathrm{id}_X$$

and therefore that

$$\frac{1}{t}.\int_0^t T(s)\,ds$$

is invertible for small enough $t > 0$. For such t, it follows from (4.1.8) that

$$A\xi = -\left(\int_0^t T(s)\,ds\right)^{-1} (T(t) - \mathrm{id}_X)\,\xi$$

for all $\xi \in D(A)$ and hence that A is bounded.
'(iv)': For this, let $\mu \in \mathbb{R}$ and $c \in [1,\infty)$ such that $\|T(t)\| \leqslant c\,e^{\mu t}$ for all $t \in [0,\infty)$. Further, let $\lambda \in \mathbb{K}$ such that $\mathrm{Re}(\lambda) < -\mu$. Then $([0,\infty) \to L(X,X), t \mapsto e^{\lambda t}.T(t))$ is strongly continuous and

$$\|e^{\lambda t}T(t)\| \leqslant c\,e^{[\mathrm{Re}(\lambda)+\mu]\cdot t}$$

for every $t \in (0,\infty)$. Hence we can define $B_\lambda \in L(X,X)$ by

$$B_\lambda := \int_0^\infty e^{\lambda t}.T(t)\,dt$$

according to Theorem 3.2.11 and in particular

$$\|B_\lambda\| \leqslant c\int_0^\infty e^{[\mathrm{Re}(\lambda)+\mu]\cdot t}\,dt = \frac{c}{|\mathrm{Re}(\lambda) + \mu|}\;.$$

Further, it follows by (4.1.4) for $\xi \in D(A)$

$$B_\lambda(A - \lambda)\xi = \int_0^\infty e^{\lambda t}.T(t)(A - \lambda)\xi\,dt = -\int_0^\infty e^{\lambda t}.[(T\xi)'(t) + \lambda.T(t)\xi]dt$$

and hence for every $\omega \in L(X, \mathbb{K})$

$$\omega\,(B_\lambda(A-\lambda)\xi) = \int_0^\infty e^{\lambda t}.\omega(T(t)(A-\lambda)\xi)\,dt$$

$$= -\int_0^\infty e^{\lambda t}.[(\omega \circ T\xi)'(t) + \lambda.(\omega \circ T\xi)(t)\xi]dt$$

$$= -\int_0^\infty \left[e^{\lambda.\mathrm{id}_{\mathbb{R}}} \cdot (\omega \circ T\xi)\right]'(t)\,dt = \omega(\xi)$$

where in the last step Lebesgue's dominated convergence theorem (along with the fundamental theorem of calculus) has been used. Hence it follows

$$\omega\,(B_\lambda(A-\lambda)\xi) = \omega(\xi)$$

and since $L(X, \mathbb{K})$ separates points on X that

$$B_\lambda(A-\lambda)\xi = \xi\,.$$

In the next step, we prove that

$$B_\lambda D(A) \subset D(A)\,. \tag{4.1.9}$$

For this, we notice that for every $t \in [0, \infty)$, because of

$$\|T(t)\xi\|_{\mathrm{A}}^2 = \|T(t)\xi\|^2 + \|AT(t)\xi\|^2 = \|T(t)\xi\|^2 + \|T(t)A\xi\|^2 \leqslant \|T(t)\|^2 \cdot \|\xi\|_{\mathrm{A}}^2$$

for every $\xi \in D(A)$, by the restriction $T_{\mathrm{A}}(t)$ of $T(t)$ in domain and in range to $D(A)$ there is given a bounded linear operator on $(D(A), \|\ \|_{\mathrm{A}})$ with bound $\|T(t)\|$. Further, as a consequence of

$$\|T_{\mathrm{A}}(t)\xi - T_{\mathrm{A}}(s)\xi\|_{\mathrm{A}}^2 = \|T(t)\xi - T(s)\xi\|_{\mathrm{A}}^2 + \|AT(t)\xi - AT(s)\xi\|_{\mathrm{A}}^2$$

$$= \|T(t)\xi - T(s)\xi\|_{\mathrm{A}}^2 + \|T(t)A\xi - T(s)A\xi\|_{\mathrm{A}}^2$$

for every $t, s \in [0, \infty)$ and $\xi \in D(A)$, it follows the strong continuity of the associated semigroup T_A. Hence by Theorem 3.2.11

$$\int_{0,\mathrm{A}}^\infty e^{\lambda t}.T_{\mathrm{A}}(t)\,dt\,,$$

where the index A in the integration symbol denotes weak integration in the Banach space $(D(A), \|\ \|_{\mathrm{A}})$, defines a bounded linear operator on $(D(A), \|\ \|_{\mathrm{A}})$. Finally, since the inclusion $\iota_{\mathrm{A}} : (D(A), \|\ \|_{\mathrm{A}}) \hookrightarrow X$ is continuous, (4.1.9) (ii) follows by

Theorem 3.2.4. Further, since A defines a bounded linear operator on $(D(A), \| \|_{\mathrm{A}})$, it follows for $\xi \in D(A)$ that

$$(A - \lambda)B_\lambda \xi = (A - \lambda) \int_{0,\mathrm{A}}^{\infty} e^{\lambda t}.T_{\mathrm{A}}(t)\xi\, dt$$

$$= \int_0^{\infty} e^{\lambda t}.(A - \lambda)T_{\mathrm{A}}(t)\xi\, dt = \int_0^{\infty} e^{\lambda t}.T_{\mathrm{A}}(t)(A - \lambda)\xi\, dt = \xi\ . \qquad (4.1.10)$$

Finally, let $\xi \in X$. Since $D(A)$ is dense in X, it follows the existence of a sequence $\xi_0, \xi_1, \ldots$ in $D(A)$ which is convergent to ξ. Hence by the boundedness of B_λ, it follows the convergence of $B_\lambda\xi_0, B_\lambda\xi_1, \ldots$ to $B_\lambda\xi$ and by (4.1.10) the convergence of $(A - \lambda)B_\lambda\xi_0, (A - \lambda)B_\lambda\xi_1, \ldots$ to ξ. Since A is closed, it follows that $B_\lambda\xi \in D(A)$ and $(A - \lambda)B_\lambda\xi = \xi$. Hence it follows that

$$R_A(\lambda) = \int_0^{\infty} e^{\lambda t}.T(t)\, dt\ .$$

The next step uses the following auxiliary result: Let $f : (0, \infty) \to \mathbb{K}$ be almost everywhere continuous and such that there are $a \in [0, \infty)$, $b \in \mathbb{R}$ satisfying

$$\|f(t)\| \leqslant a\, e^{bt}$$

for almost all $t > 0$. Then

$$\int_0^{\infty} e^{\lambda' t} f(t)\, dt = \sum_{n=0}^{\infty} \frac{(\lambda' - \lambda)^n}{n!} \cdot \int_0^{\infty} t^n e^{\lambda t} f(t)\, dt \qquad (4.1.11)$$

for all $\lambda, \lambda' \in \mathbb{K}$ satisfying $\mathrm{Re}(\lambda) < -b$ and $|\lambda' - \lambda| < |\mathrm{Re}(\lambda) + b|$. For the proof let $\lambda, \lambda' \in \mathbb{K}$ such that $\mathrm{Re}(\lambda) < -b$ and $|\lambda' - \lambda| < |\mathrm{Re}(\lambda) + b|$. Note that this implies

$$\mathrm{Re}(\lambda') + b = \mathrm{Re}(\lambda' - \lambda) + \mathrm{Re}(\lambda) + b \leqslant |\lambda' - \lambda| + \mathrm{Re}(\lambda) + b$$

$$< |\mathrm{Re}(\lambda) + b| + \mathrm{Re}(\lambda) + b = 0\ .$$

Then

$$e^{\lambda' t} f(t) = e^{(\lambda' - \lambda)t} \cdot e^{\lambda t} f(t) = \sum_{n=0}^{\infty} \frac{(\lambda' - \lambda)^n}{n!} t^n e^{\lambda t} f(t)$$

and

$$\left| \sum_{n=0}^{N} \frac{(\lambda' - \lambda)^n}{n!} t^n e^{\lambda t} f(t) \right| \leqslant \sum_{n=0}^{N} \frac{|\lambda' - \lambda|^n}{n!} |t^n e^{\lambda t} f(t)| \leqslant a\, e^{|\lambda' - \lambda| t}\, e^{\mathrm{Re}(\lambda) t}\, e^{bt}$$

$$= a\, e^{-[|\mathrm{Re}(\lambda) + b| - |\lambda' - \lambda|] \cdot t}$$

for every $t > 0$ and $N \in \mathbb{N}$. Hence (4.1.11) follows by Lebesgue's dominated convergence theorem. Note that in the particular case that $f(t) = 1$ for all $t > 0$

($a = 1, b = 0, \lambda, \lambda' \in \mathbb{K}$ such that $\text{Re}(\lambda) < 0$ and $|\lambda' - \lambda| < |\text{Re}(\lambda)|$), it follows from (4.1.11) that

$$-\frac{1}{\lambda} \cdot \sum_{n=0}^{\infty} (-1)^n \cdot \frac{(\lambda' - \lambda)^n}{\lambda^n} = -\frac{1}{\lambda} \cdot \frac{1}{1 + \frac{\lambda' - \lambda}{\lambda}} = -\frac{1}{\lambda'} = \int_0^\infty e^{\lambda' t} \, dt$$
$$= \sum_{n=0}^{\infty} \frac{(\lambda' - \lambda)^n}{n!} \cdot \int_0^\infty t^n e^{\lambda t} \, dt$$

and hence that

$$\frac{1}{n!} \cdot \int_0^\infty t^n e^{\lambda t} \, dt = \left(-\frac{1}{\lambda} \right)^{n+1} \tag{4.1.12}$$

for every $\lambda \in \mathbb{K}$ such satisfying $\text{Re}(\lambda) < 0$ and every $n \in \mathbb{N}$. Now, we continue the proof of (iv). By the foregoing along with the previous auxiliary result, it follows for every $\lambda' \in \mathbb{K}$ such that $|\lambda' - \lambda| < |\text{Re}(\lambda) + \mu|$, note that this implies

$$\text{Re}(\lambda') + \mu = \text{Re}(\lambda' - \lambda) + \text{Re}(\lambda) + \mu \leqslant |\lambda' - \lambda| + \text{Re}(\lambda) + \mu$$
$$< |\text{Re}(\lambda) + \mu| + \text{Re}(\lambda) + \mu = 0 \, ,$$

$\omega \in L(X, \mathbb{K})$ and $\xi \in X$ that

$$\omega(R_A(\lambda')\xi) = \int_0^\infty e^{\lambda' t} \, \omega(T(t)\xi) \, dt = \sum_{n=0}^{\infty} \frac{(\lambda' - \lambda)^n}{n!} \cdot \int_0^\infty t^n e^{\lambda t} \, \omega(T(t)\xi) \, dt$$
$$= \sum_{n=0}^{\infty} \frac{(\lambda' - \lambda)^n}{n!} \cdot \omega \left(\int_0^\infty t^n e^{\lambda t} \, T(t)\xi \, dt \right) .$$

and hence by the identity (3.1.6) in the proof of Theorem 3.1.11 that

$$\omega \left([R_A(\lambda)]^{n+1} \xi \right) = \frac{1}{n!} \cdot \omega \left(\int_0^\infty t^n e^{\lambda t} \, T(t)\xi \, dt \right)$$

for every $n \in \mathbb{N}$. Since $\omega \in L(X, \mathbb{K})$ and $\xi \in X$ were otherwise arbitrary and $L(X, \mathbb{K})$ separates points on X, this implies that

$$[R_A(\lambda)]^{n+1} = \frac{1}{n!} \cdot \int_0^\infty t^n e^{\lambda t} \, T(t) \, dt \, .$$

Finally, it follows by (4.1.12) that

$$\|[R_A(\lambda)]^{n+1}\xi\| \leqslant \|\xi\| \cdot \frac{c}{n!} \cdot \int_0^\infty t^n e^{(\text{Re}(\lambda)+\mu)t} \, dt = \frac{c}{|\text{Re}(\lambda) + \mu)|^{n+1}} \cdot \|\xi\|$$

for every $\xi \in X$ and hence that

$$\|[R_A(\lambda)]^{n+1}\| \leqslant \frac{c}{|\mathrm{Re}(\lambda) + \mu)|^{n+1}} .$$

'(v)': Let $t_0 \geqslant 0$, $t_1 > t_0$ and $\xi \in D(A)$. Then $u : [t_0, \infty) \to X$, defined by

$$u(t) := T(t - t_0)\xi$$

for all $t \in [t_0, \infty)$, is continuous and as a consequence of (4.1.4),(4.1.5) differentiable on (t_0, ∞) such that

$$u'(t) = -A\,u(t)$$

for all $t \in (t_0, \infty)$. On the other hand, if $u : (t_0, t_1) \to X$ is differentiable such that $\mathrm{Ran}\, u \subset D(A)$, $\lim_{t \to t_0} u(t) = \xi$ and

$$u'(t) = -A\,u(t)$$

for all $t \in (t_0, t_1)$, then we define $G : (t_0, t) \to X$ by $G(s) := T(t - s)u(s)$ for all $s \in (t_0, t)$. Then it follows for every $s \in (t_0, t)$ and $0 < h < t - s$

$$\frac{1}{h}\,[G(s+h) - G(s)] = \frac{1}{h}\,[T(t-(s+h))u(s+h) - T(t-s)u(s)]$$
$$= T(t-(s+h))\left\{\frac{1}{h}\,[u(s+h) - u(s)] - \frac{1}{h}\,[T(h) - \mathrm{id}_X]\,u(s)\right\}$$

and hence

$$\lim_{h \to 0+} \frac{1}{h}\,[G(s+h) - G(s)] = T(t-s)[u'(s) + A\,u(s)] = 0 .$$

Also, it follows for every $t_0 - s < h < 0$

$$\frac{1}{h}\,[G(s+h) - G(s)] = \frac{1}{h}\,[T(t-(s+h))u(s+h) - T(t-s)u(s)]$$
$$= T(t-s)\left\{[T(|h|) - \mathrm{id}_X]\,\frac{1}{h}\,(u(s+h) - u(s)) - \frac{1}{|h|}\,[T(|h|) - \mathrm{id}_X]\,u(s) + \frac{1}{h}\,[u(s+h) - u(s)]\right\}$$

and hence

$$\lim_{h \to 0-} \frac{1}{h}\,[G(s+h) - G(s)] = T(t-s)[u'(s) + A\,u(s)] = 0 .$$

The differentiability of G and $G'(s) = 0$ follows for all $s \in (t_0, t)$. Note that there is a continuous extension $\hat{G} : [t_0, t] \to X$ of G such that $\hat{G}(t_0) = T(t - t_0)\,\xi$ and $\hat{G}(t) = u(t)$. Hence it follows by weak integration over the interval $[t_0, t]$ and by application of Theorem 3.2.9 that $u(t) = T(t - t_0)\,\xi$.
'(vi)': The statement is a simple consequence of (v) along with the denseness of $D(A)$ in X.
'(vii)': Obviously, by (iv), (v) along with the denseness of $D(A)$ in X, it follows that $A = \bar{B}$. Further, it follows $B \supset A = \bar{B}$ and hence $B = \bar{B} = A$. □

4.2 Characterizations

Remarkably, infinitesimal generators of strongly continuous semigroups can be characterized by the conditions on the spectrum and the estimate on the powers of operators in its resolvent given in the previous Theorem 4.1.1 (iv). In the case of a quasi-contractive semigroups, that estimate is equivalent to the quasi-accretivity of the infinitesimal generator. Usually, the last is easier to prove for a given operator. The definition of quasi-accretivity is a generalization of the definition of semiboundedness (from below) for self-adjoint linear operators in Hilbert spaces. Theorem 4.2.7 states that every strongly continuous semigroup 'is' quasi-contractive if the underlying Banach space is chosen in a particular specified way. Unfortunately, the construction of that space uses the semigroup and not just the infinitesimal generator. Usually, in applications only a formal linear operator [3] is given for which a representation space along with boundary conditions have to be chosen such that the resulting operator is the infinitesimal generator of a strongly continuous semigroup. Of course, usually the generated semigroup is unknown also on a formal level.

Theorem 4.2.1. (The Hille-Yosida-Phillips theorem) Let $\mathbb{K} \in \{\mathbb{R}, \mathbb{C}\}$, $(X, \| \,\|)$ a $\mathbb{K}$-Banach space and A a densely-defined closed linear operator in X. Then A is the generator of a strongly continuous semigroup $T : [0, \infty) \to L(X, X)$ if and only if there are $c \in [1, \infty)$, $\mu \in \mathbb{R}$ such that $(-\infty, -\mu) \subset \rho(A)$ and

$$\| [R_A(\lambda)]^n \| \leqslant \frac{c}{|\lambda + \mu|^n} \tag{4.2.1}$$

for all $\lambda \in (-\infty, -\mu)$ and $n \in \mathbb{N}^*$. In this case,

$$\|T(t)\| \leqslant c\, e^{\mu t}$$

for all $t \in [0, \infty)$.

Proof. First, if A is the generator of a strongly continuous semigroup, then it follows by Theorem 4.1.1 the existence of $c \in [1, \infty)$ and $\mu \in \mathbb{R}$ such that $(-\infty, -\mu) \subset \rho(A)$ and such that (4.2.1) is valid for every $\lambda \in (-\infty, -\mu)$, $n \in \mathbb{N}^*$. On the other hand, if $c \in [1, \infty)$, $\mu \in \mathbb{R}$ are such that $(-\infty, -\mu) \subset \rho(A)$ and such that (4.2.1) is valid for every $\lambda \in (-\infty, -\mu)$ and $n \in \mathbb{N}^*$, then we conclude as follows. For this, we define the bounded linear operator

$$A_n := (n - \mu).\mathrm{id}_X - n^2.R_A(-(n + \mu))$$

for all $n \in \mathbb{N}^*$. In a first step, it follows that

$$\lim_{n \to \infty} A_n \xi = A\xi \tag{4.2.2}$$

[3] This is an operator which is linear on some linear space that is not yet embedded into a Banach space.

for every $\xi \in D(A)$. For the proof, let $\eta \in D(A)$ and $n \in \mathbb{N}^*$. Then

$$\begin{aligned}
&\|[(n-\mu).R_A(-(n+\mu)) - \mathrm{id}_X]\eta\| \\
&\quad = \|R_A(-(n+\mu))[(n-\mu).\eta - (A+n+\mu)\eta]\| \\
&\quad = \|R_A(-(n+\mu))(A\eta + 2\mu\eta)\| \leqslant \frac{c}{n}\,\|A\eta + 2\mu\eta\|
\end{aligned}$$

and hence

$$\lim_{n\to\infty} (n-\mu).R_A(-(n+\mu))\eta = \eta\ .$$

Therefore it follows because of

$$\|(n-\mu).R_A(-(n+\mu))\| \leqslant \frac{c\,|n-\mu|}{n} \leqslant c\cdot(1+|\mu|)$$

for every $n \in \mathbb{N}^*$ and the denseness of $D(A)$ in X that

$$\mathrm{s}-\lim_{n\to\infty} (n-\mu).R_A(-(n+\mu)) = \mathrm{id}_X\ . \tag{4.2.3}$$

Further, for every $n \in \mathbb{N}^*$ and $\xi \in D(A)$

$$\begin{aligned}
A_n\xi &= (n-\mu).R_A(-(n+\mu))(A+n+\mu)\xi - n^2.R_A(-(n+\mu))\xi \\
&= (n-\mu).R_A(-(n+\mu))A\xi - \mu^2 R_A(-(n+\mu))\xi
\end{aligned}$$

is satisfied and hence finally because of (4.2.3) and

$$\|R_A(-(n+\mu))\| \leqslant \frac{c}{n}$$

also the relation (4.2.2). In the next step, we define for every $n \in \mathbb{N}^*$ a corresponding $S_n : [0,\infty) \to L(X,X)$ by

$$S_n(t) := \exp(-tA_n)$$

for every $t \in [0,\infty)$. Then it follows by Theorem 3.3.1 and Theorem 4.1.1 that S_n is a strongly continuous semigroup on X with generator A_n. In addition,

$$\begin{aligned}
\|S_n(t)\| &= e^{-(n-\mu)t}\,\|\exp(n^2 t.R_A(-(n+\mu)))\| \\
&\leqslant e^{-(n-\mu)t}\sum_{k=0}^{\infty}\frac{(n^2t)^k}{k!}\,\|[R_A(-(n+\mu))]^k\| \\
&\leqslant c\,e^{-(n-\mu)t}\sum_{k=0}^{\infty}\frac{(nt)^k}{k!} = c\,e^{-(n-\mu)t}e^{nt} = c\,e^{\mu t}
\end{aligned} \tag{4.2.4}$$

for every $t \in [0,\infty)$. Further, it follows for $m, n \in \mathbb{N}^*$, $t \in [0,\infty)$ by Theorem 3.3.1 and by using that $A_m \circ A_n = A_n \circ A_m$ for $\eta \in D(A)$

$$\begin{aligned}
\|S_m(t)\eta - S_n(t)\eta\| &= \|\exp(-sA_m)\exp(-(t-s)A_n)\eta\,|_0^t\| \\
&= \left\|\int_0^t \exp(-sA_m)\exp(-(t-s)A_n)(A_n - A_m)\eta\,ds\right\|
\end{aligned}$$

$$
\begin{aligned}
&\leqslant \|A_n\eta - A_m\eta\| \int_0^t \|\exp(-sA_m)\| \cdot \|\exp(-(t-s)A_n)\|\, ds \\
&\leqslant \|A_n\eta - A_m\eta\| \int_0^t c\exp(\mu s)\, c\exp(\mu(t-s))\, ds \\
&= c^2\, t\, e^{\mu t}\, \|A_n\eta - A_m\eta\|
\end{aligned}
$$

and hence for $\xi \in X$

$$
\begin{aligned}
\|S_m(t)\xi - S_n(t)\xi\| &= \|S_m(t)(\xi-\eta) - S_n(t)(\xi-\eta) + S_m(t)\eta - S_n(t)\eta\| \\
&\leqslant c\, e^{\mu t}\, [2\,\|\xi-\eta\| + c\,t\,\|A_n\eta - A_m\eta\|\,] \\
&\leqslant c\, e^{\mu t}\, [2\,\|\xi-\eta\| + c\,t\,\|A_n\eta - A\eta\| + c\,t\,\|A_m\eta - A\eta\|\,]\,. \qquad (4.2.5)
\end{aligned}
$$

Hence it follows by (4.2.2) along with the denseness of $D(A)$ in X that $S_1(t)\xi$, $S_2(t)\xi, \dots$ is a Cauchy sequence and hence convergent to an element $S(t)\xi \in X$. Performing the limit $m \to \infty$ in (4.2.5) leads to the relation

$$
\|S_n(t)\xi - S(t)\xi\| \leqslant c\, e^{\mu t}\, [2\,\|\xi-\eta\| + c\,t\,\|A_n\eta - A\eta\|\,]\,. \qquad (4.2.6)
$$

Obviously, $S(t) := (X \to X, \xi \mapsto S(t)\xi)$ is in particular linear and by (4.2.4) bounded such that

$$
\|S(t)\| \leqslant c\, e^{\mu t}
$$

for all $t \in [0,\infty)$. Further, $S := ([0,\infty) \to L(X,X), t \mapsto S(t))$ satisfies $S(0) = \mathrm{id}_X$ and because of

$$
\begin{aligned}
\|S_n(t+s)\xi - S(t)S(s)\xi\| &= \|S_n(t)[S_n(s)\xi - S(s)\xi] + [S_n(t) - S(t)]S(s)\xi\| \\
&\leqslant c\, e^{\mu t}\|S_n(s)\xi - S(s)\xi\| + \|[S_n(t) - S(t)]S(s)\xi\|\,,
\end{aligned}
$$

for all $\xi \in X$, $S(t+s) = S(t)S(s)$ for all $t, s \in [0,\infty)$. Again by (4.2.2) along with the denseness of $D(A)$ in X, it follows from (4.2.6) for every compact subset I of $[0,\infty)$ the uniform convergence of $((I \to X, t \mapsto S_n(t)\xi))_{n\in\mathbb{N}^*}$ in $C(I,X)$ to $(I \to X, t \mapsto S(t)\xi)$ and therefore that $(I \to X, t \mapsto S(t)\xi) \in C(I,X)$ and finally that $([0,\infty) \to X, t \mapsto S(t)\xi) \in C([0,\infty),X)$. Hence S is a strongly continuous semigroup. In the final step, we show that its infinitesimal generator $\tilde{A}$ coincides with A. For this, let $\xi \in D(A)$. Then it follows by (4.1.6) from the proof of Theorem 4.1.1 that

$$
S_n(t)\xi = \xi - \int_0^t S_n(s)A_n\xi\, ds
$$

for $n \in \mathbb{N}^*$ and $t \in (0,\infty)$. Further,

$$
\begin{aligned}
\|S_n(s)A_n\xi - S(s)A\xi\| &= \|S_n(s)(A_n\xi - A\xi) + (S_n(s) - S(s))A\xi\| \\
&\leqslant c\, e^{\mu s}\, \|A_n\xi - A\xi\| + \|(S_n(s) - S(s))A\xi\|
\end{aligned}
$$

for every $s \in [0,t]$. Hence the sequence of integrable functions $([0,t] \to \mathbb{R}, s \mapsto \|S_n(s)A_n\xi - S(s)A\xi\|)_{n\in\mathbb{N}^*}$ is everywhere pointwise convergent to 0 on $[0,t]$ and is majorized by the summable function $([0,t] \to \mathbb{R}, s \mapsto c\, e^{\mu s}\|A_n\xi - A\xi\| + 2c\, e^{\mu s}\|A\xi\|)$.

Hence it follows by Lebesgue dominated convergence theorem along with (4.1.6) that

$$\frac{1}{t}.(S(t)\xi - \xi) = -\frac{1}{t}\int_0^t S(s)A\xi\,ds$$

and therefore, since by the proof of Theorem 4.1.1

$$\lim_{t\to 0+}\frac{1}{t}\int_0^t S(s)A\xi\,ds = A\xi\ ,$$

that $\xi \in D(\tilde{A})$ and $\tilde{A}\xi = A\xi$. Hence it follows $\tilde{A} \supset A$. Further, it follows from (4.2.4) by Theorem 4.1.1 that $(-\infty, -\mu) \subset \rho(\tilde{A})$. In particular, it follows for $\lambda \in (-\infty, -\mu)$, $\eta \in X$

$$(\tilde{A} - \lambda)R_A(\lambda)\eta = (A - \lambda)R_A(\lambda)\eta = \eta$$

and therefore

$$R_{\tilde{A}}(\lambda)\eta = R_A(\lambda)\eta\ .$$

Hence $D(\tilde{A}) = D(A)$ and $\tilde{A} = A$. □

Corollary 4.2.2. Let $\mathbb{K} \in \{\mathbb{R}, \mathbb{C}\}$, $(X, \| \ \|)$ a $\mathbb{K}$-Banach space, A the generator of a strongly continuous semigroup on X. Further, let $c \in [1, \infty)$, $\mu \in \mathbb{R}$ such that $\|T(t)\| \leqslant c\,e^{\mu t}$ for all $t \in [0, \infty)$. Then

$$T(t) = \mathrm{s}-\lim_{n\to\infty} e^{-(n-\mu)t}.\exp\left(n^2 t.R_A\left(-(n+\mu)\right)\right)$$

for all $t \in [0, \infty)$.

Definition 4.2.3. Let $\mathbb{K} \in \{\mathbb{R}, \mathbb{C}\}$, $(X, \| \ \|)$ a $\mathbb{K}$-Banach space.

(i) Per definition, a *normalized tangent functional* to $\xi \in X$ is an element $\omega \in L(X, \mathbb{K})$ satisfying

$$\omega(\xi) = \|\xi\|^2 \quad \text{and} \quad \|\omega\| = \|\xi\|\ .$$

Note that there is a normalized tangent functional to every $\xi \in X$ by the Hahn-Banach theorem.[4]

(ii) In addition, let A a linear operator in X. Then we call A *accretive* if for every $\xi \in D(A)$ there is a normalized tangent functional $\omega \in L(X, \mathbb{K})$ to ξ such that

$$\mathrm{Re}\,(\omega(A\xi)) \geqslant 0$$

where we define as usual $\mathrm{Re} := \mathrm{id}_{\mathbb{R}}$ in the case $\mathbb{K} = \mathbb{R}$. In addition, we call A *quasi-accretive with bound* $\mu \in \mathbb{R}$ if $A - \mu.\mathrm{id}_X$ is accretive. Obviously, this is the case if and only if for every $\xi \in X$ there is a normalized tangent functional $\omega \in L(X, \mathbb{K})$ to ξ such that

$$\mathrm{Re}\,(\omega(A\xi)) \geqslant \mu\,\|\xi\|^2\ .$$

[4] See, e.g., [179] Vol.I, Theorem III.5.

Theorem 4.2.4. Let $\mathbb{K} \in \{\mathbb{R}, \mathbb{C}\}$, $(X, \| \|)$ a $\mathbb{K}$-Banach space and A a linear operator in X. Then A is accretive if and only if

$$\|(A+\lambda)\xi\| \geqslant \lambda\|\xi\| \tag{4.2.7}$$

for all $\lambda \in (0, \infty)$ and $\xi \in D(A)$.

Proof. Let A be accretive, $\lambda \in (0, \infty)$, $\xi \in D(A)$ and $\omega \in L(X, \mathbb{K})$ a normalized tangent functional such that $\mathrm{Re}\,(\omega(A\xi)) \geqslant 0$. Then

$$\lambda\,\|\xi\|^2 \leqslant \mathrm{Re}\,(\omega\,((A+\lambda)\xi)) \leqslant |\,\omega\,((A+\lambda)\xi)\,| \leqslant \|\xi\| \cdot \|(A+\lambda)\xi\|$$

and hence (4.2.7). On the other hand, if (4.2.7) is true for all $\lambda \in (0, \infty)$ and $\xi \in D(A)$, we conclude as follows. For this, let $\lambda \in (0, \infty)$, $\xi \in D(A)$, $\omega_\lambda \in L(X, \mathbb{K})$ a normalized tangent functional to $(A+\lambda)\xi$ and $\omega'_\lambda := \omega_\lambda / \|(A+\lambda)\xi\|$ if $\|(A+\lambda)\xi\| \neq 0$, $\omega'_\lambda := \omega_\lambda$ if $(A+\lambda)\xi = 0$. Then

$$\begin{aligned}\lambda\|\xi\| \leqslant \|(A+\lambda)\xi\| &= \omega'_\lambda\,((A+\lambda)\xi) = \mathrm{Re}\,(\omega'_\lambda\,(A\xi)) + \lambda\mathrm{Re}\,(\omega'_\lambda\,(\xi)) \\ &\leqslant \mathrm{Re}\,(\omega'_\lambda\,(A\xi)) + \lambda\,|\,\omega'_\lambda\,(\xi)\,| \leqslant \mathrm{Re}\,(\omega'_\lambda\,(A\xi)) + \lambda\|\xi\|\,.\end{aligned}$$

Hence

$$\mathrm{Re}\,(\omega'_\lambda\,(A\xi)) \geqslant 0 \tag{4.2.8}$$

and

$$\mathrm{Re}\,(\omega'_\lambda\,(\xi)) \geqslant \|\xi\| - \frac{1}{\lambda}\,\mathrm{Re}\,(\omega'_\lambda\,(A\xi)) \geqslant \|\xi\| - \frac{1}{\lambda}\,|\,\omega'_\lambda\,(A\xi)\,| \geqslant \|\xi\| - \frac{1}{\lambda}\,\|A\xi\|\,. \tag{4.2.9}$$

We define

$$\mathcal{F} := \left\{ F \subset B_1(0_{L(X,\mathbb{K})}) : \bigvee_{R>0} (F \supset \{\omega'_\lambda : \lambda > R\}) \right\}.$$

Obviously, $\mathcal{F}$ is a filter on $B_1(0_{L(X,\mathbb{K})})$. Since $B_1(0_{L(X,\mathbb{K})})$ is weak*−compact, there is $\omega' \in B_1(0_{L(X,\mathbb{K})})$ which is adherent to every $F \in \mathcal{F}$. Further, since $(B_1(0_{L(X,\mathbb{K})}) \to \mathbb{K}, \omega \mapsto \omega(\eta))$ is weak*−continuous, it follows for every $\eta \in X$ that $\omega'(\eta)$ is adherent in particular to every $\{\omega'_\lambda(\eta) : \lambda > \nu\}$ for every $\nu \in \mathbb{N}^*$. Hence for every $\nu \in \mathbb{N}^*$ there is $\lambda_\nu > \nu$ such that

$$|\,\omega'_{\lambda_\nu}(\eta) - \omega'(\eta)\,| < \frac{1}{\nu}$$

and therefore

$$\lim_{\nu\to\infty} \omega'_{\lambda_\nu}(\eta) = \omega'(\eta)\,.$$

Hence it follows from (4.2.8), (4.2.9) that

$$\mathrm{Re}\,(\omega'\,(A\xi)) \geqslant 0 \quad \text{and} \quad |\omega'(\xi)| \geqslant \mathrm{Re}\,(\omega'\,(\xi)) \geqslant \|\xi\|\,.$$

Further, since $|\,\omega'(\xi)| \leqslant \|\xi\|$, it follows that $|\,\omega'(\xi)| = \|\xi\|$ and $\|\,\omega'\| = 1$. Hence $\omega := \|\xi\|\,\omega'$ is a normalized tangent functional to ξ such that $\mathrm{Re}\,(\omega\,(A\xi)) \geqslant 0$. □

Theorem 4.2.5. Let $\mathbb{K} \in \{\mathbb{R}, \mathbb{C}\}$, $(X, \| \|)$ a $\mathbb{K}$-Banach space and A a densely-defined, linear operator in X which is quasi-accretive with bound $\mu \in \mathbb{R}$. Then A is closable and $\bar{A}$ is quasi-accretive with bound $\mu \in \mathbb{R}$.

Proof. In the first step, we consider the special case $\mu = 0$ and lead the assumption that A is not closable to a contradiction. If A is not closable, $\overline{G(A)}$ is not the graph of a map. Then there exists $(0, \eta) \in \overline{G(A)}$ such that $\|\eta\| = 1$. In particular, let $\xi_0, \xi_1, \ldots$ be a sequence in $D(A)$ which is convergent to 0 and such that $A\xi_0, A\xi_1, \ldots$ is convergent to η. Further, since $D(A)$ is dense in X, there is $\xi \in D(A)$ such that $\|\xi\| = 1$ and $\|\xi - \eta\| < 1/2$.[5] Then it follows by Theorem 4.2.4 that

$$\|A\xi + c\,(\xi - A\xi_\nu) - c^2\xi_\nu\| = \|\,(A + c)\,(\xi - c\,\xi_\nu)\| \geqslant c\,\|\xi - c\,\xi_\nu\|$$

for every $\nu \in \mathbb{N}$, $c > 0$ and hence that

$$\|\xi - \eta + c^{-1}A\xi\| \geqslant \|\xi\| \ .$$

Since c can be chosen arbitrarily large, this leads to the contradiction

$$\frac{1}{2} > \|\xi - \eta\| \geqslant \|\xi\| = 1 \ . \ \lightning$$

Hence A is closable. Further, let $\lambda \in (0, \infty)$ and $\xi \in D(\bar{A})$. Then there is a sequence $\xi_0, \xi_1, \ldots$ in $D(A)$ which is convergent to ξ and such that $A\xi_0, A\xi_1, \ldots$ is convergent to $\bar{A}\xi$. In particular, by Theorem 4.2.4

$$\|(A + \lambda)\xi_\nu\| \geqslant \lambda\|\xi_\nu\|$$

and hence also

$$\|(\bar{A} + \lambda)\xi\| \geqslant \lambda\|\xi\| \ .$$

Hence it follows by Theorem 4.2.4 that $\bar{A}$ is accretive. Finally, if $\mu \neq 0$, then $A - \mu$ is a densely-defined, linear and accretive operator in X and hence closable with an accretive closure by the foregoing. Hence it follows by Theorem 3.1.3 (vi) that $A = (A - \mu) + \mu$ is closable and that $\bar{A} = \overline{A - \mu} + \mu$ is quasi-accretive with bound μ. □

Theorem 4.2.6. (Lumer-Phillips) Let $\mathbb{K} \in \{\mathbb{R}, \mathbb{C}\}$, $(X, \| \|)$ a $\mathbb{K}$-Banach space, A a densely-defined, linear operator in X and $\mu \in \mathbb{R}$. Then A is closable and $\bar{A}$ the generator of a strongly continuous semigroup T on X satisfying

$$\|T(t)\| \leqslant e^{\mu t} \ . \tag{4.2.10}$$

for all $t \in [0, \infty)$ if and only if A is quasi-accretive with bound $-\mu$ and $\mathrm{Ran}(A - \lambda)$ is dense in X for some $\lambda \in (-\infty, -\mu)$. A strongly continuous semigroup T satisfying (4.2.10) for all $t \in [0, \infty)$ is called a *quasi-contraction* and a *contraction* in the special case that $\mu = 0$.

[5] For instance choose $\xi' \in X$ such that $\|\xi' - \eta\| < 1/5$ and define $\xi := \|\xi'\|^{-1}.\xi'$.

Proof. If A is closable and $\bar{A}$ the generator of a strongly continuous semigroup T on X satisfying (4.2.10) for all $t \in [0, \infty)$, then it follows by Theorem 4.1.1 that $(-\infty, -\mu) \subset \rho(\bar{A})$ as well as

$$\|R_{\bar{A}}(\lambda)\| \leqslant \frac{1}{|\lambda + \mu|}$$

for every $\lambda \in (-\infty, -\mu)$. Hence

$$|\lambda + \mu| \cdot \|\xi\| = |\lambda + \mu| \cdot \|R_{\bar{A}}(\lambda)(A - \lambda)\xi\| \leqslant \| \left((A + \mu) + |\mu + \lambda|\right) \xi\|$$

for every $\lambda \in (-\infty, -\mu)$, $\xi \in D(A)$ and therefore $A + \mu$ is accretive by Theorem 4.2.4. As a consequence, A is quasi-accretive with bound $-\mu$. In addition, for $\lambda \in (-\infty, -\mu)$, the statement $\text{Ran}(\bar{A} - \lambda) = X$ holds. Finally, since according to Theorem 3.1.3 (vi) $\bar{A} - \lambda$ coincides with the closure of $A - \lambda$, it follows that $\text{Ran}(A - \lambda)$ is dense in X. On the other hand, if A is quasi-accretive with bound $-\mu$ and $\lambda_0 \in (-\infty, -\mu)$ is such that $\text{Ran}(A - \lambda_0)$ is dense in X, we conclude as follows. First, it follows by Theorem 4.2.5 that A is closable as well as that $\bar{A}$ is quasi-accretive with bound $-\mu$. Hence according to Theorem 4.2.4

$$\|(\bar{A} - \lambda)\xi\| = \|(\bar{A} + \mu + |\lambda + \mu|\,)\xi\| \geqslant |\lambda + \mu| \cdot \|\xi\| \tag{4.2.11}$$

for all $\lambda \in (-\infty, -\mu)$ and $\xi \in D(\bar{A})$ which implies in particular the injectivity of $\bar{A} - \lambda$ for every $\lambda \in (-\infty, -\mu)$. Also is

$$\text{Ran}\,(\bar{A} - \lambda_0) = X\,.$$

For the proof, let $\eta \in X$. Since $\text{Ran}(A - \lambda_0)$ is dense in X, there is a sequence $\xi_0, \xi_1, \dots$ in $D(A)$ such that

$$\lim_{\nu \to \infty} (A - \lambda_0)\xi_\nu = \eta\,.$$

Then it follows by (4.2.11) that $\xi_0, \xi_1, \dots$ is a Cauchy sequence in X and hence convergent to some $\xi \in X$. Since $\bar{A}$ is closed, it follows by Theorem 3.1.3 (vi) that $\xi \in D(\bar{A})$ and $(\bar{A} - \lambda_0) = \eta$. Hence $\bar{A} - \lambda_0$ is bijective and $\lambda_0 \in \rho(\bar{A})$. Now, let $\lambda \in (-\infty, -\mu)$ be such that $\bar{A} - \lambda$ is surjective and hence bijective. Then as a consequence of (4.2.11)

$$\left\|(\bar{A} - \lambda)^{-1}\right\| \leqslant \frac{1}{|\lambda + \mu|}\,. \tag{4.2.12}$$

Further, for every $\lambda' \in (\lambda - |\lambda + \mu|, -\mu)$

$$(\bar{A} - \lambda')(\bar{A} - \lambda)^{-1} = \text{id}_X - (\lambda' - \lambda)(\bar{A} - \lambda)^{-1}$$

and hence because of

$$\left\|(\lambda' - \lambda)(\bar{A} - \lambda)^{-1}\right\| < 1$$

also $\bar{A} - \lambda'$ bijective for all $\lambda' \in (\lambda - |\lambda + \mu|, -\mu)$. Hence it follows that $(-\infty, -\mu) \in \rho(\bar{A})$ and from (4.2.12) that

$$\|\,[R_{\bar{A}}(\lambda)]^n\,\| \leqslant \frac{1}{|\lambda + \mu|^n}$$

for all $\lambda \in (-\infty, -\mu)$ and $n \in \mathbb{N}^*$. Finally, it follows from Theorem 4.2.1 that $\bar{A}$ is the infinitesimal generator of a strongly continuous semigroup T on X which satisfies (4.2.10) for every $t \in [0, \infty)$. □

Theorem 4.2.7. Let $\mathbb{K} \in \{\mathbb{R}, \mathbb{C}\}$, $(X, \| \,\|)$ a $\mathbb{K}$-Banach space and T a strongly continuous semigroup on X. Then there is a norm $|||\,|||$ on X which is equivalent to $\|\,\|$ and such that T defines a strongly continuous quasi-contraction semigroup on $(X, |||\,|||)$.

Proof. Since T is a strongly continuous semigroup on X, there are $c \in [1, \infty)$ and $\mu \in \mathbb{R}$ such that

$$\|T(t)\| \leqslant c\, e^{\mu t}$$

for all $t \in [0, \infty)$. We define $|||\,||| : X \to \mathbb{R}$ by

$$|||\xi||| := \|e^{-\mu . \mathrm{id}_{\mathbb{R}}} . T\xi\|_{\infty}$$

for all $\xi \in X$. Obviously, $|||\,|||$ defines a norm on X. Because of

$$|||\xi||| \geqslant \|T(0)\xi\| = \|\xi\|$$

and

$$\|e^{-\mu t} T(t)\xi\| \leqslant c\,\|\xi\|$$

for all $\xi \in X$, it follows that

$$|||\xi||| \leqslant c\,\|\xi\|$$

for all $\xi \in X$ and hence that $\|\,\|$ and $|||\,|||$ are equivalent. Therefore, $(X, |||\,|||)$ is a $\mathbb{K}$-Banach space and T defines a strongly continuous semigroup on $(X, |||\,|||)$. Further, it follows

$$|||T(t)\xi||| = \|e^{-\mu . \mathrm{id}_{\mathbb{R}}} . T T(t)\xi\|_{\infty} = \sup\left\{e^{\mu t} e^{-\mu(s+t)} \cdot \|T(s+t)\xi\| : s \in [0, \infty)\right\}$$
$$\leqslant e^{\mu t}\, |||\xi|||$$

for all $t \in [0, \infty)$, $\xi \in X$ and hence

$$|||T(t)||| \leqslant e^{\mu t}$$

for all $t \in [0, \infty)$. Therefore, T defines in particular a quasi-contraction on $(X, |||\,|||)$. □

4.3 An Integral Representation in the Complex Case

The following theorem gives a 'weak' integral representation of any strongly continuous semigroup on a complex Banach space as integral over its resolvent along a parallel to the imaginary axis in the resolvent set. If the resolvent is known analytically, this leads to ananalytic integral representation of the solutions of the initial value problem of the associated differential equation. For an application of this theorem see [25].

Theorem 4.3.1. (An integral representation of semigroups on complex Banach spaces) Let $(X, \| \|)$ be a complex Banach space and T a strongly continuous semigroup on X with infinitesimal generator A. Further, let $c \in [1, \infty)$, $\mu \in \mathbb{R}$ be such that $\|T(t)\| \leqslant c\, e^{\mu t}$ for every $t \in [0, \infty)$. Finally, let $\mu' \in (-\infty, -\mu)$, $\xi \in X$ and $\omega \in L(X, \mathbb{C})$. Then $(\mathbb{R} \to \mathbb{C}, \varpi \mapsto \omega\,(R_A(\mu' + i\varpi)\xi)\,) \in L^2_{\mathbb{C}}(\mathbb{R})$ and, moreover,

$$e^{\mu' t}\,\omega(T(t)\xi) = \frac{1}{2\pi} \cdot \lim_{\nu \to \infty} \int_{-\nu}^{\nu} e^{-it\varpi}\,\omega\,(R_A(\mu' + i\varpi)\xi)\,d\varpi\,, \tag{4.3.1}$$

$t \in [0, \infty)$ where 'lim' denotes the limit in the mean.

Proof. First, it follows by Theorem 4.1.1 that any $\lambda \in \mathbb{C}$ such that $\mathrm{Re}(\lambda) < -\mu$ is contained in $\rho(A)$ and in particular

$$R_A(\lambda) = \int_0^\infty e^{\lambda t}\, T(t)\, dt\,.$$

Hence it follows for every $\varpi \in \mathbb{R}$ that

$$\omega(R_A(\mu' + i\varpi)\xi) = \frac{1}{\sqrt{2\pi}} \int_{\mathbb{R}} e^{i\varpi t}\, h(t)\, dt$$

where

$$h(t) := \sqrt{2\pi}\, e^{\mu' t}\, \omega(T(t)\xi)$$

for all $t \in [0, \infty)$ and $h(t) := 0$ for all $t \in (-\infty, 0)$. In particular,

$$|h(t)| \leqslant \sqrt{2\pi}\, c\, \|\omega\|\, \|\xi\| \cdot e^{-|\mu' + \mu|\, t}$$

for every $t \geqslant 0$ and hence $h \in L^2_{\mathbb{C}}(\mathbb{R})$. Hence $(\mathbb{R} \to \mathbb{C}, \varpi \mapsto \omega\,(R_A(\mu' + i\varpi)\xi)\,) \in L^2_{\mathbb{C}}(\mathbb{R})$ as well as (4.3.1) follows by the Fourier inversion theorem. □

4.4 Perturbation Theorems

The following perturbation theorems are of major importance for applications. Frequently, in applications there is given a formal linear operator[6] which is read off from a formal differential equation describing the evolution of a physical system. Often, a representation space can be found such that the closure of one of its parts can be seen to generate a strongly continuous semigroup. The theorems below give sufficient conditions on the remaining part such that the whole induced operator generates a strongly continuous semigroup.

---

[6] This is an operator which is linear on some linear space that is not yet embedded into a Banach space.

Theorem 4.4.1. (Relatively bounded perturbations of generators of quasi-contraction semigroups) Let $\mathbb{K} \in \{\mathbb{R}, \mathbb{C}\}$, $(X, \| \, \|)$ a $\mathbb{K}$-Banach space, $A : D(A) \to X$ a densely-defined, linear and closable operator whose closure is the infinitesimal generator of a strongly continuous quasi-contraction semigroup, $B : D(A) \to X$ a $\mathbb{K}$-linear operator in X such that

$$\|B\xi\| \leqslant a\|A\xi\| + b\|\xi\| \tag{4.4.1}$$

for all $\xi \in D(A)$ and some $a \in [0, 1)$, $b \in [0, \infty)$. In addition, let $A + \alpha B$ be quasi-accretive for every $\alpha \in [0, 1]$. Then $A + B$ is closable and its closure is the infinitesimal generator of a strongly continuous quasi-contraction semigroup.

Proof. The first part of the proof considers the case $a < 1/2$. By (4.4.1) it follows

$$\|B\xi\| \leqslant (a + b) \cdot \|\xi\|_{\bar{A}}$$

for every $\xi \in D(A)$ and hence by Theorem 3.1.3 (ii), (iii) that B is a densely defined bounded X-valued linear operator in $(D(\bar{A}), \| \, \|_{\bar{A}})$. Hence this operator has a unique extension to a X-valued bounded linear operator $\hat{B}$ on $(D(\bar{A}), \| \, \|_{\bar{A}})$.[7] In particular, it follows from (4.4.1) that

$$\|\hat{B}\xi\| \leqslant a\|\bar{A}\xi\| + b\|\xi\| \tag{4.4.2}$$

for all $\xi \in D(\bar{A})$. This can be seen as follows. If $\xi \in D(\bar{A})$ and $\xi_0, \xi_1, \ldots$ is a sequence in $D(A)$ converging to ξ and such that $A\xi_0, A\xi_1, \ldots$ is converging to $\bar{A}\xi$, then $\xi_0, \xi_1, \ldots$ is converging to ξ in $(D(\bar{A}), \| \, \|_{\bar{A}})$ and therefore $B\xi_0, B\xi_1, \ldots$ is converging to $\hat{B}\xi$. Note that as a consequence $(A + B)\xi_0, (A + B)\xi_1, \ldots$ is converging to $(\bar{A} + \hat{B})\xi$ and therefore $(\xi, (\bar{A} + \hat{B})\xi) \in G(\overline{A + B})$ and $\bar{A} + \hat{B} \subset \overline{A + B}$. Also note that since $A + B$ is by assumption quasi-accretive, it follows by Theorem 4.2.5 that $A + B$ is closable and that its closure is quasi-accretive. Therefore, $\bar{A} + \hat{B}$ is quasi-accretive, too. Further, since $\bar{A}$ is the infinitesimal generator of a strongly continuous quasi-contraction semigroup, there is $\mu \in \mathbb{R}$ such that $\bar{A}$ is quasi-accretive with bound $-\mu \in \mathbb{R}$ and $(-\infty, -\mu) \subset \rho(\bar{A})$. In particular, it follows for every $\lambda \in (-\infty, -\mu)$ and $\xi \in D(\bar{A})$:

$$\|\hat{B}\xi\| \leqslant a\|\bar{A}\xi\| + b\|\xi\| \leqslant a\|(\bar{A} - \lambda)\xi\| + (a\,|\lambda| + b)\,\|\xi\|$$

and hence for every $\eta \in X$:

$$\|\hat{B}R_{\bar{A}}(\lambda)\eta\| \leqslant a\|\eta\| + (a\,|\lambda| + b)\,\|R_{\bar{A}}(\lambda)\eta\| \leqslant \left[a + \frac{a\,|\lambda| + b}{|\lambda + \mu|}\right]\|\eta\|$$
$$\leqslant \left[2a + \frac{a\,|\mu| + b}{|\lambda + \mu|}\right]\|\eta\| \, .$$

Therefore,

$$(\bar{A} + \hat{B} - \lambda)R_{\bar{A}}(\lambda) = \mathrm{id}_X - \hat{B}R_{\bar{A}}(\lambda)$$

[7] See, e.g., Theorem 4 in Chapter IV, §3 of [128].

is bijective for $\lambda \in (-\infty, -\mu)$ such that

$$2a + \frac{a\,|\mu| + b}{|\lambda + \mu|} < 1 \, .$$

Hence it follows by Theorem 4.2.6 that $\bar{A} + \hat{B}$ is closable and that its closure is the infinitesimal generator of a strongly continuous quasi-contraction semigroup. In particular, it follows from the bijectivity of $\bar{A} + \hat{B} - \lambda$ for $\lambda \in (-\infty, -\mu)$ that $\bar{A} + \hat{B}$ coincides with its closure and hence is closed. Therefore, it follows $\overline{A + B} \subset \bar{A} + \hat{B}$ and finally that $\bar{A} + \hat{B} = \overline{A + B}$. In the following, we drop the condition that in $a < 1/2$. For this, let $\alpha \in [0, 1], \beta \in [0, \infty)$ and $n \in \mathbb{N}^*$. Then it follows for $\xi \in D(A)$

$$\|(A + \alpha B)\xi\| \geqslant \|A\xi\| - \alpha\|B\xi\| \geqslant \|A\xi\| - \|B\xi\| \geqslant (1 - a)\|A\xi\| - b\|\xi\| \, ,$$

hence

$$\|\beta B\xi\| \leqslant \beta a\|A\xi\| + \beta b\|\xi\| \leqslant \frac{\beta a}{1 - a} \|(A + \alpha B)\xi\| + \frac{\beta b}{1 - a} \|\xi\|$$

and by replacing α by $\alpha/n \in [0, 1]$ and choosing $\beta := 1/n$

$$\left\| \frac{1}{n} B\xi \right\| \leqslant \frac{1}{n} \frac{a}{1 - a} \left\| \left(A + \frac{\alpha}{n} B \right) \xi \right\| + \frac{1}{n} \frac{b}{1 - a} \|\xi\| \, .$$

Hence for n large enough such that $n^{-1}a/(1-a) < 1/2$ and by applying the previous result consecutively to the cases $\alpha = 0, 1, \ldots, n - 1$, it follows that $A + B$ is closable and that its closure is the infinitesimal generator of a strongly continuous quasi-contraction semigroup. □

Remark 4.4.2. Note for the Hilbert space case that (4.4.1) is a consequence of the inequality

$$\|B\xi\|^2 \leqslant a^2\|A\xi\|^2 + b^2\|\xi\|^2$$

which should be easier to prove.

Theorem 4.4.3. (Bounded perturbations) Let X be a Banach space, $A : D(A) \to X$ the infinitesimal generator of a strongly continuous semigroup $T : [0, \infty) \to L(X, X)$, $\mu \in \mathbb{R}$ and $c \in [1, \infty)$ such that $\|T(t)\| \leqslant c\, e^{\mu t}$ for all $t \in [0, \infty)$ and $B \in L(X, X)$. Then $A + B$ is the infinitesimal generator of a strongly continuous semigroup $T' : [0, \infty) \to L(X, X)$ such that

$$\|T'(t)\| \leqslant c\, e^{[\mu + c\|B\|]\cdot t} \tag{4.4.3}$$

for all $t \in [0, \infty)$.

Proof. First, obviously, it follows from the closedness of A and the continuity of B the closedness of the densely-defined linear operator $A + B$ in X. Further, $(-\infty, -\mu)$ is contained in the resolvent set of A and

$$\|[(A - \lambda)^{-1}]^n\| \leqslant \frac{c}{|\lambda + \mu|^n}$$

for all $\lambda \in (-\infty, -\mu)$ and $n \in \mathbb{N}^*$. In the following, let $\mu' := \mu + c\,\|B\|$ and $\lambda \in (\infty, -\mu')$. Then

$$A + B - \lambda = (A + B - \lambda)(A - \lambda)^{-1}(A - \lambda) = \left(\mathrm{id}_X + B\,(A - \lambda)^{-1}\right)(A - \lambda)\,.$$

Hence because of

$$\|B\,(A - \lambda)^{-1}\| \leqslant q := \frac{c\,\|B\|}{|\lambda + \mu|} < 1\,,$$

it follows the bijectivity of $\mathrm{id}_X + B\,(A - \lambda)^{-1}$ and hence also of $A + B - \lambda$ and

$$\begin{aligned}(A + B - \lambda)^{-1} &= (A - \lambda)^{-1} \sum_{k \in \mathbb{N}} \left[(-B)\,(A - \lambda)^{-1}\right]^k \\ &= \sum_{k \in \mathbb{N}} (A - \lambda)^{-1} \left[(-B)\,(A - \lambda)^{-1}\right]^k\,.\end{aligned}$$

Note that the involved families of elements of $L(X, X)$ are absolutely summable. Hence it follows also for any $n \in \mathbb{N}^*$ the absolute summability of

$$\left((A - \lambda)^{-1} \left[(-B)\,(A - \lambda)^{-1}\right]^{k_1} \ldots (A - \lambda)^{-1} \left[(-B)\,(A - \lambda)^{-1}\right]^{k_n}\right)_{k \in \mathbb{N}^n}$$

and

$$\begin{aligned}&\left[(A + B - \lambda)^{-1}\right]^n \\ &\quad = \sum_{k \in \mathbb{N}^n} (A - \lambda)^{-1} \left[(-B)\,(A - \lambda)^{-1}\right]^{k_1} \ldots (A - \lambda)^{-1} \left[(-B)\,(A - \lambda)^{-1}\right]^{k_n}\,.\end{aligned}$$

Hence it follows the absolute summability of

$$\left(\left\|(A - \lambda)^{-1} \left[(-B)\,(A - \lambda)^{-1}\right]^{k_1} \ldots (A - \lambda)^{-1} \left[(-B)\,(A - \lambda)^{-1}\right]^{k_n}\right\|\right)_{k \in \mathbb{N}^n}$$

and

$$\begin{aligned}&\left\|\left[(A + B - \lambda)^{-1}\right]^n\right\| \qquad (4.4.4)\\ &\quad \leqslant \sum_{k \in \mathbb{N}^n} \left\|(A - \lambda)^{-1} \left[(-B)\,(A - \lambda)^{-1}\right]^{k_1} \ldots (A - \lambda)^{-1} \left[(-B)\,(A - \lambda)^{-1}\right]^{k_n}\right\|\,.\end{aligned}$$

Further, for every $k \in \mathbb{N}^n$,

$$\begin{aligned}&\left\|(A - \lambda)^{-1} \left[(-B)\,(A - \lambda)^{-1}\right]^{k_1} \ldots (A - \lambda)^{-1} \left[(-B)\,(A - \lambda)^{-1}\right]^{k_n}\right\| \\ &\quad \leqslant \frac{c^{|k|+1}}{|\lambda + \mu|^{|k|+n}} \cdot \|B\|^{|k|} = \frac{c}{|\lambda + \mu|^n} \cdot q^{|k|} \qquad (4.4.5)\end{aligned}$$

where $|k| := k_1 + \cdots + k_n$. Because of $q < 1$, it follows the absolute summability of $(q^k)_{k \in \mathbb{N}}$ with sum $1/(1 - q)$ and hence also for any $n \in \mathbb{N}^*$ the absolute summability of $(q^{|k|})_{k \in \mathbb{N}^n}$ with sum $1/(1 - q)^n$. Hence it follows from (4.4.4), (4.4.5) that

$$\left\|\left[(A + B - \lambda)^{-1}\right]^n\right\| \leqslant \frac{1}{(1 - q)^n} \cdot \frac{c}{|\lambda + \mu|^n} = \frac{c}{|\lambda + \mu'|^n}\,.$$

Hence, finally, it follows by Theorem 4.2.1 that $A + B$ is the generator of a strongly continuous semigroup $T' : [0, \infty) \to L(X, X)$ and (4.4.3). □

4.5 Strongly Continuous Groups

The theorem below gives necessary and sufficient conditions which have to be satisfied in order that two strongly continuous semigroups can be joined to one strongly continuous group. This is important for hyperbolic problems which for physical reasons should generally lead on strongly continuous groups.

Theorem 4.5.1. **(Strongly continuous groups)** Let $\mathbb{K} \in \{\mathbb{R}, \mathbb{C}\}$, $(X, \| \,\|)$ a $\mathbb{K}$-Banach space and $T : \mathbb{R} \to L(X, X)$ a *strongly continuous group*, i.e., such that

$$T(0) = \mathrm{id}_X \ , \ T(t+s) = T(t)T(s)$$

for all $t, s \in \mathbb{R}$ and

$$T\xi := \big(\mathbb{R} \to X, \xi \mapsto T(t)\xi\big) \in C\big(\mathbb{R}, X\big)$$

for all $\xi \in X$. We define the *(infinitesimal) generator* $A : D(A) \to X$ of T by

$$D(A) := \left\{ \xi \in X : \lim_{t \to 0, t \neq 0} \frac{1}{t}.[T(t) - \mathrm{id}_X]\xi \text{ exists} \right\}$$

and for every $\xi \in D(A)$:

$$A\xi := - \lim_{t \to 0, t \neq 0} \frac{1}{t}.[T(t) - \mathrm{id}_X]\xi \ .$$

Finally, let $S_+, S_- : [0, \infty) \to L(X, X)$ be strongly continuous semigroups with corresponding infinitesimal generators B_+ and B_- satisfying $B_- = -B_+$.

(i) A and $-A$ are the infinitesimal generators of the strongly continuous semigroups $T|_{[0,\infty)}$ and $T \circ (-\mathrm{id}_{\mathbb{R}})|_{[0,\infty)}$, respectively.

(ii) If in addition $a, b \in \mathbb{R}$ are such that $a < b$, $t_0 \in (a, b)$ and $\xi \in D(A)$, then $u : (a, b) \to X$, defined by

$$u(t) := T(t - t_0)\xi$$

for all $t \in (a, b)$, is the uniquely determined differentiable map such that $\mathrm{Ran}\, u \subset D(A)$, $u(t_0) = \xi$ and

$$u'(t) = -A\, u(t)$$

for all $t \in (a, b)$.

(iii) $S : \mathbb{R} \to L(X, X)$ defined by $S(t) := S_+(t)$ for all $t \in [0, \infty)$ and $S(t) := S_-(-t)$ for all $t \in (-\infty, 0)$ is a strongly continuous group with infinitesimal generator B_+.

Proof. '(i)': First, since $-\mathrm{id}_{\mathbb{R}}$ is additive, it follows that $T|_{[0,\infty)}$ and $T \circ (-\mathrm{id}_{\mathbb{R}})|_{[0,\infty)}$ are strongly continuous semigroups with corresponding infinitesimal generators A_+ and A_-, respectively. In particular, it follows that $A_+ \supset A$. Further, it follows for $\varepsilon > 0$, $t \in (0, \varepsilon)$ and $\xi \in D(A_+)$

$$\begin{aligned}\left\|\frac{1}{-t}.[T(-t)-\mathrm{id}_X]\xi+A_+\xi\right\| &= \left\|T(-t)\,\frac{1}{t}.[T(t)-\mathrm{id}_X]\xi+A_+\xi\right\| \\ &= \left\|T(-t)\left\{\frac{1}{t}.[T(t)-\mathrm{id}_X]\xi+A_+\xi\right\}-[T(-t)-\mathrm{id}_X]\,A_+\xi\right\| \\ &\leqslant C\left\|\frac{1}{t}.[T(t)-\mathrm{id}_X]\xi+A_+\xi\right\|+\|[T(-t)-\mathrm{id}_X]\,A_+\xi\|\end{aligned}$$

where $C \geqslant 0$ is such $\|T(-t)\| \leqslant C$ for all $t \in [0, \varepsilon]$. Hence it follows

$$-\lim_{t\to 0, t<0}\frac{1}{t}.[T(t)-\mathrm{id}_X]\xi = A_+\xi$$

and therefore $\xi \in D(A)$ and $A\xi = A_+\xi$. As a consequence, $A \supset A_+$ and hence finally $A_+ = A$. Application of the foregoing to the strongly continuous group $T \circ (-\mathrm{id}_{\mathbb{R}})$ leads to $A_- = -A$.
'(ii)': For this, let $a, b \in \mathbb{R}$ such that $a < b$, $t_0 \in (a, b)$, $\xi \in D(A)$ and $u : (a, b) \to X$ be defined by $u(t) := T(t-t_0)\xi$. Then it follows for $s_0 \in (a, t_0)$, $t \in (s_0, b)$ that $u(t) = T(t-s_0)T(s_0-t_0)\xi$ and hence by (i), Theorem 3.1.3 (ii),(v) that $\mathrm{Ran}\,u|_{(s_0,b)} \subset D(A)$, $u(t_0) = \xi$ and that $u|_{(s_0,b)}$ is differentiable with derivative

$$u'(t) = -AT(t-s_0)T(s_0-t_0)\xi = -Au(t)$$

for all $t \in (s_0, b)$. On the other hand, if $v : (a, b) \to X$ is differentiable such that $\mathrm{Ran}\,v \subset D(A)$, $v(t_0) = \xi$ and

$$v'(t) = -Av(t)$$

for all $t \in (a, b)$, then it follows by (i) and Theorem 3.1.3 (ii),(v) that $v(t) = T(t-t_0)\xi = u(t)$ for $t \in (t_0, b)$. Further, consider $\tilde{v} := ((-t_0, -a) \to X, t \mapsto v(-t))$. Then $\mathrm{Ran}\,\tilde{v} \subset D(A)$, $\lim_{t\to -t_0}\tilde{v}(t) = \xi$ and $\tilde{v}$ is differentiable such that

$$\tilde{v}' = -(-A)\,\tilde{v}(t)$$

for all $t \in (-t_0, -a)$. Hence it follows by (i) and Theorem 3.1.3 (ii),(v) that $v(-t) = \tilde{v}(t) = T(-(t+t_0))\xi$ for $t \in (-t_0, -a)$ and hence $v(t) = T(t-t_0)\xi = u(t)$ for all $t \in (a, t_0)$. Finally, it follows $v = u$.
'(iii)': First, it follows by Theorem 3.1.3 (ii), (v) for $\xi \in D(B_+)$, $t \geqslant 0$ that $u_\xi := ((0,\infty) \to X, s \mapsto S_-(t)S_+(s)\xi)$ maps into $D(B_+)$, is differentiable with derivative $u_\xi'(s) = -S_-(t)B_+S_+(s)\xi = -B_+u_\xi(s)$ for all $s \in (0,\infty)$ and satisfies

$$\lim_{s\to 0} S_-(t)S_+(s)\xi = S_-(t)\xi \in D(B_+)\,.$$

Hence it follows by Theorem 3.1.3 (v) that $u_\xi(s) = S_+(s)S_-(t)\xi$ and hence $S_-(t)S_+(s)\xi = S_+(s)S_-(t)\xi$ for all $s \in [0,\infty)$. Since $D(B_+)$ is dense in X this implies that

$$S_-(t)S_+(s) = S_+(s)S_-(t)$$

for all $t, s \in [0, \infty)$. Further, by

$$\begin{aligned}
&S_+(t+h)S_-(t+h)\xi - S_+(t)S_-(t)\xi \\
&= [S_+(t+h) - S_+(t)]S_-(t)\xi + [S_+(t+h) - S_+(t)][S_-(t+h)\xi - S_-(t)\xi] \\
&\quad + S_+(t)[S_-(t+h)\xi - S_-(t)\xi]
\end{aligned}$$

for all $\xi \in X$, $t \in [0,\infty)$, $h \in [-t,\infty)$ it follows that $([0,\infty) \to L(X,X), t \mapsto S_+(t)S_-(t))$ is a strongly continuous semigroup on X. By

$$\begin{aligned}
\frac{1}{t}.(S_+(t)S_-(t)\xi - \xi) &= \frac{1}{t}.(S_+(t) - \mathrm{id}_X)\xi + (S_+(t) - \mathrm{id}_X)\,\frac{1}{t}.(S_-(t) - \mathrm{id}_X)\xi \\
&\quad + \frac{1}{t}.(S_-(t) - \mathrm{id}_X)\xi
\end{aligned}$$

for $t \in (0,\infty)$, $\xi \in X$, it follows that $D(B_+)$ is contained in the domain of this continuous semigroup and that its infinitesimal generators vanishes on $D(B_+)$. Hence, since this generator is in particular closed and $D(B_+)$ is dense in X, that generator coincides with 0 operator on X. Since that operator generates $([0,\infty) \to X, t \mapsto \mathrm{id}_X)$, it follows

$$S_+(t)S_-(t) = S_-(t)S_+(t) = \mathrm{id}_X$$

and therefore the bijectivity of $S_+(t)$ and

$$S_-(t) = (S_+(t))^{-1}$$

for all $t \in [0,\infty)$. In particular, it follows for $t, s \in \mathbb{R}$, if $t \geqslant 0$ and $s \geqslant 0$

$$S(t+s) = S_+(t+s) = S_+(t)S_+(s) = S(t)S(s)\ ,$$

if $t \geqslant 0$ and $-t \leqslant s < 0$

$$\begin{aligned}
S(t+s) &= S_+(t+s) = S_+(t+s)S_+(-s)(S_+(-s))^{-1} = S_+(t)S_-(-s) \\
&= S(t)S(s)\ ,
\end{aligned}$$

if $t < 0$, $s \geqslant -t$

$$\begin{aligned}
S(t+s) &= S_+(t+s) = (S_+(-t))^{-1}S_+(-t)S_+(t+s) = S_-(-t)S_+(s) \\
&= S(t)S(s)\ ,
\end{aligned}$$

if $s < -t$

$$S(t+s) = S(s+t) = S(s)S(t) = S(t)S(s)$$

and hence that S is a strongly continuous group. Finally, it follows by (i) that the infinitesimal generator of S is given by B_+. □

Corollary 4.5.2. Let $\mathbb{K} \in \{\mathbb{R}, \mathbb{C}\}$, $(X, \| \|)$ a $\mathbb{K}$-Banach space and A a linear operator in X. Then A is the infinitesimal generator of a strongly continuous group T if and only if A and $-A$ are infinitesimal generators of strongly continuous semigroups T_+ and T_-, respectively. In this case, $T(t) = T_+(t)$ for all $t \in [0,\infty)$ and $T(t) := T_-(-t)$ for all $t \in (-\infty, 0)$.

4.6 Associated Inhomogeneous Initial Value Problems

This section treats the inhomogeneous differential equation that is associated with a strongly continuous semigroup. In particular, there is given a sufficient condition on the inhomogeneity such that the corresponding initial value problem is well-posed for data from the domain of the generator.

Lemma 4.6.1. Let $\mathbb{K} \in \{\mathbb{R}, \mathbb{C}\}$, $(X, \| \,\|)$ a $\mathbb{K}$-Banach space, $A : D(A) \to X$ the infinitesimal generator of a strongly continuous semigroup $T : [0, \infty) \to L(X, X)$. Further, let $t_0 \geqslant 0$, $t_1 > t_0$, $\xi \in X$ and $f : [t_0, t_1) \to X$ be continuous. Finally, let $u : (t_0, t_1) \to X$ be differentiable with $\operatorname{Ran} u \subset D(A)$, $\lim_{t \to t_0} u(t) = \xi$ and

$$u'(t) + A\,u(t) = f(t)$$

for all $t \in (t_0, t_1)$. Then

$$u(t) = T(t - t_0)\xi + \int_{t_0}^{t} T(t - s) f(s)\, ds \tag{4.6.1}$$

for all $t \in (t_0, t_1)$.

Proof. For this, let $t \in (t_0, t_1)$. We define $G : (t_0, t) \to X$ by $G(s) := T(t - s)u(s)$ for all $s \in (t_0, t)$. Then it follows for every $s \in (t_0, t)$ and $0 < h < t - s$

$$\begin{aligned} \frac{1}{h}\,[G(s + h) - G(s)] &= \frac{1}{h}\,[T(t - (s + h))u(s + h) - T(t - s)u(s)] \\ &= T(t - (s + h))\left\{\frac{1}{h}\,[u(s + h) - u(s)] - \frac{1}{h}\,[T(h) - \mathrm{id}_X]\,u(s)\right\} \end{aligned}$$

and hence

$$\lim_{h \to 0+} \frac{1}{h}\,[G(s + h) - G(s)] = T(t - s)[u'(s) + A\,u(s)] = T(t - s)f(s)\,.$$

Also it follows for every $t_0 - s < h < 0$

$$\begin{aligned} \frac{1}{h}\,[G(s + h) - G(s)] &= \frac{1}{h}\,[T(t - (s + h))u(s + h) - T(t - s)u(s)] \\ &= T(t - s)\left\{[T(|h|) - \mathrm{id}_X]\,\frac{1}{h}\,(u(s + h) - u(s)) - \frac{1}{|h|}\,[T(|h|) - \mathrm{id}_X]\,u(s) \right. \\ &\quad \left. + \frac{1}{h}\,[u(s + h) - u(s)]\right\} \end{aligned}$$

and hence

$$\lim_{h \to 0-} \frac{1}{h}\,[G(s + h) - G(s)] = T(t - s)[u'(s) + A\,u(s)] = T(t - s)f(s)\,.$$

Altogether, it follows the differentiability of G and $G'(s) = T(t - s)f(s)$ for all $s \in (t_0, t)$. Note that there is a continuous extension $\hat{G} : [t_0, t] \to X$ of G such that

$\hat{G}(t_0) = T(t - t_0)\,\xi$ and $\hat{G}(t) = u(t)$. Finally, note that $F : [t_0, t] \to X$ defined by $F(s) := T(t - s) f(s)$ for every $s \in [t_0, t]$ is continuous, too. Hence it follows by Theorem 3.2.9

$$u(t) - T(t - t_0)\xi = \hat{G}(t) - \hat{G}(t_0) = \int_{t_0}^{t} F(s)\, ds = \int_{t_0}^{t} T(t - s) f(s)\, ds$$

and hence (4.6.1). □

Theorem 4.6.2. Let $\mathbb{K} \in \{\mathbb{R}, \mathbb{C}\}$, $(X, \| \ \|)$ a $\mathbb{K}$-Banach space, $A : D(A) \to X$ the infinitesimal generator of a strongly continuous semigroup $T : [0, \infty) \to L(X, X)$. Further, let $t_0 \geqslant 0$, $t_1 > t_0$, $\xi \in D(A)$ and $f \in C([t_0, t_1), X)$ such that $\operatorname{Ran} f \subset D(A)$ and $Af := ([t_0, t_1) \to X, t \mapsto Af(t)) \in C([t_0, t_1), X)$. Then $u : [t_0, t_1) \to X$, defined by

$$u(t) := T(t - t_0)\xi + \int_{t_0}^{t} T(t - s) f(s)\, ds \tag{4.6.2}$$

for all $t \in [t_0, t_1)$ where integration denotes weak Lebesgue integration with respect to $L(X, \mathbb{K})$, is continuous and such that $u(t_0) = \xi$, $\operatorname{Ran} u \subset D(A)$, $Au := ([t_0, t_1) \to X, t \mapsto X) \in C([t_0, t_1), X)$. Finally, u is continuously differentiable on (t_0, t_1) such that

$$u'(t) + A\, u(t) = f(t) \tag{4.6.3}$$

for all $t \in (t_0, t_1)$.

Proof. First, we introduce some notation. We define $Y := D(A)$ and $\| \ \|_Y := \| \ \|_A$. Since A is closed, $(Y, \| \ \|_Y)$ is a Banach space Further, the inclusion $\iota_{Y \hookrightarrow X}$ of Y into X is continuous, $A \in L(Y, X)$, for every $t \in [0, \infty)$ the part $T_Y(t)$ of $T(t)$ in Y is a bounded linear operator on Y and $T_Y := ([0, \infty) \to L(Y, Y), t \mapsto T_Y(t))$ is strongly continuous. Further, it follows from the continuity of f and Af because of

$$\|f(t') - f(t)\|_Y \leqslant \|f(t') - f(t)\| + \|Af(t') - Af(t)\|$$

for all $t, t' \in [t_0, t_1)$ that $f \in C([t_0, t_1), Y)$. In addition, we define $I := [a, b]$, where $a, b \in \mathbb{R}$ are such that $t_0 < a$ and $b < t_1$,

$$\triangle(I) := \{(t, s) \in \mathbb{R}^2 : a \leqslant s \leqslant t \leqslant b\}$$

and $U \in C_*(\triangle(I), L(X, X))$ by $U := (\triangle(I) \to L(X, X), (t, r) \mapsto T(t - r))$ where here and in the following the lower index $*$ denotes strong continuity. Then $U_Y := (\triangle(I) \to L(Y, Y), (t, r) \mapsto T_Y(t - r)) \in C_*(\triangle(I), L(Y, Y))$. Then the Theorem is concluded as follows. First, it follows trivially that $u(t_0) = \xi$. Further, it follows for every $t \in (t_0, t_1)$, $s \in [t_0, t]$ and $h \in [t_0 - s, t - s]$:

$$\begin{aligned}
&\|U(t, s + h) f(s + h) - U(t, s) f(s)\|_Y \\
&\quad = \|U(t, s + h) f(s + h) - U(t, s + h) f(s) + U(t, s + h) f(s) - U(t, s) f(s)\|_Y \\
&\quad \leqslant \|U\|_{\infty, Y, Y} \cdot \|f(s + h) - f(s)\|_Y + \|U(t, s + h) f(s) - U(t, s) f(s)\|_Y
\end{aligned}$$

and hence because of $U_Y \in C_*(\Delta(I), L(Y,Y))$, $f \in C([t_0, t_1), Y)$ the continuity of $([t_0, t] \to Y, s \mapsto U(t,s)f(s))$ and therefore also the continuity of $([t_0, t] \to X, s \mapsto U(t,s)f(s))$. As a consequence, it follows that $u(t) \in Y$ and

$$\int_{t_0}^{t} U(t,s)f(s)\,ds = \int_{t_0,\mathrm{Y}}^{t} U(t,s)f(s)\,ds\ ,$$

for all $t \in [t_0, t_1)$ where the index Y denotes weak integration with respect to $L(Y, \mathbb{K})$. In particular, it follows for every $t \in [t_0, t_1)$ and $h \geqslant 0$ such that $t + h \in [t_0, t_1)$

$$\begin{aligned}
&\left\| \int_{t_0,\mathrm{Y}}^{t+h} U(t+h,s)f(s)\,ds - \int_{t_0,\mathrm{Y}}^{t} U(t,s)f(s)\,ds \right\|_{\mathrm{Y}} \\
&\leqslant \left\| \int_{t_0,\mathrm{Y}}^{t} (U(t+h,s)f(s) - U(t,s)f(s))\,ds \right\|_{\mathrm{Y}} + \left\| \int_{t,\mathrm{Y}}^{t+h} U(t+h,s)f(s)\,ds \right\|_{\mathrm{Y}} \\
&\leqslant \left\| (U(t+h,t) - U(t,t)) \int_{t_0,\mathrm{Y}}^{t} U(t,s)f(s)\,ds \right\|_{\mathrm{Y}} + \|U\|_{\infty,\mathrm{Y},\mathrm{Y}} \cdot \int_{t,\mathrm{Y}}^{t+h} \|f(s)\|_Y\,ds\ ,
\end{aligned}$$

for $h \leqslant 0$ such that $t + h \in [t_0, t_1)$

$$\begin{aligned}
&\left\| \int_{t_0,\mathrm{Y}}^{t+h} U(t+h,s)f(s)\,ds - \int_{t_0,\mathrm{Y}}^{t} U(t,s)f(s)\,ds \right\|_{\mathrm{Y}} \\
&\leqslant \left\| \int_{t_0,\mathrm{Y}}^{t+h} (U(t,s)f(s) - U(t+h,s)f(s))\,ds \right\|_{\mathrm{Y}} + \left\| \int_{t+h,\mathrm{Y}}^{t} U(t,s)f(s)\,ds \right\|_{\mathrm{Y}} \\
&\leqslant \int_{t_0}^{t+h} \|U(t,s)f(s) - U(t+h,s)f(s)\|_{\mathrm{Y}}\,ds + \|U\|_{\infty,\mathrm{Y},\mathrm{Y}} \cdot \int_{t+h}^{t} \|f(s)\|_Y\,ds
\end{aligned}$$

and hence by $U_Y \in C_*(\Delta(I), L(Y,Y))$ and Lebesgue's dominated convergence theorem that $u \in C([t_0, t_1), Y)$ and hence also that $u \in C([t_0, t_1), X)$ and $Au \in C([t_0, t_1), X)$. Finally, for every $(t, r) \in \mathrm{D} := \{(t, r) : r \in [t_0, t_1) \wedge t \in [r, b]\}$ and all $h, h' \in \mathbb{R}$ such that $(t, r) + (h, h') \in \mathrm{D}$,

$$\begin{aligned}
&\|AU(t+h', r+h)f(r+h) - AU(t,r)f(r)\| \\
&\leqslant \|U(t+h', r+h)f(r+h) - U(t,r)f(r)\|_{\mathrm{Y}} \\
&= \|U(t+h', r+h)\,(f(r+h) - f(r)) + (U(t+h', r+h) - U(t,r))f(r)\|_{\mathrm{Y}} \\
&\leqslant \|U\|_{\infty,\mathrm{Y},\mathrm{Y}} \cdot \|f(r+h) - f(r)\|_{\mathrm{Y}} + \|(U(t+h', r+h) - U(t,r))f(r)\|_{\mathrm{Y}}.
\end{aligned}$$

Hence $(D \to X, (t, r) \mapsto AU(t, r)f(r))$ is continuous. As consequence, it follows by the Theorem of Fubini that

$$\begin{aligned}
&\xi + \int_{t_0}^{t} (f(s) - A\,u(s))\,ds \\
&= \xi + \int_{t_0}^{t} \left[f(s) - A\left(U(s,t_0)\xi + \int_{t_0,\mathrm{Y}}^{s} U(s,s')f(s')\,ds' \right) \right] ds
\end{aligned}$$

$$
\begin{aligned}
&= \xi + \int_{t_0}^{t} \left[f(s) - A\,U(s,t_0)\xi - \int_{t_0}^{s} A\,U(s,s')f(s')\,ds' \right] ds \\
&= \xi + \int_{t_0}^{t} f(s)\,ds - \int_{t_0}^{t} A\,U(s,t_0)\xi\,ds - \int_{t_0}^{t} \left(\int_{t_0}^{s} A\,U(s,s')f(s')\,ds' \right) ds \\
&= \xi + \int_{t_0}^{t} f(s)\,ds - (\xi - U(t,t_0)\xi) - \int_{t_0}^{t} \left(\int_{s'}^{t} A\,U(s,s')f(s')\,ds \right) ds' \\
&= U(t,t_0)\xi + \int_{t_0}^{t} f(s)\,ds - \int_{t_0}^{t} (f(s') - U(t,s')f(s'))\,ds' \\
&= U(t,t_0)\xi + \int_{t_0}^{t} U(t,s)f(s)\,ds = u(t)
\end{aligned}
$$

for all $t \in [t_0, t_1)$. Hence, finally, it follows by Lemma 3.2.8 the relation (4.6.3) for all $t \in (t_0, t_1)$. □

5

Examples of Generators of Strongly Continuous Semigroups

5.1 The Ordinary Derivative on a Bounded Interval

In the following, let a be some positive real number different from zero. In addition, we denote by I the open interval $(0, a)$ with end points 0 and a, respectively, and by $\bar{I}$ its corresponding closure in $\mathbb{R}$. For applications see Chapters 7.1, 7.2.

Definition 5.1.1. We define the densely-defined linear operator A_r in $X := L^2_{\mathbb{C}}(I)$ by

$$A_r f := f'$$

for any $f \in C^1(\bar{I}, \mathbb{C})$ satisfying in addition

$$\lim_{x \to 0+} f(x) = 0 \tag{5.1.1}$$

We give now some facts on A_r and its adjoint operator A_r^*.

Theorem 5.1.2. The closure $\bar{A}_r$ of A_r is accretive with an *empty spectrum* and a compact resolvent and generates a contractive strongly continuous semigroup $T_r : [0, \infty) \to L(X, X)$. For any $t \in [0, \infty)$ and $f \in X$ the corresponding $T_r(t)f$ is given by the restriction of

$$\hat{f}(\mathrm{id}_{\mathbb{R}} - t)$$

to I. Here $\hat{f}$ denotes the extension of f to an almost everywhere defined function on $\mathbb{R}$ assuming zero values on the complement of I.

Proof. A_r is accretive since for any $f \in C^1(\bar{I}, \mathbb{C})$ with the additional property (5.1.1) it follows by partial integration that

$$\langle f | A_r f \rangle = |f(a)|^2 - \langle A_r f | f \rangle$$

and hence that

$$Re \langle f \,|\, A_r f \rangle = |f(a)|^2 \geqslant 0 \,.$$

Hence A_r is in particular closable with an accretive closure $\bar{A}_r$. For every complex number σ and every $f \in X$ we define the $R_\sigma f : I \to \mathbb{C}$ by

$$(R_\sigma f)(x) := e^{\sigma x} \int_0^x e^{-\sigma y} f(y) dy = \int_I K_\sigma(x,y) f(y) dy \ , \ x \in I \ ,$$

for all $x \in I$ where $K_\sigma : I^2 \to \mathbb{C}$ is defined by

$$K_\sigma(x,y) := \begin{cases} e^{\sigma(x-y)} & \text{for } x \geqslant y \\ 0 & \text{for } x < y \ . \end{cases}$$

$R_\sigma f$ is in particular continuous and has a unique extension to a continuous function on $\bar{I}$. That extension vanishes at 0. Hence by $R_\sigma(f) := R_\sigma f$ for all $f \in X$ there is defined a mapping $R_\sigma : X \to X$ which, obviously, is in particular linear. Moreover, R_σ is Hilbert-Schmidt and hence also compact since the kernel K_σ is an element of $L^2_{\mathbb{C}}(I^2)$. Moreover, by partial integration it follows for all $f \in C^1(\bar{I}, \mathbb{C})$ with the additional property (5.1.1) that

$$R_\sigma (A_r - \sigma) f = f$$

and for all $g \in C_0(I, \mathbb{C})$ that

$$(\bar{A}_r - \sigma) R_\sigma g = g \ .$$

Hence it follows also

$$R_\sigma (\bar{A}_r - \sigma) f = f$$

for all $f \in D(\bar{A}_0)$ and since $C_0(I, \mathbb{C})$ is dense in X and R_σ is in particular continuous that

$$(\bar{A}_r - \sigma) R_\sigma g = g$$

for all $g \in X$. Hence $\bar{A}_r - \sigma$ is bijective with a compact inverse and the spectrum of $\bar{A}_r$ is empty. Further, since $\bar{A}_r$ is accretive and, as a consequence of the foregoing for e.g., the range of $\bar{A}_r + 1$ is dense in X, it follows that $\bar{A}_r$ generates a contractive strongly continuous semigroup $T_r : [0, \infty) \to L(X, X)$.

In the following, let σ be some element of $(-\infty, 0) \times \mathbb{R}$ and $f \in X$. Then by

$$\tau_t f := \hat{f}(\mathrm{id}_{\mathbb{R}} - t)\big|_I$$

for all $t \in [0, \infty)$ there is given a bounded uniformly continuous function $\tau f : [0, \infty) \to X$ (see e.g. [185]). Note that for any $t \in [0, \infty)$ the corresponding $\tau_t f$ has its support in $[t, a]$ and hence that the support of τf is contained in $[0, a]$. Hence for any $g \in X$

$$\begin{aligned} \int_0^\infty e^{\sigma t} \langle g | \tau_t f \rangle dt &= \int_0^a g^*(x) \left[\int_0^x e^{\sigma t} f(x-t) dt \right] dx \\ &= \int_0^a g^*(x) \left[e^{\sigma x} \int_0^x e^{-\sigma y} f(y) dy \right] dx = \langle g | R_\sigma f \rangle \\ &= \int_0^\infty e^{\sigma t} \langle g | T_r(t) f \rangle dt \ . \end{aligned}$$

Since this is true for any $\sigma \in (-\infty, 0) \times \mathbb{R}$, we get from the injectivity of the Fourier transform on $L^1_{\mathbb{C}}(\mathbb{R})$ and the continuity of the involved functions:

$$\langle g|T_r(t)f\rangle = \langle g|\tau_t f\rangle \quad ,$$

for all $t \in [0, \infty)$ and hence

$$T_r(t)f = \tau_t f$$

for all $t \in [0, \infty)$. Since this true for any $f \in X$, finally, the theorem follows. □

As a corollary, it follows for the densely-defined linear operator A_l in X defined by

Definition 5.1.3.

$$A_l f := -f'$$

for any $f \in C^1(\bar{I}, \mathbb{C})$ satisfying in addition

$$\lim_{x \to a-} f(x) = 0 \ .$$

Corollary 5.1.4. The closure $\bar{A}_l$ of A_l is accretive with an *empty spectrum* and a compact resolvent and generates a contractive strongly continuous semigroup $T_l : [0, \infty) \to L(X, X)$. For any $t \in [0, \infty)$ and $f \in X$ the corresponding $T_l(t)f$ is given by the restriction of

$$\hat{f}(\mathrm{id}_{\mathbb{R}} + t)$$

to I.

Proof. The proof is a straightforward consequence of the identity

$$A_l = UA_rU$$

where $U : X \to X$ is the unitary linear map defined by

$$Uf := f \circ (a - \mathrm{id}_I)$$

for all $f \in X$. □

Corollary 5.1.5. The operator $\bar{A}_l$ is the adjoint operator of A_r

$$\bar{A}_l = A_r^* \ .$$

Proof. By using

$$A_r^* = (\bar{A}_r)^* \ ,$$

it follows that the spectrum of A_r^* is equal to the complex-conjugate of the spectrum of $\bar{A}_r$ and hence by Theorem 5.1.2 empty and that

$$(A_r^*)^{-1} = (\bar{A}_r^{-1})^* \ .$$

Now by the proof of Theorem 5.1.2 we have:

$$\left(\bar{A}_r^{-1} f\right)(x) = \int_0^x f(y)\,dy$$

for all $x \in I$ and $f \in X$ and hence

$$\left((A_r^*)^{-1} f\right)(x) = \int_x^a f(y)\,dy$$

for all $x \in I$ and $f \in X$. Further, by the proof of Corollary 5.1.4

$$\bar{A}_l^{-1} = U \bar{A}_r^{-1} U$$

and hence

$$\left(\bar{A}_l^{-1} f\right)(x) = \left(\bar{A}_r^{-1} U f\right)(a - x) = \int_0^{a-x} f(a - y)\,dy$$

$$= \int_x^a f(y')\,dy' = \left((A_r^*)^{-1} f\right)(x)$$

for all $x \in I$ and $f \in X$. From this follows

$$(A_r^*)^{-1} = \bar{A}_l^{-1}$$

and finally the corollary. □

5.2 Linear Stability of Ideal Rotating Couette Flows

In the following we consider Couette flow [43, 51], i.e., an incompressible ideal fluid contained between two infinitely long concentric cylinders of radii R_1, R_2 ($R_2 > R_1 > 0$) rotating about there common axis, which is assumed to be the z-axis of a Cartesian coordinate system, with angular velocities Ω_1 and Ω_2, respectively. Stationary divergence free axial solutions of Euler's equation with vanishing normal components at the cylinders are given by

$$v_0(x, y, z) := \left(-y \cdot \Omega\left(\sqrt{x^2 + y^2}\right), x \cdot \Omega\left(\sqrt{x^2 + y^2}\right), 0\right)$$

$$p_0(x, y, z) := \rho \int_{R_1}^{\sqrt{x^2+y^2}} r^2\, \Omega^2(r)\,dr$$

for all $(x, y, z) \in U \times \mathbb{R}$ where

$$U := \{(x, y) \in \mathbb{R}^2 : R_1^2 < x^2 + y^2 < R_2^2\} ,$$

$\rho > 0$ is the (constant) density of the fluid and Ω is some element of $C^2(I, \mathbb{R}) \cap C(\bar{I}, \mathbb{R})$. Here $I := (R_1, R_2)$. Note that

$$v_0 = \nabla \times (0, 0, \psi_0)$$

where

$$\psi_0(x, y, z) := -\int_{R_1}^{\sqrt{x^2+y^2}} r\,\Omega(r)\,dr$$

for all $(x, y, z) \in U \times \mathbb{R}$ and that the vorticity $\omega_0 := \nabla \times v_0$ is given by

$$\omega_0(x, y, z) = (0, 0, -(\triangle\psi_0)(x, y, z)) = \left(0, 0, \omega_{0z}\left(\sqrt{x^2 + y^2}\right)\right)$$

for all $(x, y, z) \in U \times \mathbb{R}$. Here $\omega_{0z} : I \to \mathbb{R}$ is defined by

$$\omega_{0z}(r) := r\,\Omega'(r) + 2\,\Omega(r)$$

for all $r \in I$. In the vorticity formulation the governing equation for reduced small axial variations of such a ω_0 of the form

$$(0, 0, \omega(r, z)\exp(im\varphi)),$$

$(r, z) \in I \times \mathbb{R}, \varphi \in (-\pi, \pi)$, where $m \in \mathbb{Z}$ and polar coordinates have been introduced in the (x, y)-plane, is given by

$$\frac{\partial\omega}{\partial t} = -A_m\,\omega\,.$$

Here

$$(A_m\omega)(r, z) := im\left[\Omega(r)\,\omega(r, z) + \frac{\omega_{0z}'(r)}{r^2}\int_{R_1}^{R_2} G_m(r, r')\,\omega(r', z)\,dr'\right] \tag{5.2.1}$$

where the kernel G_m is defined by

$$G_m(r, r') := \begin{cases} -f_{m2}(r)\,f_{m1}(r') & \text{for } r' < r \\ -f_{m1}(r)\,f_{m2}(r') & \text{for } r' \geqslant r \end{cases}$$

for all $(r, r') \in I^2$ and

$$f_{01} := -\frac{r}{\ln\eta}\ln\left(\frac{r}{R_1}\right),$$

$$f_{02} := r\ln\left(\frac{r}{R_2}\right)$$

for $m = 0$,

$$f_{m1} := \frac{\eta^m}{2m(1-\eta^{2m})}\left(R_1^m r^{-(m-1)} - R_1^{-m} r^{m+1}\right),$$

$$f_{m2} := R_2^m r^{-(m-1)} - R_2^{-m} r^{m+1}\,.$$

for $m \neq 0$ for all $r \in I$ and

$$\eta := \frac{R_1}{R_2} .$$

Note that we use the same symbol for functions in Cartesian coordinates and their corresponding transformations in cylindrical coordinates. Obviously, for every $m \in \mathbb{Z}$ the corresponding $G_m : I^2 \to \mathbb{R}$ can be extended to a continuous and symmetric function on $\bar{I}^2$. Hence it follows

$$G_m \in L^2_{\mathbb{C}}(I^2)$$

and that by

$$[\text{Int}(G_m)f](r) := \int_{R_1}^{R_2} G_m(r, r')\, f(r')\, dr'$$

for all $r \in I$ and every $f \in L^2_{\mathbb{C}}(I)$, there is defined a linear self-adjoint Hilbert-Schmidt and hence also *compact* operator on $L^2_{\mathbb{C}}(I)$.

Theorem 5.2.1. Let

$$\omega'_{0z} \in L^{\infty}(I) .$$

Then it follows for every $m \in \mathbb{Z}$:

(i)

$$A_m := im.\left(\mathrm{T}_{\Omega} + \mathrm{T}_{\omega'_{0z}/id_{\mathbb{R}}}\text{Int}(G_m)\right) ,$$

where for every a.e. on I defined complex-valued and measurable function f the symbol T_f denotes the maximal multiplication operator in $L^2_{\mathbb{C}}(I)$, is a *bounded* linear operator on $L^2_{\mathbb{C}}(I)$ and hence the generator of a strongly continuous semigroup on $L^2_{\mathbb{C}}(I)$.

(ii) The essential spectrum $\sigma_{\text{ess}}(A_m)$ of A_m, defined by its spectrum minus its isolated eigenvalues of finite multiplicity, is given by

$$\sigma_{\text{ess}}(A_m) = \overline{\text{Ran}(im\Omega)} . \tag{5.2.2}$$

(iii) If ω'_{0z} is not a.e. zero as well as either a.e. positive or a.e. negative on I:

$$\sigma(A_m) \subset i.\mathbb{R} .$$

Proof. For this, let $m \in \mathbb{Z}$. '(i)': Since Ω and $\omega'_{0z}/id_{\mathbb{R}}$ are bounded, it follows that $\mathrm{T}_{\Omega}, \mathrm{T}_{\omega'_{0z}/id_{\mathbb{R}}}$ are bounded linear operators on $L^2_{\mathbb{C}}(I)$ and hence also that A_m is a bounded linear operator on $L^2_{\mathbb{C}}(I)$.
'(ii)': Since $\text{Int}(G_m)$ is in particular compact, it follows also the compactness of $\mathrm{T}_{\omega'_{0z}id_{\mathbb{R}}} \circ \text{Int}(G_m)$ and therefore by 'Weyl's essential spectrum theorem', for e.g., see Corollary 2 of Theorem XIII.14 in Vol. IV of [179], that $\mathrm{T}_{im\Omega}$ and A_m have the same essential spectrum. Since Ω is continuous, the spectrum $\sigma(\mathrm{T}_{im\Omega})$ of $\mathrm{T}_{im\Omega}$ is given by

$$\sigma(\mathrm{T}_{im\Omega}) = \overline{\text{Ran}(im\Omega)} .$$

Now, let $\lambda \in \sigma(\mathrm{T}_{im\Omega})$ be in particular an eigenvalue of finite multiplicity of $\mathrm{T}_{im\Omega}$. Then $(im\Omega)^{-1}(\lambda)$ is not a set of measure zero. Since $im\Omega$ is continuous, it follows that $(im\Omega)^{-1}(\lambda)$ is bounded as well as measurable and hence also summable and of measure > 0. Hence there exists for any $n \in \mathbb{N}^*$ a decomposition of $(im\Omega)^{-1}(\lambda)$ into disjoint subsets of positive measure. This leads to the contradiction that $\mathrm{Ker}(\mathrm{T}_{im\Omega})$ is not finite dimensional. As a consequence, $\mathrm{T}_{im\Omega}$ has no eigenvalues of finite multiplicity and it follows (5.2.2).
'(iii)': For this, let ω_{0z}' be not a.e. zero as well as either a.e. positive or a.e. negative on I. Further, let $\lambda \in \mathbb{C}$ be an eigenvalue of finite multiplicity of A_m and $f \in L^2_{\mathbb{C}}(I)$ be a corresponding eigenvector. Then

$$m.\mathrm{Int}(G_m)f = -\frac{id_{\mathbb{R}}}{\omega_{0z}'}(m\Omega + i\lambda)f$$

and hence

$$m\,\langle f\,|\,\mathrm{Int}(G_m)f\rangle = -\int_I \frac{id_{\mathbb{R}}}{\omega_{0z}'}(m\Omega + i\lambda)\,|f|^2\,dv^1\,.$$

Since $\mathrm{Int}(G_m)$ is in particular self-adjoint, this implies that

$$\mathrm{Re}(\lambda)\cdot\int_I \frac{id_{\mathbb{R}}}{\omega_{0z}'}\,|f|^2\,dv^1 = 0$$

and hence that $\mathrm{Re}(\lambda) = 0$. □

The evolution of 'small' perturbations of ideal rotating Couette flows is one of the rare cases in applications where the governing operator is a *bounded* linear operator.[1] From a physical point of view, the example is interesting because of the occurrence of a continuous part in the oscillation spectrum of a *finitely extended* system. On the other hand, in Quantum Theory such occurrence would be very surprising because continuous parts in the spectrum of a Schrödinger operator are usually associated to scattering states reaching infinity. See [26] for a similar example of an occurrence of a continuous part in the oscillation spectrum of *rigidly* rotating general relativistic stars. See [18] for the proof of the existence of unstable inviscid *plane* Couette flows under the assumption of periodic boundary conditions. For further results on the stability of Couette flows, see [34, 184].

5.3 Outgoing Boundary Conditions

In the following, we give a well-posed formulation of the initial-boundary value problem for

$$\frac{\partial^2 u}{\partial t^2} - \frac{\partial^2 u}{\partial x^2} + Vu = 0 \qquad (5.3.1)$$

[1] Note that instead of $L^2_{\mathbb{C}}(I)$ a large number of other choices of the representation space of the formal operator A_m in (5.2.1) is possible.

on the interval $I = (0, a)$, where $a > 0$, for standard Sommerfeld outgoing boundary conditions

$$\left(\frac{\partial u}{\partial t} - \frac{\partial u}{\partial x}\right)\Big|_{x=0} = 0\,, \quad \left(\frac{\partial u}{\partial t} + \frac{\partial u}{\partial x}\right)\Big|_{x=a} = 0$$

and Engquist-Majda [58] outgoing boundary conditions

$$\left(\frac{\partial^2 u}{\partial x \partial t} - \frac{\partial^2 u}{\partial x^2} + \frac{V}{2}u\right)(0) = 0\,, \left(\frac{\partial^2 u}{\partial x \partial t} + \frac{\partial^2 u}{\partial x^2} - \frac{V}{2}u\right)(a) = 0\,.$$

The choice of the used function spaces is suggested by formulations of (5.3.1) as a first order system given in Chapters 7.1, 7.2. The motivation of considering such initial boundary value problems comes from numerical evolutions. Frequently, in such evolutions it is performed an artificial space cut off because of the necessarily finite extension of computational domains. Therefore, boundary conditions have to be posed in such a way that the numerical solution of the cut off system approximates as best as possible the solution of the original problem on infinite space. Ideally, the boundaries of the cutoff problem should be 'transparent', i.e., let energy dissipate through the boundaries. In particular, reflections from the boundaries have to be minimized. If the data have compact support inside the computational domain, then there should be no incoming waves entering that domain.[2] For this reason, such conditions are named 'outgoing' boundary conditions. The formulation of such boundary conditions for Einstein's evolution equations for the gravitational field is an important current problem in numerical relativity. See, for instance, [181, 192, 211] and for a review on the Cauchy problem for Einstein's equations [78]. For a review of the formulation of outgoing boundary conditions for wave equations, see [85]. In this connection, see also [3, 16, 92]. For an approach to the difference method in the numerics of PDEs which is related to the semigroup method, see [123–125]. In this connection, see [55] for a major generalization of Kreiss' condition. Theorem 5.3.4 below is a new result.

Definition 5.3.1.

(i) We define for every $k \in \mathbb{N}$ the densely-defined linear operator

$$D_I^k : C_0^\infty(I, \mathbb{C}) \to L^2_{\mathbb{C}}(I)$$

in $L^2_{\mathbb{C}}(I)$ by

$$D_I^k f := f^{(k)}\,,$$

for every $f \in C_0^\infty(I, \mathbb{C})$.

(ii) For any $n \in \mathbb{N}^*$,

$$W^n_{\mathbb{C}}(I) := \bigcap_{k=0}^{n} D\left(D_I^{k*}\right)\,.$$

[2] In this strictness, this is reasonable only for linear constant coefficient systems.

Equipped with the scalar product

$$\langle , \rangle_n \; : (W^n_{\mathbb{C}}(I))^2 \to \mathbb{C}$$

defined by

$$\langle f, g \rangle_n := \sum_{k=0}^{n} \langle D_I^{k*} f | D_I^{k*} g \rangle_2$$

for all $f, g \in W^n_{\mathbb{C}}(I)$, $W^n_{\mathbb{C}}(I)$ is a complex Hilbert space.

(iii) Let $V \in C(\bar{I}, \mathbb{R})$ be such that $V \geqslant \varepsilon$ for some $\varepsilon > 0$. Then, obviously, $W^1_{\mathbb{C}}(I)$ equipped with the scalar product

$$\langle f | g \rangle_1 := \langle D_I^{1*} f | D_I^{1*} g \rangle + \langle V^{1/2} f | V^{1/2} g \rangle$$

for all $f, g \in W^1_{\mathbb{C}}(I)$ is a Hilbert space with induced topology being equivalent to the standard topology on $W^1_{\mathbb{C}}(I)$.

(iv) Finally, we define the linear operator

$$A_V : D(A_V) \to W^1_{\mathbb{C}}(I) \times L^2_{\mathbb{C}}(I)$$

in $X := W^1_{\mathbb{C}}(I) \times L^2_{\mathbb{C}}(I)$ by

$$D(A_V) := \{(f, g) \in C^2(\bar{I}, \mathbb{C}) \times C^1(\bar{I}, \mathbb{C}) : f'_0 - g_0 = f'_a + g_a = 0\}$$

and

$$A_V(f, g) := (-g, -f'' + Vf)$$

for all $(f, g) \in D(A_V)$.

Theorem 5.3.2.

(i) A_V is a densely-defined, linear and accretive operator in X.

(ii) $\bar{A}_V$ generates a contractive strongly continuous semigroup $T_V \; : \; [0, \infty) \to L(X, X)$.

Proof. Obviously, by A_V there is defined a linear operator in X. That $D(A_V)$ is dense in X can be proved as follows. For this, let (f, g) be some element of X and ε some nonvanishing positive real number. Since $C^2(\bar{I}, \mathbb{C})$ is dense in $W^1_{\mathbb{C}}(I)$, there is some $u \in C^2(\bar{I}, \mathbb{C})$ such that

$$\|u - f\|_1 \leqslant \sqrt{\varepsilon^2/2} \, .$$

Further, since $C_0^\infty(I, \mathbb{C})$ is dense in $L^2_{\mathbb{C}}(I)$, there is some $v_1 \in C_0^\infty(I, \mathbb{C})$ such that

$$\|v_1 - g\| \leqslant \sqrt{\varepsilon^2/8} \, .$$

Obviously, in addition, there is some $v_2 \in C^1(\bar{I}, \mathbb{C})$ such that

$$v_{20} = u'_0 \, , \; v_{2a} = -u'_a$$

and at the same time such that

$$\|v_2\| \leqslant \sqrt{\varepsilon^2/8} \, .$$

Then by construction $(u, v_1 + v_2) \in D(A_V)$ and

$$\begin{aligned}\|(f,g) - (u, v_1 + v_2)\|^2 &= \|f - u\|_1^2 + \|g - (v_1 + v_2)\|^2 \\ &\leqslant \|f - u\|_1^2 + 2\|g - v_1\|^2 + 2\|v_2\|^2 \leqslant \varepsilon^2 \, .\end{aligned}$$

Since ε and (f,g) were otherwise arbitrary, from this it follows that $D(A_V)$ is dense in X.

For the proof that A_V is accretive, let (u, v) be some otherwise arbitrary element from $D(A_V)$. Moreover, denote by $(\,|\,)$ the scalar product of X. Then we conclude by partial integration and by using the definition of $D(A_V)$ that

$$\mathrm{Re}\,((u,v)\,|\,A_V(u,v)) = \mathrm{Re} \int_0^a [u'^*(-v') + Vu^*(-v) + v^*(-u'' + Vu)]\, dx$$

$$= -\frac{1}{2}\int_0^a (v^*u' + vu'^*)'\, dx = |v_a|^2 + |v_0|^2 \geqslant 0$$

and hence that A_V is accretive. Finally, the statement (i) follows.

The statement of (ii) follows from that of (i) if it can be shown that $\mathrm{Ran}(A_V + \lambda)$ is dense in X for some real $\lambda > 0$. For this, let λ be such a real number and let f, g be some elements of $C^1(\bar{I}, \mathbb{C})$ and $C_0(I, \mathbb{C})$, respectively. In addition, we define

$$\alpha := \frac{f_a - f_0}{\lambda a + 2} \, , \; \beta := \frac{1}{\lambda}\frac{f_a + f_0 + \lambda a f_0}{\lambda a + 2}$$

and $u_s \in C^\infty(\bar{I}, \mathbb{C})$ by

$$u_s(x) := \alpha x + \beta \, , \; x \in I \, .$$

Finally, consider the operator $B : D(B) \to L^2_{\mathbb{C}}(I)$ defined by

$$D(B) := \{u \in C^2(\bar{I}, \mathbb{C}) : u_0' - \lambda u_0 = u_a' + \lambda u_a = 0\}$$

and

$$Bu := -u'' + (V + \lambda^2)u \, , \; u \in D(B) \, .$$

By the theory of regular Sturm-Liouville operators, B is a densely-defined, linear, symmetric and essentially self-adjoint operator in $L^2_{\mathbb{C}}(I)$. By partial integration, it follows for every $u \in D(B)$ that

$$\langle u|Bu\rangle = \lambda\,(|u_a|^2 + |u_0|^2) + \int_0^a [\,|u'|^2 + (V + \lambda^2)|u|^2\,]\, dx \geqslant \lambda^2\|u\|^2$$

and hence that $\bar{B}$ is bijective. Again, by the theory of regular Sturm-Liouville operators, from this follows the existence of $u \in D(B)$ such that

$$-u'' + (V + \lambda^2)u = \lambda f + g - (V + \lambda^2)u_s \, .$$

Hence $u + u_s$ is an element of $C^2(\bar{I}, \mathbb{C})$ such that

$$- (u + u_s)'' + (V + \lambda^2)(u + u_s) = \lambda f + g$$
$$(u + u_s)'_0 - \lambda(u + u_s)_0 + f_0 = (u + u_s)'_a + \lambda(u + u_s)_a - f_a = 0$$

and, finally,

$$(u + u_s, \lambda(u + u_s) - f) \in D(A_v)$$

and

$$(A_V + \lambda)(u + u_s, \lambda(u + u_s) - f) = (f, g) .$$

Since f, g were otherwise arbitrary and $C^1(\bar{I}, \mathbb{C}) \times C_0(I, \mathbb{C})$ is dense in X, from this follows that $\mathrm{Ran}(A_V + \lambda)$ is dense in X and hence finally (ii) and the theorem. □

Definition 5.3.3.

(i) Let V be some element of $C^1(\bar{I}, \mathbb{R})$ such that $V \geqslant \varepsilon$ for some $\varepsilon > 0$. Then we define the positive definite Hermitian sesquilinear forms $\langle | \rangle_1, \langle | \rangle_2$ by

$$\langle f|g\rangle_1 := 4 \langle D_I^{1*} f | D_I^{1*} g\rangle + 2\langle V^{1/2} f | V^{1/2} g\rangle$$

for all $f, g \in W^1_{\mathbb{C}}(I)$ and

$$\langle f|g\rangle_2 := \langle (2\, D_I^{2*} - V) f | (2\, D_I^{2*} - V) g\rangle + 2\, \langle V^{1/2} D_I^{1*} f | V^{1/2} D_I^{1*} g\rangle + \langle V f | V g\rangle$$

for all $f, g \in W^2_{\mathbb{C}}(I)$. By the inequalities

$$\|(2D_I^{2*} - V)f\|^2 \leqslant 6\|D_I^{*2} f\|^2 + 3\|Vf\|^2 \ , \ \|(2D_I^{2*} - V)f\|^2 \geqslant 2\|D_I^{2*} f\|^2 - \|Vf\|^2$$

for all $f \in W^2_{\mathbb{C}}(I)$, it follows that the induced topologies are equivalent to the standard topologies on $W^1_{\mathbb{C}}(I)$ and $W^2_{\mathbb{C}}(I)$, respectively.

(ii) Finally, we define the linear operator

$$A_{V,1} : D(A_{V,1}) \to W^2_{\mathbb{C}}(I) \times W^1_{\mathbb{C}}(I)$$

in $X_1 := W^2_{\mathbb{C}}(I) \times W^1_{\mathbb{C}}(I)$ by

$$D(A_{V,1}) : = \{(f, g) \in C^3(\bar{I}, \mathbb{C}) \times C^2(\bar{I}, \mathbb{C}) : 2f''_0 - V_0 f_0 - 2g'_0 = 2f''_a - V_a f_a + 2g'_a = 0\}$$

and

$$A_{V,1}(f, g) := (-g, -f'' + Vf)$$

for all $(f, g) \in D(A_{V,1})$.

Theorem 5.3.4.

(i) $A_{V,1}$ is a densely-defined, linear and quasi-accretive operator in X_1 with left bound

$$-\|(\ln(V))'\|_\infty .$$

(ii) $\bar{A}_{V,1}$ generates a strongly continuous semigroup $T_{V,1} : [0,\infty) \to L(X_1, X_1)$ satisfying

$$\|T_{V,1}(t)\| \leqslant \exp(\|(\ln(V))'\|_\infty \cdot t)$$

for all $t \in [0,\infty)$.

Proof. Obviously, by $A_{V,1}$ there is defined a linear operator in X_1. That $D(A_{V,1})$ is dense in X_1 can be proved as follows. For this, let (f, g) be some element of X_1 and ε some nonvanishing positive real number. Since $C^3(\bar{I},\mathbb{C})$ is dense in $W^2_{\mathbb{C}}(I)$, there is some $u \in C^3(\bar{I},\mathbb{C})$ such that

$$\|u - f\|_2 \leqslant \sqrt{\varepsilon^2/2}\ .$$

Further, since $C^2(\bar{I},\mathbb{C})$ is dense in $W^1_{\mathbb{C}}(I)$, there is some $v_1 \in C^2(\bar{I},\mathbb{C})$ such that

$$\|v_1 - g\|_1 \leqslant \sqrt{\varepsilon^2/8}\ .$$

Finally, define

$$s_0 := u_0'' - \frac{1}{2}V_0u_0 - v_{10}'\ ,\ \ s_a := -u_a'' + \frac{1}{2}V_au_a - v_{1a}'$$

and for every $\delta \in (0, a/2)$

$$v_2(x) := \begin{cases} (s_0/(3\delta^2)) \cdot (x-\delta)^3 & \text{for } 0 < x < \delta \\ 0 & \text{for } \delta \leqslant x \leqslant a - \delta \\ (s_a/(3\delta^2)) \cdot (x+\delta-a)^3 & \text{for } a-\delta < x < a\ . \end{cases}$$

Then v_2 is an element of $C^2(\bar{I},\mathbb{C})$ such that

$$v_{20}' = s_0\ ,\ \ v_{2a}' = s_a\ .$$

Moreover, some calculation gives

$$\int_0^a (|v_2'|^2 + |v_2|^2)\,dx = (|s_a|^2 + |s_0|^2)\left(\frac{1}{5} + \frac{\delta^2}{63}\right)\delta\ .$$

Hence there is some $\delta \in (0, a/2)$ such that

$$\|v_2\|_1 \leqslant \sqrt{\varepsilon^2/8}\ .$$

By construction, for such a δ $(u, v_1 + v_2) \in D(A_{V,1})$ and

$$\begin{aligned}\|(f,g) - (u, v_1+v_2)\|^2 &= \|f-u\|_2^2 + \|g-(v_1+v_2)\|_1^2 \\ &\leqslant \|f-u\|_2^2 + 2\|g-v_1\|_1^2 + 2\|v_2\|_1^2 \leqslant \varepsilon^2\ .\end{aligned}$$

Since ε and (f, g) were otherwise arbitrary, from this follows that $D(A_{V,1})$ is dense in X_1.

For the proof that $A_{V,1}$ is accretive, let (u,v) be some otherwise arbitrary element from $D(A_{V,1})$. Moreover, denote by $(\,|\,)$ the scalar product of X_1. Then we conclude by partial integration, the Hölder inequality for $L^2_{\mathbb{C}}(I)$ and by using the definition of $D(A_{V,1})$ that

$$\begin{aligned}
&\operatorname{Re}((u,v)\,|\,A_{V,1}(u,v)) \\
&= \operatorname{Re}\int_0^a \big[-(2u''-Vu)^*(2u''-Vu) - 2Vu'^*v' - V^2u^*v + 4V'(-u''+Vu)' \\
&\qquad +2Vv^*(-u''+Vu)\big]\,dx \\
&= 4(|v_a'|^2+|v_0'|^2) + 2\operatorname{Re}\int_0^a V'v'^*u\,dx \geqslant -\|(\ln(V))'\|_\infty\,\|(u,v)\|^2
\end{aligned}$$

and hence that $A_{V,1}$ is accretive. Finally, the statement (i) follows.

The statement of (ii) follows from that of (i) if it can be shown that $\operatorname{Ran}(A_{V,1}+\lambda)$ is dense in X_1 for some real $\lambda > 0$. For this let λ be such a real number and let f,g be some elements of $C^2(\bar{I},\mathbb{C})$ and $C^1(\bar{I},\mathbb{C})$, respectively. In addition, let α,β be complex numbers such that $u_s \in C^\infty(\bar{I},\mathbb{C})$ defined by

$$u_s(x) := \alpha x + \beta\ ,\ x \in I\ .$$

satisfies

$$\begin{aligned}
u'_{s0} - \left(\lambda + \frac{V_0}{2\lambda}\right)u_{s0} &= \frac{1}{\lambda}\,(f_0' - \lambda f_0 - g_0) \\
u'_{sa} + \left(\lambda + \frac{V_a}{2\lambda}\right)u_{sa} &= \frac{1}{\lambda}\,(f_a' + \lambda f_a + g_a)\ .
\end{aligned}$$

Obviously, such α,β do exist. Finally, consider the operator $B : D(B) \to L^2_{\mathbb{C}}(I)$ defined by

$$D(B) := \{u \in C^2(\bar{I},\mathbb{C}) : u_0' - [\lambda + (V_0/(2\lambda))]u_0 = u_a' + [\lambda + (V_a/(2\lambda))]u_a = 0\}$$

and

$$Bu := -u'' + (V+\lambda^2)u\ ,\ u \in D(B)\ .$$

By the theory of regular Sturm-Liouville operators, B is a densely-defined, linear, symmetric and essentially self-adjoint operator in $L^2_{\mathbb{C}}(I)$. By partial integration, it follows for every $u \in D(B)$ that

$$\begin{aligned}
\langle u|Bu\rangle &= \left(\lambda + \frac{V_a}{2\lambda}\right)|u_a|^2 + \left(\lambda + \frac{V_0}{2\lambda}\right)|u_0|^2 \\
&\quad + \int_0^a [\,|u'|^2 + (V+\lambda^2)|u|^2\,]\,dx \geqslant \lambda^2\|u\|^2
\end{aligned}$$

and hence that $\bar{B}$ is bijective. Again by the theory of regular Sturm-Liouville operators, from this follows the existence of $u \in D(B) \cap C^3(\bar{I}, \mathbb{C})$ such that

$$-u'' + (V + \lambda^2)u = \lambda f + g - (V + \lambda^2)u_s \, .$$

Hence $u + u_s$ is an element of $C^3(\bar{I}, \mathbb{C})$ such that

$$- (u + u_s)'' + (V + \lambda^2)(u + u_s) = \lambda f + g$$
$$(u + u_s)'_0 - \left(\lambda + \frac{V_0}{2\lambda}\right)(u + u_s)_0 = \frac{1}{\lambda}(f'_0 - \lambda f_0 - g_0)$$
$$(u + u_s)'_a + \left(\lambda + \frac{V_a}{2\lambda}\right)(u + u_s)_a = \frac{1}{\lambda}(f'_a + \lambda f_a + g_a)$$

and finally

$$(u + u_s, \lambda(u + u_s) - f) \in D(A_{V,1})$$

and

$$(A_{V,1} + \lambda)(u + u_s, \lambda(u + u_s) - f) = (f, g) \, .$$

Since f, g were otherwise arbitrary and $C^2(\bar{I}, \mathbb{C}) \times C^1(\bar{I}, \mathbb{C})$ is dense in X_1, from this follows that $\mathrm{Ran}(A_{V,1} + \lambda)$ is dense in X_1 and hence finally (ii) and the theorem. □

5.4 Damped Wave Equations

In the following, we consider wave equations of the form

$$(u')'(t) + iBu'(t) + (A + C)u(t) = 0$$

for all $t \in \mathbb{R}$ where A, B and C are suitable linear operators. Such equations occur in particular in the description of rotating systems in General Relativity and Astrophysics. The main stress of this section is on the development of sufficient conditions for the stability of the solutions of such equations, i.e., the absence of solutions that grow exponentially with time. For applications of the results in this section, see [27, 28, 54]. For work on the stability of the solutions of the wave equation and Dirac equations on the background of a rotating Kerr black hole using related methods, see [62–65].

In addition, there is a large literature on damped wave equations. See, for instance, [56, 89, 91, 122, 126, 133, 154, 156, 162, 189, 190, 218, 223, 226]. For additional references, see [57], Section VI.3.

Assumption 5.4.1. In the following, let $(X, \langle | \rangle)$ be a non trivial complex Hilbert space. Denote by $\| \, \|$ the norm induced on X by $\langle | \rangle$. Further, let $A : D(A) \to X$ be a densely-defined, linear self-adjoint operator in X for which there is an $\varepsilon \in (0, \infty)$ such that

$$\langle \xi | A\xi \rangle \geqslant \varepsilon \langle \xi | \xi \rangle \tag{5.4.1}$$

for all $\xi \in D(A)$. Denote by $A^{1/2}$ the square root of A with domain $D(A^{1/2})$. Further, let $B : D(A^{1/2}) \to X$ be a linear operator in X such that for some $a \in [0, 1)$ and $b \in \mathbb{R}$

$$\|B\xi\|^2 \leqslant a^2 \|A^{1/2}\xi\|^2 + b^2 \|\xi\|^2 \tag{5.4.2}$$

for all $\xi \in D(A^{1/2})$. Finally, let $C : D(A^{1/2}) \to X$ be linear and such that for some real numbers c and d

$$\|C\xi\|^2 \leqslant c^2 \|A^{1/2}\xi\|^2 + d^2 \|\xi\|^2 \tag{5.4.3}$$

for all $\xi \in D(A^{1/2})$.

Note that as a consequence of (5.4.1) the spectrum of A is contained in the interval $[\varepsilon, \infty)$. Hence A is in particular positive and bijective, and there is a uniquely defined linear and positive self-adjoint operator $A^{1/2} : D(A^{1/2}) \to X$ such that $(A^{1/2})^2 = A$. That operator is the so called *square root of* A. Further, note that from its definition and the bijectivity of A it follows that $A^{1/2}$ is in particular bijective. This can be concluded for instance as follows. By using the fact that $A^{1/2}$ commutes with A, it easy to see that for every $\lambda \in [0, \varepsilon^{1/2})$ by $(A^{1/2} + \lambda)(A - \lambda^2)^{-1}$ there is given the inverse to $A^{1/2} - \lambda$. Hence the spectrum of $A^{1/2}$ is contained in the interval $[\varepsilon^{1/2}, \infty)$. All these facts will be used later on.

Definition 5.4.2. We define

$$Y := D(A^{1/2}) \times X \tag{5.4.4}$$

and $(\,|\,) : Y^2 \to \mathbb{C}$ by

$$(\xi|\eta) := \langle A^{1/2}\xi_1 | A^{1/2}\eta_1 \rangle + \langle \xi_2 | \eta_2 \rangle$$

for all $\xi = (\xi_1, \xi_2), \eta = (\eta_1, \eta_2) \in Y$.

Then we have the following

Theorem 5.4.3.

(i) $(Y, (\,|\,))$ is a complex Hilbert space.

(ii) The operator $H : D(A) \times D(A^{1/2}) \to Y$ in Y defined by

$$H\xi := (-i\xi_2, iA\xi_1)$$

for all $\xi = (\xi_1, \xi_2) \in D(A) \times D(A^{1/2})$ is densely-defined, linear and self-adjoint.

(iii) The operator $\hat{B} : D(H) \to Y$ defined by

$$\hat{B}\xi := (0, -B\xi_2)$$

for all $\xi = (\xi_1, \xi_2) \in D(H)$ is linear. If B is symmetric, then $\hat{B}$ is symmetric, too. If B is bounded, then $\hat{B}$ is bounded, too, and the corresponding operator norms $\|B\|$ and $|\hat{B}|$ satisfy

$$|\hat{B}| \leqslant \|B\| \,. \tag{5.4.5}$$

(iv) The sum $H + \hat{B}$ is closed. If B is symmetric, then $H + \hat{B}$ is self-adjoint.

(v) The operator $V : Y \to Y$ defined by

$$V\xi := (0, iC\xi_1)$$

for all $\xi = (\xi_1, \xi_2) \in Y$ is linear and bounded. The operator norm $|V|$ of V satisfies

$$|V| \leqslant (c^2 + d^2/\epsilon)^{1/2} .$$

Proof. '(i)': Obviously, $(\,|\,)$ defines a Hermitian sesquilinear form on Y^2. That $(\,|\,)$ is in addition positive definite follows from the positive definiteness of $\langle\,|\,\rangle$ and the injectivity of $A^{1/2}$. Finally, the completeness of $(Y, |\,|)$, where $|\,|$ denotes the norm on Y induced by $(\,|\,)$, follows from the completeness of $(X, \|\,\|)$ together with the fact that $A^{1/2}$ has a *bounded* inverse. Here it is essentially used that 0 is not contained in the spectrum of A.
'(ii)': That $D(A) \times D(A^{1/2})$ is dense in Y is an obvious consequence of the facts that $D(A)$ is a core for $A^{1/2}$ (see e.g. Theorem 3.24 in Chapter V.3 of [106]) and that $D(A^{1/2})$ is dense in X. The linearity of H is obvious. Also the symmetry of H follows straightforwardly from the symmetry of $A^{1/2}$. By that symmetry, one gets for every $\xi = (\xi_1, \xi_2) \in D(H^*)$ and any $\eta = (\eta_1, \eta_2) \in D(H)$:

$$\begin{aligned}(H^*\xi|\eta) &= \langle (H^*\xi)_1 | A\eta_1 \rangle + \langle (H^*\xi)_2 | \eta_2 \rangle \\ &= (\xi|H\eta) = \langle -i\xi_2 | A\eta_1 \rangle + \langle iA^{1/2}\xi_1 | A^{1/2}\eta_2 \rangle ,\end{aligned}$$

and from this by using that A is bijective and $A^{1/2}$ is self-adjoint that $\xi_1 \in D(A)$ and

$$(H^*\xi)_1 = -i\xi_2 , \ (H^*\xi)_2 = iA\xi_1 .$$

Hence H is an extension of H^* and thus $H = H^*$.
'(iii)': The linearity of $\hat{B}$ is obvious. Also it is straightforward to see that $\hat{B}$ is symmetric if B is symmetric. If B is bounded, then

$$|\hat{B}\xi|^2 = \|B\xi_2\|^2 \leqslant \|B\|^2 \|\xi_2\|^2 \leqslant \|B\|^2 |\xi|^2$$

for all $\xi = (\xi_1, \xi_2) \in D(H)$. Hence $\hat{B}$ is also bounded and $|\hat{B}|$, $\|B\|$ satisfy the claimed inequality.
'(iv)': Obviously, (5.4.2) implies

$$|\hat{B}\xi|^2 \leqslant a^2 |H\xi|^2 + b^2 |\xi|^2 \tag{5.4.6}$$

for all $\xi \in D(H)$. From this, it is easily seen that $H + \hat{B}$ is closed (see, e.g., [86], Lemma V.3.5). Moreover, in the case that B (and hence by (iii) also $\hat{B}$) is symmetric, (5.4.6) implies according to the *Kato-Rellich* Theorem (see, e.g., Theorem X.12 in [179] Vol. II) that $H + \hat{B}$ is self-adjoint. For the application of these theorems, the assumption $a < 1$ made above is essential.
'(v)': The linearity of V is obvious. For every $\xi = (\xi_1, \xi_2) \in Y$ one has

$$\begin{aligned}|V\xi|^2 &= \|iC\xi_1\|^2 \leqslant c^2 \|A^{1/2}\xi_1\|^2 + d^2 \|\xi_1\|^2 \\ &= c^2 \|A^{1/2}\xi_1\|^2 + d^2 \|(A^{1/2})^{-1} A^{1/2}\xi_1\|^2 \leqslant (c^2 + d^2/\epsilon)\,|\xi|^2 .\end{aligned} \tag{5.4.7}$$

In the last step, it has been used that

$$\|(A^{1/2})^{-1}\| \leqslant 1/\sqrt{\varepsilon}\ .$$

This follows by an application of the spectral theorem (see, e.g. Theorem VIII.5 in [179] Vol. I) to $A^{1/2}$. Since ξ is otherwise arbitrary, from (5.4.7) it follows the boundedness of V and the claimed inequality. □

Assumption 5.4.4. In the following, we assume in addition that B is symmetric or bounded.

Note that condition (5.4.2) is trivially satisfied if B is bounded. We define:

Definition 5.4.5.

$$G_+ := -i(H + \hat{B} + V)\ ,\ G_- := i(H + \hat{B} + V)\ .$$

Then

Lemma 5.4.6. The operators G_+ and G_- are closed and quasi-accretive. In particular,

$$\mathrm{Re}(\xi|G\xi) \geqslant -(\mu_B + |V|)\,(\xi|\xi)$$

for $G \in \{G_+, G_-\}$ and all $\xi \in D(H)$. Here Re denotes the real part and

$$\mu_B := \begin{cases} 0 & \text{if } B \text{ is symmetric} \\ \|B\| & \text{if } B \text{ is bounded} \end{cases}\ .$$

Proof. That G_+ and G_- are closed is an obvious consequence of (iv) and (v) of the previous theorem. Further, if B is symmetric, one has because of (iv) and (v) of the preceding theorem

$$\mathrm{Re}(\xi|G_\pm\xi) = \mp\,\mathrm{Re}(\xi|iV\xi) \geqslant -|(\xi|iV\xi)| \geqslant -|V|\,(\xi|\xi)$$

for all $\xi \in D(H)$. Similarly, if B is bounded, one has because of (ii), (iii), (iv), (5.4.5)

$$\mathrm{Re}(\xi|G_\pm\xi) = \mp\,\mathrm{Re}(\xi|i(\hat{B} + V)\xi) \geqslant -|(\xi|i(\hat{B} + V)\xi)| \geqslant -(\|B\| + |V|)\,(\xi|\xi)$$

for all $\xi \in D(H)$. Hence in both cases G_+ and G_- are quasi-accretive. □

Theorem 5.4.7. The operators G_+ and G_- are infinitesimal generators of strongly continuous semigroups $T_+ : [0,\infty) \to L(Y,Y)$ and $T_- : [0,\infty) \to L(Y,Y)$, respectively. If $\mu_\pm \in \mathbb{R}$ are such that

$$\mathrm{Re}(\xi|G_\pm\xi) \geqslant -\mu_\pm\,(\xi|\xi)$$

for all $\xi \in D(H)$, then the spectra of G_+ and G_- are contained in the half-plane $[-\mu_+,\infty) \times \mathbb{R}$ and $[-\mu_-,\infty) \times \mathbb{R}$, respectively, and

$$|T_+(t)| \leqslant \exp(\mu_+ t)\ ,\ |T_-(t)| \leqslant \exp(\mu_- t)$$

for all $t \in [0,\infty)$.

Proof. Obviously, by Theorem 4.2.6 and the preceding lemma, the theorem follows if we can show that there is a real number $\lambda < \min\{-\mu_+, -\mu_-\}$ such that $G_\pm - \lambda$ has a dense range in Y. For the proof, let ξ be some element of $D(H)$ and λ be some real number such that $|\lambda| \geqslant |V|^2$. Then we get from the symmetry of H

$$|(H - i\lambda)\xi|^2 = |H\xi|^2 + \lambda^2|\xi|^2$$

and

$$|(H - i\lambda)\xi| \geqslant \max\{|H\xi|, |\lambda|^{1/2}|V\xi|\} .$$

Using these identities together with (5.4.2)

$$\begin{aligned}
|(\hat{B} + V)\xi|^2 &\leqslant |\hat{B}\xi|^2 + 2|\hat{B}\xi|\,|V\xi| + |V\xi|^2 \\
&\leqslant a^2|H\xi|^2 + 2|\hat{B}\xi|\,|V\xi| + (b^2 + |V|^2)|\xi|^2 \\
&\leqslant a^2|H\xi|^2 + 2a|H\xi|\,|V\xi| + (b + |V|)^2|\xi|^2 \\
&\leqslant a^2|(H - i\lambda)\xi|^2 + 2a|(H - i\lambda)\xi|\,|V\xi| + [(b + |V|)^2 - a^2\lambda^2]|\xi|^2 \\
&\leqslant a(a + 2|\lambda|^{-1/2})|(H - i\lambda)\xi|^2 + [(b + |V|)^2 - a^2\lambda^2]|\xi|^2 .
\end{aligned}$$

Hence for any real λ with

$$|\lambda| > \max\{|V|^2, 4(1 - a)^{-2}, (b + |V|)/a, |\mu_+|, |\mu_-|\} ,$$

where we assume without restriction that $a > 0$, we get

$$|(\hat{B} + V)\xi| \leqslant a'|(H - i\lambda)\xi|$$

where a' is some real number from $[0, 1)$. Since $\xi \in D(H)$ is otherwise arbitrary, we conclude that

$$(\hat{B} + V)(H - i\lambda)^{-1}$$

defines a bounded linear operator on Y with operator norm smaller than 1. Since

$$H + \hat{B} + V - i\lambda = \left(1 + (\hat{B} + V)(H - i\lambda)^{-1}\right)(H - i\lambda) ,$$

we conclude that $H + \hat{B} + V - i\lambda$ is bijective and hence also that $G_+ - \lambda$ and $G_- - \lambda$ are both bijective. Hence the theorem follows. □

We note that Assumption 5.4.4 has been used only to conclude that G_+ and G_- are *both* quasi-accretive. Now it is easy to see that if B is in addition such that iB is quasi-accretive (but not necessarily bounded or antisymmetric), then $-i\hat{B}$ and hence also G_+ are quasi-accretive, too. As a consequence, we have the following

Corollary 5.4.8. Instead of Assumption 5.4.4 let B be such that iB is quasi-accretive. Then G_+ is the infinitesimal generator of a strongly continuous semigroup $T_+ : [0, \infty) \to L(Y, Y)$. If $\mu_+ \in \mathbb{R}$ is such that

$$\mathrm{Re}(\xi|G_+\xi) \geqslant -\mu_+ \, (\xi|\xi)$$

for all $\xi \in D(H)$, then the spectrum of G_+ is contained in the half-plane $[-\mu_+, \infty) \times \mathbb{R}$ and

$$|T_+(t)| \leqslant \exp(\mu_+ t)$$

for all $t \in [0, \infty)$.

In addition, it follows

Corollary 5.4.9.

(i) By

$$T(t) := \begin{cases} T_+(t) & \text{for } t \geqslant 0 \\ T_-(-t) & \text{for } t < 0 \end{cases}$$

for all $t \in \mathbb{R}$ there is defined a strongly continuous group $T : \mathbb{R} \to L(Y, Y)$.

(ii) For every $t_0 \in \mathbb{R}$ and every $\xi \in D(G_+)$, there is a uniquely determined differentiable map $u : \mathbb{R} \to Y$ such that

$$u(t_0) = \xi$$

and

$$u'(t) = -G_+ u(t) \tag{5.4.8}$$

for all $t \in \mathbb{R}$. Here $'$ denotes differentiation of functions assuming values in Y.

(iii) The function $(u|u) : \mathbb{R} \to \mathbb{R}$ defined by

$$(u|u)(t) := (\, u(t)|u(t)\,) \ , t \in \mathbb{R}$$

is differentiable and

$$(u|u)'(t) = -2\,\mathrm{Re}\,(\, u(t)|G_+ u(t)\,)$$

for all $t \in \mathbb{R}$.

Proof. Parts (i) and (ii) follow from the previous theorem by Theorem 4.5.1. Part (iii) is an obvious consequence of (ii). □

Note in particular the *the special case*[3] that there is a non trivial element η in the kernel of $A + C$ for which there is $\xi \in D(A)$ such that

$$(A + C)\xi = -iB\eta \ .$$

Then by

$$u(t) := (\xi + t\eta, \eta) \ , \ t \in \mathbb{R}$$

there is given a growing solution of (5.4.8).

[3] Such cases are easy to construct.

The following lemma is needed in the formulation of the subsequent theorem.

Lemma 5.4.10. By

$$\|\xi\|_{A^{1/2}} := \|A^{1/2}\xi\| \ , \ \xi \in D(A^{1/2})$$

there is defined a norm $\| \ \|_{A^{1/2}}$ on $D(A^{1/2})$. Moreover,

$$W_1 := (D(A^{1/2}), \| \ \|_{A^{1/2}})$$

is complete.

Proof. The lemma is a trivial consequence of the completeness of X and the bijectivity of $A^{1/2}$. □

Theorem 5.4.11. Let $t_0 \in \mathbb{R}$, $\xi \in D(A)$ and $\eta \in D(A^{1/2})$. Then there is a uniquely determined differentiable map $u : \mathbb{R} \to W_1$ satisfying

$$u(t_0) = \xi \text{ and } u'(t_0) = \eta \tag{5.4.9}$$

and such that $u' : \mathbb{R} \to X$ is differentiable with

$$(u')'(t) + iBu'(t) + (A + C)u(t) = 0 \tag{5.4.10}$$

for all $t \in \mathbb{R}$.

Proof. For this, let $v = (v_1, v_2) : \mathbb{R} \to Y$ be such that

$$v(t_0) = (\xi, \eta) \tag{5.4.11}$$

and

$$v'(t) = -G_+ v(t) \ , t \in \mathbb{R} \ . \tag{5.4.12}$$

Such v exists according to Corollary 5.4.9 (ii). Using the continuity of the canonical projections of Y onto W_1 and X, it is easy to see that $u := v_1$ is a differentiable map into W_1 such that $u' : \mathbb{R} \to X$ is differentiable and such that (5.4.9), (5.4.10) are both satisfied. On the other hand, if $u : \mathbb{R} \to W_1$ has the properties stated in the corollary, it follows by the continuity of the canonical imbeddings of W_1, X into Y that $w := (u, u')$ satisfies both equations (5.4.11) and (5.4.12). Then $u = v_1$ follows by Corollary 5.4.9 (ii). □

Corollary 5.4.12. In addition to the assumptions, let C be in particular bounded.[4] Further, let $u : \mathbb{R} \to W_1$ be differentiable with a differentiable derivative $u' : \mathbb{R} \to X$ and such that (5.4.10) holds. Finally, define $E_u : \mathbb{R} \to \mathbb{R}$ by

$$E_u(t) := \frac{1}{2} \left(\langle u'(t) | u'(t) \rangle + \langle u(t) | (A + \mathrm{Re}(C)) u(t) \rangle \right) \ . \tag{5.4.13}$$

[4] Note that in this case (5.4.3) is trivially satisfied.

Then E_u is differentiable and

$$E_u'(t) = \begin{cases} -\operatorname{Im}\langle u(t)|\operatorname{Im}(C)u'(t)\rangle & \text{for symmetric } B \\ \langle u'(t)|\operatorname{Im}(B)u'(t)\rangle - \operatorname{Im}\langle u(t)|\operatorname{Im}(C)u'(t)\rangle & \text{for bounded } B \end{cases} \tag{5.4.14}$$

for all $t \in \mathbb{R}$ where for any bounded linear operator F on X:

$$\operatorname{Re}(F) := \frac{1}{2}(F + F^*) \ , \ \operatorname{Im}(F) := \frac{1}{2i}(F - F^*) \ .$$

Proof. For this, define $v := (u, u')$. Then according to the preceding proof v satisfies (5.4.12). *For a symmetric B* it follows by Corollary 5.4.9 and Theorem 5.4.3 (iv) that

$$\begin{aligned} (v|v)'(t) &= 2\operatorname{Re}(\, v(t)|iVv(t)\,) \\ &= -\langle u'(t)|Cu(t)\rangle - \langle Cu(t)|u'(t)\rangle \\ &= -\langle u|\operatorname{Re}(C)u\rangle'(t) - 2\operatorname{Im}\langle u(t)|\operatorname{Im}(C)u'(t)\rangle \end{aligned} \tag{5.4.15}$$

for all $t \in \mathbb{R}$. In the last step, it has been used that u is also differentiable with the same derivative viewed as map with values in X. This follows from the fact the canonical imbedding of W_1 into X is continuous since $A^{1/2}$ is bijective. Further, the definition

$$\langle u|\operatorname{Re}(C)u\rangle(t) := \langle u(t)|\operatorname{Re}(C)u(t)\rangle \ , \ t \in \mathbb{R}$$

for the map $\langle u|\operatorname{Re}(C)u\rangle : \mathbb{R} \to \mathbb{R}$ has been used. Obviously, (5.4.14) follows from (5.4.15) by using definition (5.4.13). In this step also the symmetry of $A^{1/2}$ is used together with the fact that u assumes values in $D(A)$. *For a bounded B* by Corollary 5.4.9 and Theorem 5.4.3 (ii), it follows that

$$\begin{aligned} (v|v)'(t) &= 2\operatorname{Re}(\, v(t)|i(\hat{B} + V)v(t)) \\ &= 2\operatorname{Im}\langle u'(t)|Bu'(t)\rangle \\ &\quad - \langle u'(t)|Cu(t)\rangle - \langle Cu(t)|u'(t)\rangle \\ &= 2\langle u'(t)|\operatorname{Im}(B)u'(t)\rangle \\ &\quad - \langle u|\operatorname{Re}(C)u\rangle'(t) - 2\operatorname{Im}\langle u(t)|\operatorname{Im}(C)u'(t)\rangle \end{aligned}$$

for all $t \in \mathbb{R}$. Obviously, (5.4.14) follows from (5.4.15) by using definition (5.4.13). □

The next theorem relates the spectrum of G_+ to the spectrum of the so called *operator polynomial* $A + C - \lambda B - \lambda^2$ where λ runs through the complex numbers [137, 183].

Theorem 5.4.13. Let λ be some complex number.

(i) Then $H + \hat{B} + V - \lambda$ is not injective if and only if $A + C - \lambda B - \lambda^2$ is not injective. If $H + \hat{B} + V - \lambda$ is not injective, then

$$\ker(H + \hat{B} + V - \lambda) = \{(\xi, i\lambda\xi) : \xi \in \ker(A + C - \lambda B - \lambda^2)\} \ .$$

(ii) Further, $H+\hat{B}+V-\lambda$ is bijective if and only if $A+C-\lambda B-\lambda^2$ is bijective. If $H+\hat{B}+V-\lambda$ is bijective, then for all $\eta=(\eta_1,\eta_2)\in Y$:

$$(H+\hat{B}+V-\lambda)^{-1}\eta=(\xi,i(\lambda\xi+\eta_1))$$

where

$$\xi=(A+C-\lambda B-\lambda^2)^{-1}[(B+\lambda)\eta_1-i\eta_2]\ . \tag{5.4.16}$$

Proof. '(i)': If $H+\hat{B}+V-\lambda$ is not injective and $\xi=(\xi_1,\xi_2)\in\ker(H+\hat{B}+V-\lambda)$, it follows from the definitions in Theorem 5.4.3 that

$$\xi_2=i\lambda\xi_1\ ,\ (A+C-\lambda B-\lambda^2)\xi_1=0$$

and hence also that $A+C-\lambda B-\lambda^2$ is not injective. If $A+C-\lambda B-\lambda^2$ is not injective, it follows again from the definitions in Theorem 5.4.3 that

$$(H+\hat{B}+V-\lambda)(\xi,i\lambda\xi)=0$$

and hence also that $H+\hat{B}+V-\lambda$ is not injective.
'(ii)': If $H+\hat{B}+V-\lambda$ is bijective, it follows by (i) that $A+C-\lambda B-\lambda^2$ is injective. For $\eta\in X$ and $\xi=(\xi_1,\xi_2):=(H+\hat{B}+V-\lambda)^{-1}(0,i\eta)$, it follows from the definitions in Theorem 5.4.3 that

$$(A+C-\lambda B-\lambda^2)\xi_1=\eta$$

and hence that $A+C-\lambda B-\lambda^2$ is also surjective. If $A+C-\lambda B-\lambda^2$ is bijective, it follows by (i) that $H+\hat{B}+V-\lambda$ is injective. Further, if $\eta=(\eta_1,\eta_2)\in Y$ and ξ is defined by (5.4.16), it follows from the definitions in Theorem 5.4.3 that

$$(H+\hat{B}+V-\lambda)(\xi,i(\lambda\xi+\eta_1))=\eta$$

and hence that $H+\hat{B}+V-\lambda$ is also surjective. □

Lemma 5.4.14. Let $\varepsilon'<\varepsilon$ and

$$A':=A-\varepsilon'\ ,\ C':=C+\varepsilon'\ . \tag{5.4.17}$$

Then

(i)

$$D(A'^{1/2})=D(A^{1/2}) \tag{5.4.18}$$

and for all $\xi\in D(A^{1/2})$

$$\|A^{1/2}\xi\|^2=\|A'^{1/2}\xi\|^2+\varepsilon'\|\xi\|^2\ . \tag{5.4.19}$$

(ii) The operators A',B and C' satisfy

$$\langle\xi|A'\xi\rangle\geqslant(\varepsilon-\varepsilon')\langle\xi|\xi\rangle$$
$$\|B\xi\|^2\leqslant a^2\|A'^{1/2}\xi\|^2+(a^2\varepsilon'+b^2)\|\xi\|^2$$

$$\|C'\xi\|^2 \leqslant |c| \left[|c| + 2|\varepsilon'| (\varepsilon - \varepsilon')^{-1/2} \right] \|A'^{1/2}\xi\|^2 + \left[|\varepsilon'| + (c^2|\varepsilon'| + d^2)^{1/2} \right]^2 \|\xi\|^2$$

for all $\xi \in D(A^{1/2})$.

Proof. '(i)': First, since $\varepsilon' < \varepsilon$, by (5.4.17) there is defined a linear self-adjoint and positive operator A' in X. Obviously, using the symmetry of $A^{1/2}$ and $A'^{1/2}$, (5.4.19) follows for all elements of $D(A)$. From this (5.4.18) and (5.4.19) follow straightforwardly by using the facts that $D(A)$ is a core for $A^{1/2}$ and $A'^{1/2}$ (see e.g. Theorem 3.24 in chapter V.3 of [106]), that X is complete and that $A^{1/2}$ and $A'^{1/2}$ are both closed.
(ii) The first two inequalities are obvious consequences of the corresponding ones in Assumption 5.4.1 1, the definition (5.4.17) and of (5.4.19). For the proof of the third, we notice that from the first inequality along with an application of the spectral theorem (see, e.g. Theorem VIII.5 in [179] Vol. I) to $A'^{1/2}$ follows that

$$\|(A'^{1/2})^{-1}\| \leqslant 1/\sqrt{\varepsilon - \varepsilon'} \,. \tag{5.4.20}$$

Further, from Assumption 5.4.1 and (5.4.19) one gets

$$\|C\xi\|^2 \leqslant c^2 \|A'^{1/2}\xi\|^2 + (c^2|\varepsilon'| + d^2)\|\xi\|^2$$

for all $\xi \in D(A^{1/2})$. From these inequalities, we get

$$\begin{aligned} \|C'\xi\|^2 &\leqslant \|C\xi\|^2 + 2|\varepsilon'| \, \|C\xi\| \, \|\xi\| + \varepsilon'^2 \|\xi\|^2 \\ &\leqslant c^2 \|A'^{1/2}\xi\|^2 + (\varepsilon'^2 + c^2|\varepsilon'| + d^2)\|\xi\|^2 + 2|\varepsilon'| \, \|C\xi\| \, \|\xi\| \\ &\leqslant c^2 \|A'^{1/2}\xi\|^2 + \left[|\varepsilon'| + (c^2|\varepsilon'| + d^2)^{1/2} \right]^2 \|\xi\|^2 + 2|\varepsilon'| \, |c| \, \|A'^{1/2}\xi\| \, \|\xi\| \\ &\leqslant |c| \left[|c| + 2|\varepsilon'| (\varepsilon - \varepsilon')^{-1/2} \right] \|A'^{1/2}\xi\|^2 + \left[|\varepsilon'| + (c^2|\varepsilon'| + d^2)^{1/2} \right]^2 \|\xi\|^2 \end{aligned} \tag{5.4.21}$$

for all $\xi \in D(A^{1/2})$ and hence the third inequality. □

As a consequence of (ii), the sequence X, A', B, C' satisfies Assumption 5.4.1. The corresponding Y given by Definition 5.4.2 is because of (i) again given by (5.4.4). Moreover, the corresponding norm $|\,|'$ on Y turns out to be equivalent to $|\,|$. More precisely, one has for every $\varepsilon' \in (0, \varepsilon)$

Lemma 5.4.15.

$$|\,|' \leqslant |\,| \leqslant \varepsilon^{1/2} (\varepsilon - \varepsilon')^{-1/2} \, |\,|'$$

and for every bounded linear operator F on Y:

$$\varepsilon^{-1/2} (\varepsilon - \varepsilon')^{1/2} \, |F|' \leqslant |F| \leqslant \varepsilon^{1/2} (\varepsilon - \varepsilon')^{-1/2} \, |F|' \,.$$

Proof. The first inequality is a straightforward consequence of (5.4.19) and (5.4.20). The second inequality is a straightforward implication of the first. □

Note that the $G_{\pm}$ corresponding to the the sequence X, A', B, C' are the same for all ε' since ε' drops out of the definition. Moreover, as a consequence of the preceding lemma, the topologies induced on Y are equivalent. Hence the generated groups are the same, too. This will be used in the following important special case.

Theorem 5.4.16. Let $A = A_0 + \varepsilon$ where A_0 is a densely defined linear positive self-adjoint operator and let $C = -\varepsilon$. Then

$$|T_{\pm}(t)| \leqslant e\,\varepsilon^{1/2}\,t\,\exp(\mu_B t) \tag{5.4.22}$$

for all $t \geqslant \varepsilon^{-1/2}$.

Proof. For this, let $\varepsilon' \in [0, \varepsilon)$ and define A' and C' as in Lemma 5.4.14. Hence

$$A' = A_0 + \varepsilon - \varepsilon' \;,\; C' = -(\varepsilon - \varepsilon') \;.$$

Then from Theorem 5.4.3 (v), Lemma 5.4.6 and Theorem 5.4.7, we conclude that

$$|T_{\pm}(t)|' \leqslant \exp\left([\mu_B + (\varepsilon - \varepsilon')^{1/2}]\,t \right)$$

and hence by the previous Lemma that

$$|T_{\pm}(t)| \leqslant \varepsilon^{1/2}\,(\varepsilon - \varepsilon')^{-1/2}\,\exp\left([\mu_B + (\varepsilon - \varepsilon')^{1/2}]\,t \right)$$

for all $t \geqslant 0$. For $t \geqslant \varepsilon^{-1/2}$, we get from this (5.4.22) by choosing

$$\varepsilon' := \varepsilon - t^{-2} \;.$$

□

Note that in this special case (5.4.13) is conserved and positive if B is in particular symmetric.

We are now giving stability criteria.

Theorem 5.4.17. In addition, let B and C be both symmetric.

(i) Let A, B and C be such that

$$\langle \xi | (A + C)\xi \rangle + \frac{1}{4}\langle \xi | B\xi \rangle^2 \;\geqslant 0 \tag{5.4.23}$$

for all $\xi \in D(A)$ with $\|\xi\| = 1$. Then the spectrum of iG_+ is real.

(ii) In addition, let B and C be both bounded and let $A + C + (b/2)B - (b^2/4)$ be positive for some $b \in \mathbb{R}$. Then the spectrum of iG_+ is real and there are $K \geqslant 0$ and $t_0 \geqslant 0$ such that

$$|T(t)| \leqslant K|t| \tag{5.4.24}$$

for all $|t| \geqslant t_0$.

Proof. '(i)': First, from Assumption 5.4.1 and the assumed symmetry of B and C, it follows that by $A^{-1/2}BA^{-1/2}$ and $A^{-1/2}CA^{-1/2}$ there are given bounded symmetric and hence by Theorem 3.1.7 (viii) also self-adjoint linear operators on X. Hence

$$A(\lambda) := \lambda^2 A^{-1} + \lambda A^{-1/2}BA^{-1/2} - \left(1 + A^{-1/2}CA^{-1/2}\right) \, , \, \lambda \in \mathbb{C} \tag{5.4.25}$$

defines a self-adjoint operator polynomial in $L(X,X)$. In addition, one has $A^{-1} \geqslant 1/\varepsilon$. Further, for every $\xi \in D(A^{1/2})$ and $\lambda \in \mathbb{C}$

$$\langle \xi | A(\lambda)\xi \rangle = \langle \eta | A^{1/2}A(\lambda)A^{1/2}\eta \rangle = -\langle \eta | (A + C - \lambda B - \lambda^2)\eta \rangle \tag{5.4.26}$$

where $\eta := A^{-1/2}\xi \in D(A)$. Now (5.4.23) implies that the roots of the polynomial $\langle \eta | (A + C - \lambda B - \lambda^2)\eta \rangle, \lambda \in \mathbb{C}$ are real. Hence by (5.4.26) the roots of $\langle \xi | A(\lambda)\xi \rangle, \lambda \in \mathbb{C}$ are real, too. Since $\xi \in D(A^{1/2})$ is otherwise arbitrary and $D(A^{1/2})$ is dense in X, this implies also that $\langle \xi | A(\lambda)\xi \rangle$ has only real roots for all $\xi \in X$. Hence (see [137], Lemma 31.1) the polynomial $A(\lambda), \lambda \in \mathbb{C}$ is weakly hyperbolic and has therefore a real spectrum. As a consequence, $A(\lambda)$ is bijective for all non real λ. Now for any such λ

$$A + C - \lambda B - \lambda^2 = -A^{1/2}A_r(\lambda)A_r^{1/2} \tag{5.4.27}$$

where $A_r^{1/2}$ denotes the restriction of $A^{1/2}$ to $D(A)$ in domain and $D(A^{1/2})$ in range, and $A_r(\lambda)$ denotes the restriction of $A(\lambda)$ to $D(A^{1/2})$ in domain and in range. For this, note that $A(\lambda)$ leaves $D(A^{1/2})$ invariant. Further, from the bijectivity of $A^{1/2}$, $A(\lambda)$ and (5.4.25), it follows the bijectivity of $A_r^{1/2}$ and $A_r(\lambda)$, respectively and hence by (5.4.27) that $A + C - \lambda B - \lambda^2$ is bijective. This is true for all non real λ, and hence it follows by Theorem 5.4.13 that the spectrum of iG_+ is real.

'(ii)': Let B and C be both bounded and let $A + C + (b/2)B - (b^2/4)$ be positive for some $b \in \mathbb{R}$. In addition, let ϵ be some real number greater than zero and define

$$A' := A + C + (b/2)B - (b^2/4) + \varepsilon \, , \, C' := -\varepsilon \, , \, B' := B - b \, .$$

First, it is observed that

$$D(A'^{1/2}) = D(A^{1/2}) \tag{5.4.28}$$

and that there exist nonvanishing real constants K_1 and K_2 such that

$$K_1^2 \|A^{1/2}\xi\|^2 \leqslant \|A'^{1/2}\xi\|^2 \leqslant K_2^2 \|A^{1/2}\xi\|^2 \tag{5.4.29}$$

for every $\xi \in D(A^{1/2})$. This can be proved as follows. Obviously, by the symmetry of $A^{1/2}$ and $A'^{1/2}$, the Cauchy-Schwarz inequality, the boundedness of B, C, $A^{-1/2}$ and $A'^{-1/2}$, it follows the existence of nonvanishing real constants K_1 and K_2 such that (5.4.29) is valid for all $\xi \in D(A)$. Since $D(A)$ is a core for both $A^{1/2}$ and $A'^{1/2}$ (see e.g. Theorem 3.24 in chapter V.3 of [106]), from that inequality follows (5.4.28) and (5.4.29) for all $\xi \in D(A^{1/2})$. Note that in this conclusion it is used that X is complete and that $A^{1/2}$ and $A'^{1/2}$ are both closed.

Obviously, from the assumptions made it follows that also A', B' and C' instead of A, B and C, respectively, satisfy Assumption 5.4.1 and Assumption 5.4.4. Hence by Theorem 5.4.16 it follows that

$$|T'_{\pm}(t)|' \leqslant e\,\varepsilon^{1/2}\,t \tag{5.4.30}$$

for all $t \geqslant \varepsilon^{-1/2}$ where primes indicate quantities whose definition uses one or more of the operators A', B' and C' instead of A, B and C. In addition, (5.4.28) and (5.4.29) imply $Y = Y'$ as well as the equivalence of the norms $|\,|$ and $|\,|'$. Now define the auxiliary transformation $S_0 : Y' \to Y$ by

$$S_0\xi := (\xi_1, \xi_2 - i(b/2)\xi_1)$$

for all $\xi = (\xi_1, \xi_2) \in Y'$. Obviously, S_0 is bijective and bounded with the bounded inverse S_0^{-1} given by $S_0^{-1}\xi := (\xi_1, \xi_2 + i(b/2)\xi_1)$ for all $\xi = (\xi_1, \xi_2) \in Y$. In addition, we define $S_{\pm} : [0, \infty) \to L(Y, Y)$ by

$$S_{\pm}(t) := \exp(\mp ibt/2) S_0 T'_{\pm}(t) S_0^{-1}\,, \tag{5.4.31}$$

for all $t \in [0, \infty)$. Obviously, $S_{\pm}$ defines a strongly continuous semigroup with the corresponding generator

$$S_0 G'_{\pm} S_0^{-1} \pm i\frac{b}{2} = G_{\pm}\,.$$

This implies $S_{\pm} = T_{\pm}$ and by (5.4.30) and (5.4.31) the existence of $K \geqslant 0$ and $t_0 \geqslant 0$ such that (5.4.24) is valid for all $|t| \geqslant t_0$. Finally, from this follows by Theorem 4.2.1 that the spectrum of iG_+ is real. □

Lemma 5.4.18. Let D be a core for A. Further, let $B_0 : D \to X$ be a linear operator in X such that for some real numbers a_0 and b_0

$$\|B_0\xi\|^2 \leqslant a_0^2\,\langle\xi|A\xi\rangle + b_0^2\,\|\xi\|^2 \tag{5.4.32}$$

for all $\xi \in D$. Then there is a uniquely determined linear extension $\bar{B}_0 : D(A^{1/2}) \to X$ of B_0 such that

$$\|\bar{B}_0\xi\|^2 \leqslant a_0^2\,\|A^{1/2}\xi\|^2 + b_0^2\,\|\xi\|^2 \tag{5.4.33}$$

for all $\xi \in D(A^{1/2})$. If B_0 is in addition symmetric, $\bar{B}_0$ is symmetric, too.

Proof. First, we notice that D is a core for $A^{1/2}$, too. Obviously, since $D(A)$ is a core for $A^{1/2}$ (see e.g. Theorem 3.24 in chapter V.3 of [106]), this follows if we can show that the closure of the restriction of $A^{1/2}$ to D extends the restriction of $A^{1/2}$ to $D(A)$. To prove this let ξ be some element of $D(A)$. Since D is a core for A, there is a sequence $\xi_0, \xi_1 \dots$ of elements of D converging to ξ and at the same time such that $A\xi_0, A\xi_1 \dots$ converges to $A\xi$. Since $A^{1/2}$ has a bounded inverse, it follows from this that $A^{1/2}\xi_0, A^{1/2}\xi_1 \dots$ converges to $A^{1/2}\xi$. Since ξ can be chosen otherwise arbitrarily, it follows that the closure of the restriction of $A^{1/2}$ to D extends the restriction of $A^{1/2}$ to $D(A)$ and hence that D is a core for $A^{1/2}$. Hence for any $\xi \in D(A^{1/2})$ there is a a sequence $\xi_0, \xi_1 \dots$ in D converging to ξ and at the same time such that $A^{1/2}\xi_0, A^{1/2}\xi_1 \dots$ is converging to $A^{1/2}\xi$. Hence by (5.4.32) along with the completeness of X follows the convergence of the sequence $B_0\xi_1, B_0\xi_2 \dots$ to some element $\bar{B}\xi$ of X and

$$\|\bar{B}\xi\|^2 \leqslant a_0^2 \, \|A^{1/2}\xi\|^2 + b_0^2 \, \|\xi\|^2 \, .$$

Moreover, if $\xi_0', \xi_1' \dots$ is another sequence having the same properties as $\xi_0, \xi_1 \dots$, it follows by (5.4.32) that

$$\bar{B}\xi = \lim_{n\to\infty} B_0\xi_n = \lim_{n\to\infty} B_0\xi_n' \, .$$

From this it easily seen that by defining

$$\bar{B} := (D(A^{1/2}) \to X \, , \, \xi \mapsto \bar{B}\xi)$$

there is also given a linear map. Hence the existence of a linear extension of B_0 satisfying (5.4.33) is shown. Moreover, from the definition it is obvious that $\bar{B}$ is symmetric if B_0 is in addition symmetric. If on the other hand $\bar{B}_0$ is a linear extension of B_0 satisfying (5.4.33) and ξ and ξ_1, ξ_2 are as above from (5.4.33), it follows that

$$\bar{B}_0\xi = \lim_{n\to\infty} B_0\xi_n \, .$$

Finally, since ξ can be chosen otherwise arbitrarily, from this follows $\hat{B}_0 = \hat{B}$. □

5.5 Autonomous Linear Hermitian Hyperbolic Systems

In addition to wave equations, Hermitian hyperbolic systems are another important prototype of hyperbolic equations. Simple examples are given by the evolutional part of Maxwell's equations and Dirac's equations in flat space-time. The treatment of hyperbolic systems in this chapter *includes the case of singular coefficients*, i.e., the vanishing of determinants of matrices multiplying the highest order derivatives in the governing operator. As a consequence, in general, the domains of the associated operators depend on the coefficients. This generality is also considered in the subsequent study of non-autonomous linear Hermitian hyperbolic systems and quasi-linear Hermitian hyperbolic systems in Chapters 10.2, 12.2 which are based on the results in this section. In particular, the Theorems 9.0.6, 11.0.7 on the well-posedness of linear and quasi-linear evolution equations are strong enough to deal with such a situation. For similar treatments of autonomous linear Hermitian hyperbolic systems see [60, 180]. For additional material on initial boundary value problems for such systems, see, for instance, [130, 169, 176, 198].

In the following, we define for every $n \in \mathbb{N}^*$ and every multi-index $\alpha \in \mathbb{N}^n$ the densely-defined linear operator ∂^α in $L^2_{\mathbb{C}}(\mathbb{R}^n)$ by

$$\partial^\alpha := (-1)^{|\alpha|} . \left(C_0^\infty(\mathbb{R}^n, \mathbb{C}) \to L^2_{\mathbb{C}}(\mathbb{R}^n), f \mapsto \frac{\partial^\alpha f}{\partial x^\alpha} \right)^*$$

where

$$|\alpha| := \sum_{j=1}^{n} \alpha_j \, .$$

Moreover, we define as usual for any $k \in \mathbb{N}$

$$W^k_{\mathbb{C}}(\mathbb{R}^n) := \bigcap_{\alpha \in \mathbb{N}^n, |\alpha| \leqslant k} D(\partial^\alpha) \, .$$

Equipped with the scalar product

$$\langle \, , \rangle_k \; : \; (W^k_{\mathbb{C}}(\mathbb{R}^n))^2 \to \mathbb{C}$$

defined by

$$\langle f, g \rangle_k := \sum_{\alpha \in \mathbb{N}^n, |\alpha| \leqslant k} \langle \partial^\alpha f | \partial^\alpha g \rangle_2$$

for all $f, g \in W^k_{\mathbb{C}}(\mathbb{R}^n)$, $W^k_{\mathbb{C}}(\mathbb{R}^n)$ becomes a Hilbert space.

In the following Friedrichs' mollifiers will be used.

Lemma 5.5.1. Let $h \in C_0^\infty(\mathbb{R}^n)$. In addition, let h be positive with support contained in the closed unit ball and such that $h(x) = h(-x)$ for all $x \in \mathbb{R}^n$ and $\|h\|_1 = 1$. Moreover, define for any $\nu \in \mathbb{N}^*$ the corresponding $h_\nu \in C_0^\infty(\mathbb{R}^n)$ by

$$h_\nu(x) := \nu^n h(\nu x)$$

for all $x \in \mathbb{R}^n$. Finally, define for every $\nu \in \mathbb{N}^*$ and every $f \in L^2_{\mathbb{C}}(\mathbb{R}^n)$

$$H_\nu f := h_\nu * f$$

where '$*$' denotes the convolution product. Then

(i) for every $\nu \in \mathbb{N}^*$ the corresponding H_ν defines a bounded self-adjoint linear operator on $L^2_{\mathbb{C}}(\mathbb{R}^n)$ with operator norm $\|H_\nu\| \leqslant 1$.

(ii) for every $\nu \in \mathbb{N}^*$ the range of H_ν is part of $W^k_{\mathbb{C}}(\mathbb{R}^n)$ for all $k \in \mathbb{N}$. Moreover, for every multi-index $\alpha \in \mathbb{N}^n$

$$H^\alpha_\nu f := \partial^\alpha (H_\nu f) = \frac{\partial^\alpha h_\nu}{\partial x^\alpha} * f$$

for all $f \in L^2_{\mathbb{C}}(\mathbb{R}^n)$. Moreover, H^α_ν defines a bounded linear operator on $L^2_{\mathbb{C}}(\mathbb{R}^n)$ with operator norm

$$\|H^\alpha_\nu\| \leqslant \|\partial^\alpha h_\nu / \partial x^\alpha\|_1 \, .$$

(iii)

$$\text{s-}\lim_{\nu \to \infty} H_\nu = \mathrm{id}_{L^2_{\mathbb{C}}(\mathbb{R}^n)} \, .$$

Proof. For this, let ν be some element of $\mathbb{N}^*$. Moreover, define $K_\nu \in C^\infty(\mathbb{R}^n \times \mathbb{R}^n)$ by

$$K_\nu(x, y) := h_\nu(x - y)$$

for all $x, y \in \mathbb{R}^n$. Then K_ν is in particular measurable and such that $K_\nu(x, \cdot), K_\nu(\cdot, y) \in L^1(\mathbb{R}^n)$ and

$$\|K_\nu(x, \cdot)\|_1 = 1 \, , \; \|K_\nu(\cdot, y)\|_1 = 1$$

for all $x, y \in \mathbb{R}^n$. Hence to K_ν there is associated a bounded linear integral operator $\text{Int}(K_\nu) = H_\nu$ on $L^2_{\mathbb{C}}(\mathbb{R}^n)$ with operator norm equal or smaller than 1. Finally, this operator is self-adjoint since it follows from the assumptions on K that $K^*(y - x) = K(x - y)$ for all $x, y \in \mathbb{R}^n$. Further, for any multi-index $\alpha \in \mathbb{N}^n$ and $f \in L^2_{\mathbb{C}}(\mathbb{R}^n)$ it follows

$$\frac{\partial^\alpha h_\nu}{\partial x^\alpha} \in C_0^\infty(\mathbb{R}^n, \mathbb{C}) \subset L^2_{\mathbb{C}}(\mathbb{R}^n)$$

and hence also

$$h_\nu * f \in C^\infty(\mathbb{R}^n, \mathbb{C})$$

and

$$\frac{\partial^\alpha}{\partial x^\alpha}(h_\nu * f) = \frac{\partial^\alpha h_\nu}{\partial x^\alpha} * f \, .$$

Now define $K_\nu^\alpha \in C^\infty(\mathbb{R}^n \times \mathbb{R}^n)$ by

$$K_\nu^\alpha(x, y) := \frac{\partial^\alpha h_\nu}{\partial x^\alpha}(x - y)$$

for all $x, y \in \mathbb{R}^n$. Then K_ν^α is in particular measurable and such that $K_\nu^\alpha(x, \cdot)$, $K_\nu^\alpha(\cdot, y) \in L^1(\mathbb{R}^n)$ and

$$\|K_\nu^\alpha(x, \cdot)\|_1 = \|\partial^\alpha h_\nu / \partial x^\alpha\|_1 \, , \; \|K_\nu^\alpha(\cdot, y)\|_1 = \|\partial^\alpha h_\nu / \partial x^\alpha\|_1$$

for all $x, y \in \mathbb{R}^n$. Hence to K_ν^α there is associated a bounded linear integral operator $\text{Int}(K_\nu^\alpha)$ on $L^2_{\mathbb{C}}(\mathbb{R}^n)$ with operator norm equal or smaller than $\|\partial^\alpha h_\nu / \partial x^\alpha\|_1$. Hence it follows

$$\frac{\partial^\alpha}{\partial x^\alpha}(h_\nu * f) \in L^2_{\mathbb{C}}(\mathbb{R}^n) \, .$$

Since this is true for any $\alpha \in \mathbb{N}^n$, also

$$h_\nu * f \in W^k_{\mathbb{C}}(\mathbb{R}^n)$$

for all $k \in \mathbb{N}$ and, obviously,

$$\partial^\alpha(h_\nu * f) = \frac{\partial^\alpha h_\nu}{\partial x^\alpha} * f \, .$$

Further, for $f \in C_0(\mathbb{R}^n, \mathbb{C})$, it follows that

$$\text{supp}(H_\nu f - f) \subset \text{supp}(f) + B_1(0)$$

and

$$|H_\nu f - f|(x) \leqslant \int_{\mathbb{R}^n} |f(x - \mathrm{id}_{\mathbb{R}^n}) - f(x)|\, h_\nu \, dv^n$$

for all $x \in \mathbb{R}^n$. Since f is in particular uniformly continuous, it follows for every $\varepsilon > 0$ the existence of a $\delta > 0$ such that for all $x \in \mathbb{R}^n, y \in U_\delta(0)$

$$|f(x-y) - f(x)| \leqslant [v^n(\mathrm{supp}(f) + B_1(0))]^{-1/2}\, \varepsilon^{1/2} .$$

As a consequence, for all $\nu \in \mathbb{N}^*$ such that $\nu > 1/\delta$

$$\|H_\nu f - f\|_2 \leqslant \varepsilon$$

holds. Hence it follows for every $f \in C_0(\mathbb{R}^n, \mathbb{C})$ that

$$\lim_{\nu \to \infty} \|H_\nu f - f\|_2 = 0 .$$

Since $C_0(\mathbb{R}^n, \mathbb{C})$ is dense in $L^2_{\mathbb{C}}(\mathbb{R}^n)$ and $H_1, H_2, \ldots$ is in particular uniformly bounded, this implies also that

$$\mathrm{s-}\lim_{\nu \to \infty} H_\nu = \mathrm{id}_{L^2_{\mathbb{C}}(\mathbb{R}^n)} .$$

Finally, the statement of the lemma follows. □

The basis for Hermitian hyperbolic systems is provided by the following theorem of Friedrichs [79]

Theorem 5.5.2. Let $n, p \in \mathbb{N}^*$ and $A^1, \ldots, A^n$ be elements of $C^1(\mathbb{R}^n, M(p \times p, \mathbb{C}))$ such that $(A^j)_{kl}, (A^j_{,m})_{kl}$ are bounded for all $k, l \in \{1, \ldots, p\}$ and $j, m \in \{1, \ldots, n\}$. Finally, define the Hilbert space

$$X := (L^2_{\mathbb{C}}(\mathbb{R}^n))^p$$

and the linear operators $A, A' : (C^1_0(\mathbb{R}^n, \mathbb{C}))^p \to X$ by

$$Au := \sum_{j=1}^n A^j u_{,j} \, , \; A'u := -\sum_{j=1}^n (A^{j*} u)_{,j} = -\sum_{j=1}^n (A^{j*} u_{,j} + A^{j*}_{,j} u)$$

for all $u \in (C^1_0(\mathbb{R}^n, \mathbb{C}))^p$.[5] Then

$$A'^* = \bar{A} .$$

Moreover,

$$D(\bar{A}) \supset \left(W^1_{\mathbb{C}}(\mathbb{R}^n)\right)^p$$

and

$$\bar{A}u = \sum_{j=1}^n A^j \partial^j u$$

for all $u \in \left(W^1_{\mathbb{C}}(\mathbb{R}^n)\right)^p$.

[5] Note that in this elements of $\mathbb{C}^p$ are considered as column vectors, matrix multiplication is used and partial derivatives and weak derivatives are defined component-wise. In connection with matrices * denotes Hermitian conjugation.

Proof. First, it follows by partial integration that

$$\langle v|Au\rangle = \sum_{j=1}^{n}\sum_{k,l=1}^{p}\int_{\mathbb{R}^n} v_k^* A_{kl}^j u_{l,j}\,dv^n = -\sum_{j=1}^{n}\sum_{k,l=1}^{p}\int_{\mathbb{R}^n} (v_k^* A_{kl}^j)_{,j}\,u_l\,dv^n$$
$$= -\sum_{j=1}^{n}\sum_{k,l=1}^{p}\int_{\mathbb{R}^n} ((A^{j*})_{lk}v_k)^*_{,j}\,u_l\,dv^n = \langle A'v|u\rangle$$

for all $u, v \in (C_0^1(\mathbb{R}^n,\mathbb{C}))^p$ and hence

$$A' \subset A^*$$

and

$$A^{**} = \bar{A} \subset A'^* . \tag{5.5.1}$$

It remains to be proved that

$$\bar{A} \supset A'^* . \tag{5.5.2}$$

For this, we notice in a second step that since by (using obvious notation)

$$\left((W_{\mathbb{C}}^1(\mathbb{R}^n))^p \to (L_{\mathbb{C}}^2(\mathbb{R}^n))^p , u \mapsto \sum_{j=1}^{n} A^j \partial^j u \right)$$

there is given a continuous linear map and since $\left(C_0^1(\mathbb{R}^n,\mathbb{C})\right)^p$ is dense in $\left(W_{\mathbb{C}}^1(\mathbb{R}^n)\right)^p$ that

$$\left(W_{\mathbb{C}}^1(\mathbb{R}^n)\right)^p \subset D(\bar{A})$$

and

$$\bar{A}u = \sum_{j=1}^{n} A^j \partial^j u$$

for all $u \in \left(W_{\mathbb{C}}^1(\mathbb{R}^n)\right)^p$. Analogously, it follows that

$$\left(W_{\mathbb{C}}^1(\mathbb{R}^n)\right)^p \subset D(\overline{A'})$$

and

$$\overline{A'}\,u = -\sum_{j=1}^{n}(A^{j*}\,\partial^j u + A^{j*}_{,j} u)$$

for all $u \in \left(W_{\mathbb{C}}^1(\mathbb{R}^n)\right)^p$. In the following, the notation of the previous lemma will be used. For this, let h be as specified in that lemma. (Obviously, such a function h satisfying the prescribed properties in that lemma exists.) In particular, we define for $\nu \in \mathbb{N}^*$ the corresponding $H_\nu^p \in L(X, X)$ by

$$(H_\nu^p u)_k := H_\nu u_k$$

for all $u \in X$ and $k \in \{1,\dots,p\}$. In the following, we consider the sequence of operators

$$(H^p_\nu A'^* - \bar{A}\, H^p_\nu)_{\nu\in\mathbb{N}^*} \,.$$

First, because of (5.5.1), we have for $u \in D(A)$ and $\nu \in \mathbb{N}^*$ (notice that u and hence also $H^p_\nu u$ has a compact support)

$$(H^p_\nu A'^* - \bar{A}\, H^p_\nu)\, u = H^p_\nu Au - A\, H^p_\nu u = H^p_\nu Au - \sum_{j=1}^{n} A^j H^p_\nu u_{,j}$$

and hence because of

$$\text{s-}\lim_{\nu\to\infty} H^p_\nu = \mathrm{id}_X$$

that

$$\lim_{\nu\to\infty} (H^p_\nu A'^* - \bar{A}\, H^p_\nu)\, u = 0 \,. \tag{5.5.3}$$

Further, it follows for $u \in X$ (using obvious notation)

$$\bar{A}\, H^p_\nu u = \sum_{j=1}^{n} A^j \,\mathrm{Int}(K^j_\nu)\, u \;,\; \overline{A'}\, H^p_\nu u = -\sum_{j=1}^{n} \left[A^{j*}\, \mathrm{Int}(K^j_\nu) + A^{j*}_{,j}\, \mathrm{Int}(K_\nu) \right] u \,.$$

Note that $\bar{A}H^p_\nu$ and $\overline{A'}H^p_\nu$ define bounded linear integral operators on X. For $u \in D(A'^*)$ and $v \in X$, it follows

$$\langle v|H^p_\nu A'^* u\rangle = \langle H^p_\nu v|A'^* u\rangle = \langle \overline{A'}\, H^p_\nu v|u\rangle = \langle v|(\overline{A'}\, H^p_\nu)^* u\rangle$$

and hence

$$H^p_\nu A'^* u = (\overline{A'}\, H^p_\nu)^* u \,.$$

Then a short calculation leads to

$$[(H^p_\nu A'^* - \bar{A}\, H^p_\nu)\, u]_k = \sum_{j=1}^{n}\sum_{l=1}^{p} \mathrm{Int}(K^j_{kl}) u_l - \sum_{j=1}^{n} (H^p_\nu A^j_{,j} u)_k \;,\; k \in \{1,\dots,p\}$$

where

$$K^j_{kl}(x,y) := -\nu^{n+1} \frac{\partial h}{\partial x^j}(\nu(x-y)) \left(A^j_{kl}(x) - A^j_{kl}(y) \right)$$

for $j \in \{1,\dots,n\}$ and $k,l \in \{1,\dots,p\}$. Now by the mean value theorem, it follows from the assumptions on $A_{,j}, j \in \{1,\dots,n\}$ the existence of $M > 0$ such that

$$|A^j_{kl}(x) - A^j_{kl}(y)| \leqslant M|\, x - y|$$

for all $x, y \in \mathbb{R}^n$ and hence

$$\|\mathrm{Int}(K^j_{kl})\| \leqslant M \,\|\, |\, \mathrm{id}_{\mathbb{R}^n}|\, \frac{\partial h}{\partial x^j}\, \|_1 \,.$$

The sequence consisting of the continuous extensions of $H_1^p A'^* - \bar{A} H_1^p, H_2^p A'^* - \bar{A} H_2^p, \dots$ to bounded linear operators on X is therefore uniformly bounded and hence it follows because of (5.5.3) also that

$$\lim_{\nu \to \infty} (H_\nu^p A'^* - \bar{A} H_\nu^p)\, u = 0 \, .$$

Altogether, we have

$$\lim_{\nu \to \infty} H_\nu^p u = u \, , \ \lim_{\nu \to \infty} \bar{A} H_\nu^p u = \lim_{\nu \to \infty} H_\nu^p A'^* u = A'^* u$$

and hence

$$u \in D(\bar{A}) \quad \text{and} \quad \bar{A} u = A'^* u \, .$$

Since this true for all $u \in D(A'^*)$, this implies (5.5.2) and, finally, the theorem. □

The previous theorem has the following immediate corollary on Hermitian hyperbolic systems.

Corollary 5.5.3. In addition, let $A^1(x), \dots, A^n(x)$ be Hermitian for all $x \in \mathbb{R}^n$. Then the bounded linear operator $B \in L(X, X)$ defined by

$$Bu := \left(\sum_{j=1}^{n} A^j_{\ ,j} \right) u$$

for all $u \in X$ is in particular self-adjoint. Then

(i) the operator $i\,(A + (1/2)B)$ is essentially self-adjoint. Moreover,

$$D(\overline{i\,(A + (1/2)B)}\,) \supset \left(W^1_{\mathbb{C}}(\mathbb{R}^n)\right)^p$$

and

$$\overline{i\,(A + (1/2)B)}\, u = i \sum_{j=1}^{n} \left(A^j \partial^j + \frac{1}{2} A^j_{\ ,j} \right) u$$

for all $u \in \left(W^1_{\mathbb{C}}(\mathbb{R}^n)\right)^p$.

(ii) the operator $\bar{A}$ is the generator of a strongly continuous group on X.

Note that this has applications for the treatment of Dirac equations.

6

Intertwining Relations, Operator Homomorphisms

It is a common feature of partial differential equations describing the evolution of physical systems that the order of differentiability in the weak sense ('regularity') of the initial data is preserved by the evolution. Theorem 6.1.1 below and its corollary describe this phenomenon in terms of semigroups of linear operators. There the 'regularity' of the data is quantified by its containedness in the domain of a power of the infinitesimal generator. We give only a very brief discussion. For a more in depth analysis see Section II.5 of [57]. From Theorem 6.1.6 on, the theoretical part in the remainder of the course essentially follows [114]. That part contained in this chapter mainly serves the purpose to define and study 'closed' invariant subspaces ('admissible spaces') of semigroups of linear operators. In this, the notion of 'invariance' is more general than the usual for one-parameter groups of unitary transformations on Hilbert spaces. A sufficient condition for the existence of such an admissible space is the existence of an linear operator that 'intertwines' between the semigroup and a further semigroup. In the examples of the next chapter, it can be seen that such intertwining operators include constraint operators imposed on evolution equations. In these cases, the additional semigroup is associated to the so called 'constraint evolution system'.

6.1 Semigroups and Their Restrictions

Theorem 6.1.1. (Regularity) Let $\mathbb{K} \in \{\mathbb{R}, \mathbb{C}\}$, $(X, \| \,\|)$ a $\mathbb{K}$-Banach space, $A : D(A) \to X$ the infinitesimal generator of a strongly continuous semigroup $T : [0, \infty) \to L(X, X)$. Then $Y := D(A)$ is left invariant by $T(t)$ for every $t \in [0, \infty)$ and the family of restrictions $T_Y(t)$ of $T(t)$ in domain and in range to Y, $t \in [0, \infty)$, define a strongly continuous semigroup $T_Y : [0, \infty) \to L((Y, \| \,\|_A), (Y, \| \,\|_A))$ with infinitesimal generator $\left(D(A^2) \to Y, \xi \mapsto A\xi\right)$. In particular, $D(A^2)$ is a core for A.

Proof. First, it follows by Theorem 4.1.1 (ii) that Y is invariant under $T(t)$ for every $t \in [0, \infty)$. In the following, we denote for every $t \in [0, \infty)$ by $T_Y(t)$ the restriction of $T(t)$ in domain and in range to Y. Then

$$\|T(t)\xi\|_A^2 = \|T(t)\xi\|^2 + \|AT(t)\xi\|^2 = \|T(t)\xi\|^2 + \|T(t)A\xi\|^2 \leqslant \|T(t)\|^2 \cdot \|\xi\|_A^2$$

for every $\xi \in Y$ and hence $T_Y(t)$ defines a bounded linear operator on $(Y, \| \|_A)$ with bound $\|T(t)\|$ for every $t \in [0, \infty)$. As a consequence, $T_Y := ([0, \infty) \to L(Y, Y), t \mapsto T_Y(t))$ is a semigroup. Further, by

$$\begin{aligned}\|T_Y(t)\xi - T_Y(s)\xi\|_A^2 &= \|T(t)\xi - T(s)\xi\|^2 + \|AT(t)\xi - AT(s)\xi\|^2 \\ &= \|T(t)\xi - T(s)\xi\|^2 + \|T(t)A\xi - T(s)A\xi\|^2\end{aligned}$$

for every $t, s \in [0, \infty)$ and $\xi \in Y$, it follows the strong continuity of T_Y. We denote by A_Y the infinitesimal generator of T_Y. For every $\xi \in D(A_Y)$, it follows

$$-\lim_{t \to 0, t > 0} \frac{1}{t}.[T(t) - \mathrm{id}_X]\xi = A_Y \xi$$

where the limit is performed in $(Y, \| \|_A)$. Hence it follows by the continuity of the inclusion $\iota_{Y \hookrightarrow X}$ of $(Y, \| \|_A)$ into X that $A_Y \xi = A\xi \in Y$ and hence that $\xi \in D(A^2)$. In addition, it follows for every $\xi \in D(A^2)$ that

$$-\lim_{t \to 0, t > 0} \frac{1}{t}.[T(t) - \mathrm{id}_X]\xi = A\xi \, ,$$

$$-\lim_{t \to 0, t > 0} A\,\frac{1}{t}.[T(t) - \mathrm{id}_X]\xi = -\lim_{t \to 0, t > 0} \frac{1}{t}.[T(t) - \mathrm{id}_X]A\xi = A^2\xi \, ,$$

hence

$$\lim_{t \to 0, t > 0} \left\| \frac{1}{t}.[T(t) - \mathrm{id}_X]\xi + A\xi \right\|_A = 0$$

and therefore $\xi \in D(A_Y)$. Hence it follows that $A_Y = (D(A^2) \to Y, \xi \mapsto A\xi)$. As a consequence, $D(A^2)$ is dense in $(Y, \| \|_A)$, and therefore $D(A^2)$ is a core for A. □

Corollary 6.1.2. (Regularity, continued) Let $\mathbb{K} \in \{\mathbb{R}, \mathbb{C}\}$, $(X, \| \|)$ a $\mathbb{K}$-Banach space, $A : D(A) \to X$ the infinitesimal generator of a strongly continuous semigroup $T : [0, \infty) \to L(X, X)$, $n \in \mathbb{N}^*$. Then $Y := D(A^n)$ is left invariant by $T(t)$ for every $t \in [0, \infty)$ and the family of restrictions $T_Y(t)$ of $T(t)$ in domain and range to Y, $t \in [0, \infty)$, define a strongly continuous semigroup $T_Y : [0, \infty) \to L((Y, \||\ \||_n), (Y, \||\ \||_n))$ with infinitesimal generator $(D(A^{n+1}) \to Y, \xi \mapsto A\xi)$, on the Banach space $(Y, \||\ \||_n)$ where

$$\||\xi\||_n := \sqrt{\sum_{k=0}^{n} \binom{n}{k} \|A^k \xi\|^2}$$

for all $\xi \in Y$ and $A^0 := \mathrm{id}_X$. In particular, $D(A^{n+1})$ is a core for A.

Proof. The proof proceeds by induction over n. The statement is true for $n = 1$ as a consequence of Theorem 6.1.1. Assume that the statement is true for $n \in \mathbb{N}^*$. Then $Y := D(A^n)$ is left invariant by $T(t)$ for every $t \in [0, \infty)$ and the family of restrictions $T_Y(t)$ of $T(t)$ in domain and in range to Y, $t \in [0, \infty)$, define a strongly continuous semigroup $T_Y : [0,\infty) \to L((Y, \| \ \|_n), (Y, \| \ \|_n))$ with infinitesimal generator $\tilde{A} := (D(A^{n+1}) \to Y, \xi \mapsto A\xi)$, on the Banach space $(Y, \| \ \|_n)$. Also is $D(A^{n+1})$ a core for A. Then it follows by Theorem 6.1.1 that $\bar{Y} := D(A^{n+1})$ is left invariant by $T_Y(t)$ for every $t \in [0,\infty)$, that the family of restrictions $T_{\bar{Y}}(t)$ of $T(t)$ in domain and in range to $\bar{Y}$, $t \in [0,\infty)$, define a strongly continuous semigroup $T_{\bar{Y}} : [0,\infty) \to L((\bar{Y}, \| \ \|_{\tilde{A}}), (\bar{Y}, \| \ \|_{\tilde{A}}))$ with infinitesimal generator $(D(\tilde{A}^2) \to \bar{Y}, \xi \mapsto \tilde{A}\xi = A\xi)$. It follows for every $\xi \in \bar{Y}$

$$\|\xi\|_{\tilde{A}} = \sqrt{\|\xi\|_n^2 + \|\tilde{A}\xi\|_n^2} = \sqrt{\sum_{k=0}^{n} \binom{n}{k} \|A^k\xi\|^2 + \sum_{k=0}^{n} \binom{n}{k} \|A^{k+1}\xi\|^2}$$

$$= \sqrt{\|\xi\|^2 + \sum_{k=1}^{n} \binom{n}{k} \|A^k\xi\|^2 + \sum_{k=1}^{n} \binom{n}{k-1} \|A^k\xi\|^2 + \|A^{n+1}\xi\|^2}$$

$$= \sqrt{\|\xi\|^2 + \sum_{k=1}^{n} \binom{n+1}{k} \|A^k\xi\|^2 + \|A^{n+1}\xi\|^2} = \|\xi\|_{n+1} \, .$$

Further, it follows $D(\tilde{A}^2) = \{\xi \in D(A^{n+1}) : A\xi \in D(A^{n+1})\} = D(A^{n+2})$. Hence $D(A^{n+2})$ is in particular dense in $(D(A^{n+1}), \| \ \|_{n+1})$. Since, as a consequence of

$$\|\xi\|_{n+1} = \sqrt{\sum_{k=0}^{n+1} \binom{n+1}{k} \|A^k\xi\|^2} \geqslant \|\xi\|_A$$

for all $\xi \in D(A^{n+1})$, the inclusion $\iota_{D(A^{n+1}) \hookrightarrow D(A)}$ of $(D(A^{n+1}), \| \ \|_{n+1})$ into $(D(A), \| \ \|_A)$ is continuous and $D(A^{n+1})$ is dense in $(D(A), \| \ \|_A)$, it follows that $D(A^{n+2})$ is dense in $(D(A), \| \ \|_A)$ and therefore also that $D(A^{n+2})$ is a core for A. □

Corollary 6.1.3. Let $\mathbb{K} \in \{\mathbb{R}, \mathbb{C}\}$, $(X, \| \ \|)$ a $\mathbb{K}$-Banach space and $A : D(A) \to X$ the infinitesimal generator of a strongly continuous semigroup $T : [0,\infty) \to L(X,X)$. Further, let D be a dense subspace of X which is contained in $D(A)$ and invariant under $T(t)$:

$$T(t)D \subset D$$

for all $t \geqslant 0$. Then D is a core for A.

Proof. For this, let $\bar{D}$ be the closure of D in $(D(A), \| \ \|_A)$. Then $(\bar{D}, \| \ \|_A|_{\bar{D}})$ is a closed subspace of $(D(A), \| \ \|_A)$ and hence a $\mathbb{K}$-Banach space. In particular, the inclusion of $\bar{D}$ into $D(A)$ is a continuous linear map from $(\bar{D}, \| \ \|_A|_{\bar{D}})$ to $(D(A), \| \ \|_A)$. Further, let $\xi \in D(A)$ and $t \geqslant 0$. Since D is dense in X, there is a sequence $\xi_1, \xi_2, \ldots$ in D converging to ξ. By Theorem 6.1.1, it follows the continuity of

$$([0,t] \to (\bar{D}, \| \ \|_A|_{\bar{D}}), t \mapsto T(t)\xi_\nu) \, .$$

As a consequence,

$$\int_{0,A}^{t} T(s)\xi_\nu \, ds \in \bar{D}$$

where the A in the integral sign denotes weak integration in $(D(A), \| \ \|_A)$ and $\nu \in \mathbb{N}$. In addition, it follows by (4.1.6) that

$$\begin{aligned}
\left\| \int_{0,A}^{t} T(s)\xi_\nu \, ds - \int_{0,A}^{t} T(s)\xi \, ds \right\|_A^2 &= \left\| \int_0^t T(s)\xi_\nu \, ds - \int_0^t T(s)\xi \, ds \right\|^2 \\
+ \left\| \int_0^t T(s)A\xi_\nu \, ds - \int_0^t T(s)A\xi \, ds \right\|^2 &= \left\| \int_0^t T(s)\xi_\nu \, ds - \int_0^t T(s)\xi \, ds \right\|^2 \\
+ \|T(t)(\xi - \xi_\nu) + \xi_\nu - \xi\|^2 \ .
\end{aligned}$$

As a consequence,

$$\int_{0,A}^{t} T(s)\xi \, ds \in \bar{D} \ .$$

Further, by Theorem 6.1.1 and (4.1.3), it follows that

$$\lim_{t \to 0+} \left\| \frac{1}{t} \cdot \int_{0,A}^{t} T(s)\xi \, ds - \xi \right\|_A = 0 \ .$$

Hence $\xi \in \bar{D}$ and $\bar{D} = D(A)$. This implies that D is a core of A. □

Theorem 6.1.4. **(Mazur)** Let $(X, \| \ \|)$ be a normed vector space, $\xi \in X, \xi_1, \xi_2, \dots$ a sequence in X such that

$$\text{w-}\lim_{\nu \to \infty} \xi_\nu = \xi$$

and $\varepsilon > 0$. Here 'w−lim' denotes the weak limit. Then there are elements $n \in \mathbb{N}^*$, $\alpha_1, \dots, \alpha_n \in [0, \infty)$ such that $\sum_{\nu=1}^{n} \alpha_\nu = 1$ and $\| \sum_{\nu=1}^{n} \alpha_\nu . \xi_\nu - \xi \| \leqslant \varepsilon$.

Proof. For this, define the set A consisting of all convex linear combinations of members of the sequence $\xi_1, \xi_2, \dots$ by

$$A := \left\{ \sum_{\nu=1}^{n} \alpha_\nu . \xi_\nu : n \in \mathbb{N}^*, \alpha_1, \dots, \alpha_n \in [0, \infty) \text{ such that } \sum_{\nu=1}^{n} \alpha_\nu = 1 \right\} .$$

Obviously, A and hence also its closure $\bar{A}$ in $(X, \| \|)$ are convex. The statement of this lemma follows if $x \in \bar{A}$. The proof of this is indirect. Assume on the contrary that $\bar{A}$ and $B := \{\xi\}$ are disjoint. Since B is in particular compact, these sets can be strictly separated by a hyperplane, i.e., there is a continuous real linear map $l : X \to \mathbb{R}$ along with $\gamma_1, \gamma_2 \in \mathbb{R}$ such that $\gamma_1 < \gamma_2$ and $l(A) \subset (-\infty, \gamma_1)$ and $l(B) \subset (\gamma_2, \infty)$.[1] If $\mathbb{K} = \mathbb{R}$, this implies that $\lim_{\nu \to \infty} l(\xi_\nu) \neq l(\xi) \lightning$. If $\mathbb{K} = \mathbb{C}$, by $\lambda := (X \to \mathbb{C}, \eta \mapsto l(\eta) - il(i.\eta)$ there is defined an element of X' such $\mathrm{Re}(\lambda) = l$. Hence it follows that $\lim_{\nu \to \infty} \lambda(\xi_\nu) \neq \lambda(\xi) \lightning$. □

[1] See, e.g., Theorem V.4 in the first volume of [179].

Lemma 6.1.5. Let $\mathbb{K} \in \{\mathbb{R}, \mathbb{C}\}$, $(X, \| \, \|)$ a $\mathbb{K}$-Banach space and $A : D(A) \to X$ be the infinitesimal generator of a strongly continuous semigroup. Then

$$\text{s-}\lim_{\lambda \to -\infty} -\lambda.(A - \lambda)^{-1} = \text{id}_X .$$

Proof. By Theorem 4.2.1, there are $\mu \in \mathbb{R}$ and $c \in [1, \infty)$ such that $(-\infty, -\mu)$ is part of the resolvent set of A and in particular such that

$$\|(A - \lambda)^{-1}\| \leqslant \frac{c}{|\lambda + \mu|} \tag{6.1.1}$$

for all $\lambda \in (-\infty, -\mu)$. Let $\lambda_1, \lambda_2, \dots$ be some a sequence in $(-\infty, -\mu)$ which is diverging to $-\infty$. Then it follows by (6.1.1) that

$$\| - \lambda_\nu.(A - \lambda_\nu)^{-1}\| \leqslant \frac{c \cdot |\lambda_\nu|}{|\lambda_\nu + \mu|} = c + \frac{|\mu|}{|\lambda_\nu| - |\mu|}$$

and hence the boundedness of the sequence $-\lambda_\nu.(A - \lambda_\nu)^{-1}$ in $L(X, X)$. Further, it follows for $\xi \in D(A)$ that

$$\begin{aligned} \| - \lambda_\nu.(A - \lambda_\nu)^{-1}\xi - \xi\| &= \| - \lambda_\nu.(A - \lambda_\nu)^{-1}\xi - (A - \lambda_\nu)^{-1}(A - \lambda_\nu)\xi\| \\ &= \|(A - \lambda_\nu)^{-1}A\xi\| \leqslant \frac{c}{|\lambda_\nu| - |\mu|} \cdot \|A\xi\| \end{aligned}$$

and hence that

$$\lim_{\nu \to \infty} -\lambda_\nu.(A - \lambda_\nu)^{-1}\xi = \xi .$$

Since $D(A)$ is dense in X and by using the boundedness of $-\lambda_\nu.(A-\lambda_\nu)^{-1}$ in $L(X, X)$, from this follows that

$$\text{s-}\lim_{\nu \to \infty} -\lambda_\nu.(A - \lambda_\nu)^{-1} = \text{id}_X .$$

□

Theorem 6.1.6. (Admissible spaces) Let $\mathbb{K} \in \{\mathbb{R}, \mathbb{C}\}$, $(X, \| \, \|_X)$ a $\mathbb{K}$-Banach space, and $A : D(A) \to X$ the infinitesimal generator of a strongly continuous semigroup $T : [0, \infty) \to L(X, X)$. Further, let Y be a subspace of X and $\| \, \|_Y : Y^2 \to \mathbb{C}$ a norm on Y such that $(Y, \| \, \|_Y)$ is a Banach space and such that the inclusion $\iota_{Y \hookrightarrow X}$ of $(Y, \| \, \|_Y)$ into X is continuous.

(i) Then Y is 'A-admissible', i.e, Y is left invariant by $T(t)$ for all $t \in [0, \infty)$ and the family $T_Y(t)$ of restrictions of $T(t)$ to Y in domain and in image, $t \in [0, \infty)$, defines a strongly continuous semigroup $T_Y : [0, \infty) \to L(Y, Y)$ on $(Y, \| \, \|_Y)$, if and only if there are $\mu \in \mathbb{R}$ and $c \in [1, \infty)$ such that $(-\infty, -\mu)$ is part of the resolvent set of A, for every $\lambda \in (-\infty, -\mu)$ the corresponding operator $(A-\lambda)^{-1}$ leaves Y invariant and its part in Y, i.e., its restriction in domain and in image to Y, $[(A - \lambda)^{-1}]_Y$ satisfies

$$\|[(A-\lambda)^{-1}]_Y^n\|_{\text{Op,Y}} \leqslant \frac{c}{|\lambda+\mu|^n} \tag{6.1.2}$$

for all $n \in \mathbb{N}^*$ and $(A-\lambda)^{-1}Y$ is dense in $(Y, \| \ \|_Y)$. If $(Y, \| \ \|_Y)$ is reflexive, the last condition is redundant.

(ii) If Y is A-admissible and $\mu \in \mathbb{R}$ and $c \in [1, \infty)$ are as described in (ii), then the part $A_Y := (\{\xi \in D(A) \cap Y : A\xi \in Y\} \to Y, \xi \mapsto A\xi)$ of A in Y is the infinitesimal generator of T_Y and

$$\|T_Y(t)\|_{\text{Op,Y}} \leqslant c\, e^{\mu t} \tag{6.1.3}$$

for all $t \in [0, \infty)$.

Proof. If Y is A-admissible, by Theorem 4.2.1 there are $\mu \in \mathbb{R}$ and $c \in [1, \infty)$ such that $(-\infty, -\mu)$ is part of the resolvent sets of A and the infinitesimal generator $\tilde{A}$ of T_Y and in particular such that for $\lambda \in (-\infty, -\mu)$

$$\|[(A-\lambda)^{-1}]^n\|_{\text{Op}} \leqslant \frac{c}{|\lambda+\mu|^n} \tag{6.1.4}$$

and

$$\|[(\tilde{A}-\lambda)^{-1}]^n\|_{\text{Op,Y}} \leqslant \frac{c}{|\lambda+\mu|^n} \tag{6.1.5}$$

for all $n \in \mathbb{N}^*$. Hence by Theorem 4.2.1 and Theorem 4.1.1, it follows (6.1.3) and for $\eta \in Y$

$$(A-\lambda)^{-1}\eta = \int_0^\infty e^{\lambda t}\, T(t)\, \eta\, dt\,,$$

where integration is weak Lebesgue integration with respect to $L(X, \mathbb{K})$, and

$$(\tilde{A}-\lambda)^{-1}\eta = \int_0^\infty e^{\lambda t}\, T(t)\, \eta\, dt$$

where integration is weak Lebesgue integration with respect to $L((Y, \| \ \|_Y), \mathbb{K})$. Since $\iota_{Y \hookrightarrow X}$ is continuous, it follows that $\iota^*_{Y \hookrightarrow X}(L(X, \mathbb{C})) \subset L((Y, \| \ \|_Y), \mathbb{K})$ and hence by Theorem 3.2.4 that

$$(A-\lambda)^{-1}\eta = (\tilde{A}-\lambda)^{-1}\eta \tag{6.1.6}$$

for all $\eta \in Y$ and hence that

$$(\tilde{A}-\lambda)^{-1} = [(A-\lambda)^{-1}]_Y \tag{6.1.7}$$

where $[(A-\lambda)^{-1}]_Y$ denotes the restriction of $(A-\lambda)^{-1}$ in domain and in image to Y. Hence (6.1.5) implies (6.1.2). (This implies in particular that $(A-\lambda)^{-1}Y$ is dense in $(Y, \| \ \|_Y)$.)

From (6.1.6), it also follows that $\tilde{A}$ is equal to the part A_Y of A in Y. "$A_Y \supset \tilde{A}$": If $\eta \in D(\tilde{A})$, then by (6.1.6)

$$\eta = (\tilde{A}-\lambda)^{-1}(\tilde{A}-\lambda)\eta = (A-\lambda)^{-1}(\tilde{A}-\lambda)\eta \in D(A) \cap Y$$

and

$$(A-\lambda)\eta = (\tilde{A}-\lambda)\eta$$

which implies that

$$A\eta = \tilde{A}\eta \in Y \ .$$

'$A_Y \subset \tilde{A}$': If $\eta \in D(A) \cap Y$ is such that $A\eta \in Y$, then by (6.1.6)

$$\eta = (A-\lambda)^{-1}(A-\lambda)\eta = (\tilde{A}-\lambda)^{-1}(\tilde{A}-\lambda)\eta \in D(\tilde{A}) \ .$$

Summarizing, so far we proved (ii) and the direction '$\Rightarrow$' of (i). The proof of the direction '$\Leftarrow$' of (i) proceeds as follows. Let $\mu \in \mathbb{R}$ and $c \in [1,\infty)$ be such that $(-\infty,-\mu)$ is part of the resolvent set of A, that for every $\lambda \in (-\infty,-\mu)$ the estimate (6.1.4) is true as well such that $(A-\lambda)^{-1}$ leaves Y invariant and that its part in Y, i.e., its restriction in domain and in image to Y, $[(A-\lambda)^{-1}]_Y$ satisfies (6.1.2). In particular, let $\lambda \in (-\infty,-\mu)$ and let $\tilde{A}$ be the part of A in Y. Then it follows (6.1.7) where $[(A-\lambda)^{-1}]_Y$ denotes the restriction of $(A-\lambda)^{-1}$ to Y in domain and in image to Y: For $\xi \in D(\tilde{A})$, it follows $\xi \in D(A) \cap Y$ and $\tilde{A}\xi = A\xi \in Y$. Hence $(\tilde{A}-\lambda)\xi = (A-\lambda)\xi \in Y$ and

$$(A-\lambda)^{-1}(\tilde{A}-\lambda)\xi = \xi \ .$$

For $\eta \in Y$, it follows $(A-\lambda)^{-1}\eta \in D(A) \cap Y$ and

$$A(A-\lambda)^{-1}\eta = \eta + \lambda(A-\lambda)^{-1}\eta \in Y \ .$$

Hence $(A-\lambda)^{-1}\eta \in D(\tilde{A})$ and

$$(\tilde{A}-\lambda)(A-\lambda)^{-1}\eta = (A-\lambda)(A-\lambda)^{-1}\eta = \eta \ .$$

From this follows the bijectivity of $\tilde{A}-\lambda$ and the validness of (6.1.7). In case that $(Y, \| \ \|_Y)$ is reflexive, it follows that the domain $D(\tilde{A})$ of $\tilde{A}$ is dense in $(Y, \| \ \|_Y)$: For this, let $\eta \in Y$ and $\lambda_\nu := -\mu - \nu$ for $\nu \in \mathbb{N}^*$. Then by (6.1.2), it follows

$$\| -\lambda_\nu.(A-\lambda_\nu)^{-1}\eta\|_Y \leqslant c \cdot (|\mu|+1) \cdot \|\eta\|_Y$$

for every $\nu \in \mathbb{N}^*$. Since bounded subsets of $(Y, \| \ \|_Y)$ are sequentially compact,[2] there is an increasing sequence ν_1, ν_2 in $\mathbb{N}^*$ such that

$$\mathrm{w}_Y - \lim_{\mu\to\infty} \left(-\lambda_{\nu_\mu}.(A-\lambda_{\nu_\mu})^{-1}\eta\right) = \eta'$$

where $\eta' \in Y$. Further, by Lemma 6.1.5 and the continuity of the inclusion $\iota_{Y\hookrightarrow X}$, it follows that $\eta' = \eta$ and hence by Lemma 6.1.4 that η is in the $\| \ \|_Y$-closure of $D(\tilde{A})$. Finally, $\tilde{A}$ is closed: If $\eta_1, \eta_2, \ldots$ is a sequence in $D(\tilde{A})$ converging in $(Y, \| \ \|_Y)$ to some element $\eta \in Y$ and further such that $\tilde{A}\eta_1, \tilde{A}\eta_2, \ldots$ is converging in $(Y, \| \ \|_Y)$ to some element $\eta' \in Y$, then it follows by (6.1.7) and (6.1.2), where $\lambda \in (-\infty,-\mu)$, that

$$\eta = (\tilde{A}-\lambda)^{-1}(\eta' - \lambda\eta) \in D(\tilde{A})$$

[2] See, e.g., [52] II.3.28.

and hence also that $\tilde{A}\eta = \eta'$. Therefore, it follows by Theorem 4.2.1 that $\tilde{A}$ is the infinitesimal generator of a strongly continuous semigroup $T_Y : [0,\infty) \to L(Y,Y)$ on $(Y, \langle | \rangle_Y)$. Finally, by Corollary 4.2.2 and (6.1.7), it follows

$$T(t) = \text{s}_X - \lim_{\nu\to\infty} e^{(\mu-\nu)t} . \exp(\nu^2 t.(A+\mu+\nu)^{-1})$$

and

$$T_Y(t) = \text{s}_Y - \lim_{\nu\to\infty} e^{(\mu-\nu)t} . \exp(\nu^2 t.[(A+\mu+\nu)^{-1}]_Y)$$

for every $t \in [0,\infty)$. Hence by the continuity of the inclusion $\iota_{Y\hookrightarrow X}$, it follows that Y is left invariant by $T(t)$ and that $T_Y(t)$ coincides with the restriction of $T(t)$ in domain and in range to Y, for every $t \in [0,\infty)$. □

Theorem 6.1.7. Let $\mathbb{K} \in \{\mathbb{R}, \mathbb{C}\}$, $(X, \| \|_X)$, $(Z, \| \|_Z)$ be $\mathbb{K}$-Banach spaces. Further, let $T : [0,\infty) \to L(X,X)$, $\hat{T} : [0,\infty) \to L(Z,Z)$ be strongly continuous semigroups with corresponding infinitesimal generators $A : D(A) \to X$ and $\hat{A} : D(\hat{A}) \to Z$, respectively. Finally, let S be a closed linear map from some dense subspace Y of X into Z such that the following 'intertwining' relation holds

$$S\,T(t) \supset \hat{T}(t)\,S \tag{6.1.8}$$

for all $t \in [0,\infty)$. Note that this implies that $T(t)$ leaves Y invariant for all $t \in [0,\infty)$. Then the family $T_Y(t)$ of restrictions of $T(t)$ in domain and in image to Y, $t \in [0,\infty)$, defines a strongly continuous semigroup $T_Y : [0,\infty) \to L(Y,Y)$ on $(Y, \| \|_Y)$. Its infinitesimal generator is given by the part $A_Y := (\{\xi \in D(A) \cap Y : A\xi \in Y\} \to Y, \xi \mapsto A\xi)$ of A in Y. Finally, $D(A) \cap Y$ is a core for A.

Proof. For this, denote by $\iota_{Y\hookrightarrow X}$ the inclusion of $(Y, \| \|_Y)$ into X where $\| \|_Y := \| \|_S$. In addition, let $T_Y(t)$ be the restriction of $T(t)$ to Y in domain and image, for every $t \in [0,\infty)$. Further, let $(c,\mu) \in [1,\infty) \times \mathbb{R}$, $(\hat{c},\hat{\mu}) \in [1,\infty) \times \mathbb{R}$ such that

$$\|T(t)\|_{\text{Op},X} \leqslant c\, e^{\mu t} \,,\ \|\hat{T}(t)\|_{\text{Op},Z} \leqslant \hat{c}\, e^{\hat{\mu} t}$$

for all $t \in [0,\infty)$. Then it follows by (6.1.8) for $t \in [0,\infty)$

$$\begin{aligned}\|T_Y(t)\xi\|_Y^2 &= \|T(t)\xi\|_X^2 + \|S\,T(t)\xi\|_Z^2 = \|T(t)\xi\|_X^2 + \|\hat{T}(t)S\xi\|_Z^2 \\ &\leqslant c^2 e^{2\mu t}\|\xi\|_X^2 + \hat{c}^2 e^{2\hat{\mu} t}\|S\xi\|_Z^2 \leqslant (\max\{c,\hat{c}\})^2 \cdot e^{2\max\{\mu,\hat{\mu}\}\,t}\|\xi\|_Y^2\end{aligned}$$

for all $\xi \in Y$ and hence that $T_Y(t)$ is a bounded linear operator on $(Y, \langle | \rangle_Y)$ and

$$\|T_Y(t)\|_{\text{Op},Y} \leqslant \max\{c,\hat{c}\} \cdot e^{\max\{\mu,\hat{\mu}\}\,t} .$$

Hence, obviously by $T_Y : [0,\infty) \to L(Y,Y)$, associating to each $t \in [0,\infty)$ the operator $T_Y(t)$, there is defined a semigroup on $(Y, \langle | \rangle_Y)$. Further, $t \in [0,\infty)$, $(t_\nu)_{\nu\in\mathbb{N}} \in [0,\infty)^{\mathbb{N}}$ is convergent to t and $\xi \in Y$. Therefore, it follows by (6.1.8) for every $\nu \in \mathbb{N}$

$$\begin{aligned}\|T_Y(t_\nu)\xi - T_Y(t)\xi\|_Y^2 &= \|T(t_\nu)\xi - T(t)\xi\|_X^2 + \|S\,T(t_\nu)\xi - S\,T(t)\xi\|_Z^2 \\ &= \|T(t_\nu)\xi - T(t)\xi\|_X^2 + \|\hat{T}(t_\nu)S\xi - \hat{T}(t)S\xi\|_Z^2\end{aligned}$$

and hence by the strong continuity of T, $\hat{T}$ that

$$\lim_{\nu\to\infty} \|T_Y(t_\nu)\xi - T_Y(t)\xi\|_Y = 0 \, .$$

Since $\xi, (t_\nu)_{\nu\in\mathbb{N}}$ and $t \in [0,\infty)$ were otherwise arbitrary, from this it follows the strong continuity of T_Y. Hence Y is A-admissible, and it follows by Theorem 6.1.6 that the infinitesimal generator of T_Y is given by the part A_Y of A in Y. Also, since $D(A_Y)$ is dense in $(Y, \| \, \|_Y)$, it follows that $D(A) \cap Y$ is dense in $(Y, \| \, \|_Y)$. Hence, since the inclusion $\iota_{Y\hookrightarrow X}$ is continuous, $D(A) \cap Y$ is dense in $(Y, \| \, \|_X)$ and, since Y is dense in X, also dense in X. Therefore $D(A) \cap Y$ is a dense subspace of X which is invariant under $T(t)$ for all $t \in [0,\infty)$ and therefore according to Corollary 6.1.3 also a core for A. □

Theorem 6.1.8. Let $\mathbb{K} \in \{\mathbb{R}, \mathbb{C}\}$, $(X, \| \, \|)$ be a $\mathbb{K}$-Banach space. Further, let $T : [0,\infty) \to L(X,X)$ be a strongly continuous contraction semigroup. Finally, let S be a densely-defined closed linear operator in X such that the intertwining relation

$$S\,T(t) \supset T(t)\,S$$

holds for all $t \in [0,\infty)$. Then

$$\|S\,T(t)\xi\| \leqslant \|S\xi\| \tag{6.1.9}$$

for all $\xi \in D(S)$ and $t \in [0,\infty)$.

Proof. For this, let $t \in [0,\infty)$ and $\xi \in D(S)$. Then

$$\|S\,T(t)\xi\| = \|T(t)\,S\xi\| \leqslant \|S\xi\|$$

since $T(t)$ is a contraction by assumption. □

Remark 6.1.9. Equality holds in (6.1.9) if T consists of isometries. The corresponding statement is often called Emmi Noether's theorem. It plays an important role in classical field theories and their quantization.[3] Note that the formulation of the initial-value problem for wave equations, Maxwell's equations and a subclass of autonomous Hermitian hyperbolic systems can be chosen in such a way that time evolution is unitary. See Corollary 2.2.3, Theorem 7.3.1 and Corollary 5.5.3. This is of course obvious for equations from Quantum Theory, like Schrödinger and Dirac equations. Also note that in all these cases polarization identities lead to further 'conserved quantities' under time evolution.

---

[3] For e.g., see [32]. Usually, that theorem is derived from a variational principle.

6.2 Intertwining Relations

Theorem 6.2.1. Let $(X, \| \|_X)$, $(Z, \| \|_Z)$ be $\mathbb{K}$-Banach spaces. Further, let $T : [0,\infty) \to L(X,X)$, $\hat{T} : [0,\infty) \to L(Z,Z)$ be strongly continuous semigroups with corresponding infinitesimal generators $A : D(A) \to X$ and $\hat{A} : D(\hat{A}) \to Z$, respectively. Further, let $(c,\mu) \in [1,\infty) \times \mathbb{R}$ be such that $\|T(t)\|_X \leqslant c\,e^{\mu t}$ and $\|\hat{T}(t)\|_Z \leqslant c\,e^{\mu t}$ for all $t \in [0,\infty)$. Finally, let S be a closed linear map from some dense subspace Y of X into Z. Then the following statements are equivalent:

(i)
$$S\,T(t) \supset \hat{T}(t)\,S \tag{6.2.1}$$
for all $t \in [0,\infty)$.

(ii) There is $\lambda \in (-\infty, -\mu)$ such that
$$S\,(A-\lambda) \subset (\hat{A}-\lambda)S\,. \tag{6.2.2}$$

(iii) There is $\lambda \in (-\infty, -\mu)$ such that
$$S\,(A-\lambda)^{-1} \supset (\hat{A}-\lambda)^{-1}S\,. \tag{6.2.3}$$

Proof. For this, define $\| \|_Y := \| \|_S$.
'(i) $\Rightarrow$ (ii)': By Theorem 6.1.7 (ii), Theorem 6.1.6, the $\mathbb{K}$-Banach space $(Y, \| \|_Y)$ is A-admissible and there is some $\lambda \in (-\infty, -\mu)$ such that $(A-\lambda)^{-1}$ leaves Y invariant. Let ξ be some element of $D(S\,(A-\lambda))$. Then $\xi \in D(A)$ and $(A-\lambda)\xi \in Y$. Hence it follows that $\xi = (A-\lambda)^{-1}(A-\lambda)\xi \in Y$ and that ξ is part of the domain of the part of A in Y. Therefore, it follows for every null sequence $(t_n)_{n\in\mathbb{N}} \in (0,\infty)^{\mathbb{N}}$ by the continuity of $S : (Y, \| \|_Y) \to Z$ and (6.2.1)

$$\begin{aligned}
SA\xi &= S\left(\lim_{n\to\infty,Y} -\frac{1}{t_n}\,(T(t_n)\xi - \xi)\right) = \lim_{n\to\infty} S\left(-\frac{1}{t_n}\,(T(t_n)\xi - \xi)\right) \\
&= \lim_{n\to\infty} -\frac{1}{t_n}\,(\hat{T}(t_n)S\xi - S\xi)
\end{aligned}$$

and hence that $S\xi \in D(\hat{A})$ and that $SA\xi = \hat{A}S\xi$. From the last, it follows that $S\,(A-\lambda)\xi = (\hat{A}-\lambda)S\xi$. '(ii) $\Rightarrow$ (iii)': For this, let $\lambda \in (-\infty,-\mu)$ be such that $S\,(A-\lambda) \subset (\hat{A}-\lambda)S$. Then it follows for every $\eta \in Y$ that $(A-\lambda)^{-1}\eta \in D(S\,(A-\lambda))$ and hence that $(A-\lambda)^{-1}\eta$ is part of $D((\hat{A}-\lambda)S)$ as well as

$$\begin{aligned}
(\hat{A}-\lambda)^{-1}S\eta &= (\hat{A}-\lambda)^{-1}S\,(A-\lambda)(A-\lambda)^{-1}\eta \\
&= (\hat{A}-\lambda)^{-1}(\hat{A}-\lambda)S\,(A-\lambda)^{-1}\eta = S\,(A-\lambda)^{-1}\eta\,.
\end{aligned}$$

'(iii) $\Rightarrow$ (i)': For this, let $\lambda_0 \in (-\infty,-\mu)$ be such that

$$S\,(A-\lambda_0)^{-1} \supset (\hat{A}-\lambda_0)^{-1}S\,.$$

Obviously, it follows from this by induction that

$$S\left[(A-\lambda_0)^{-1}\right]^n \supset \left[(\hat{A}-\lambda_0)^{-1}\right]^n S \tag{6.2.4}$$

for all $n \in \mathbb{N}$. Further, by Theorem 4.1.1 (iv)

$$\|(A-\lambda_0)^{-1}\|_{\text{Op,X}} \leqslant \frac{c}{-\mu-\lambda_0}\,,\ \|(\hat{A}-\lambda_0)^{-1}\|_{\text{Op,Z}} \leqslant \frac{c}{-\mu-\lambda_0}\,.$$

and hence by (3.1.5)

$$(A-\lambda)^{-1} = \sum_{n=0}^{\infty} (\lambda-\lambda_0)^n.\left[(A-\lambda_0)^{-1}\right]^{n+1}$$

$$(\hat{A}-\lambda)^{-1} = \sum_{n=0}^{\infty} (\lambda-\lambda_0)^n.\left[(\hat{A}-\lambda_0)^{-1}\right]^{n+1}$$

for $\lambda \in (\lambda_0 - (-\mu-\lambda_0)/c, \lambda_0]$. From this follows by (6.2.4) and the closedness of S that also

$$S\,(A-\lambda)^{-1} \supset (\hat{A}-\lambda)^{-1}S\ .$$

for $\lambda \in (\lambda_0 - (-\mu-\lambda_0)/c, \lambda_0]$. By repeating this reasoning, it follows inductively that

$$S\left[(A-\lambda)^{-1}\right]^n \supset \left[(\hat{A}-\lambda)^{-1}\right]^n S \tag{6.2.5}$$

for all $\lambda \in (-\infty, \lambda_0]$. Further, by Corollary 4.2.2 and (6.1.7), it follows for $t \in [0,\infty)$

$$T(t) = \mathrm{s_X}-\lim_{\nu\to\infty} e^{(\mu-\nu)t}.\exp(\nu^2 t.(A+\mu+\nu)^{-1}) \tag{6.2.6}$$

and

$$\hat{T}(t) = \mathrm{s_Z}-\lim_{\nu\to\infty} e^{(\mu-\nu)t}.\exp(\nu^2 t.(\hat{A}+\mu+\nu)^{-1})\ . \tag{6.2.7}$$

By (6.2.5) and the closedness of S, it follows

$$S\,e^{(\mu-\nu)t}.\exp(\nu^2 t.(A+\mu+\nu)^{-1}) \supset e^{(\mu-\nu)t}.\exp(\nu^2 t.(\hat{A}+\mu+\nu)^{-1})S$$

for $\nu \in [-\lambda_0, \infty)$. Finally, by this along with (6.2.6), (6.2.7) and the closedness of S, it follows (6.2.1). □

Theorem 6.2.2. (Sufficient conditions) Let $(X, \| \ \|_X)$, $(Z, \| \ \|_Z)$ be $\mathbb{K}$-Banach spaces. Further, let $T : [0,\infty) \to L(X,X)$, $\hat{T} : [0,\infty) \to L(Z,Z)$ be strongly continuous semigroups with corresponding infinitesimal generators $A : D(A) \to X$ and $\hat{A} : D(\hat{A}) \to Z$, respectively. In addition, let $(c,\mu) \in [1,\infty) \times \mathbb{R}$ be such that $\|T(t)\|_X \leqslant c\,e^{\mu t}$ and $\|\hat{T}(t)\|_Z \leqslant c\,e^{\mu t}$ for all $t \in [0,\infty)$. Finally, let S be a closed linear map from some dense subspace Y of X into Z and $\tilde{D} \leqslant D(SA) \cap D(\hat{A}S)$ such that

$$SA\xi = \hat{A}S\xi$$

for all $\xi \in \tilde{D}$. Note that this implies that every $\xi \in \tilde{D}$ satisfies $\xi \in D(A) \cap Y$, $A\xi \in Y$ and $S\xi \in D(\hat{A})$. Then each of the conditions (i), (ii) below implies that

$$S\,T(t) \supset \hat{T}(t)\,S \tag{6.2.8}$$

for all $t \in [0, \infty)$:

(i) $(A - \lambda)\tilde{D}$ is dense in $(Y, \|\,\|_Y)$ for some $\lambda \in (-\infty, -\mu)$ where $\|\,\|_Y := \|\,\|_S$.
(ii) S is bijective and $(\hat{A} - \lambda)S\tilde{D}$ is dense in Z for some $\lambda \in (-\infty, -\mu)$. In this case, it also follows that $\hat{A} = SAS^{-1}$.

Proof. '(i)': Let $\lambda \in (-\infty, -\mu)$ be such that $(A-\lambda)\tilde{D}$ is dense in $(Y, \|\,\|_Y)$. In addition, let $\xi \in D(A)$ be such that $(A - \lambda)\xi \in Y$. Then there is $(\xi_\nu)_{\nu \in \mathbb{N}} \in \tilde{D}^{\mathbb{N}}$ such that

$$\lim_{\nu \to \infty, Y} (A - \lambda)\xi_\nu = (A - \lambda)\xi \ . \tag{6.2.9}$$

By the continuity of the inclusion $\iota_{Y \hookrightarrow X}$ and the continuity of $(A - \lambda)^{-1}$, this implies $\lim_{\nu \to \infty, X} \xi_\nu = \xi$. Further, since $S : (Y, \|\,\|_Y) \to (Z, \|\,\|_Z)$ is continuous, it follows from (6.2.9)

$$\lim_{\nu \to \infty} (\hat{A} - \lambda)S\xi_\nu = \lim_{\nu \to \infty} S(A - \lambda)\xi_\nu = S(A - \lambda)\xi$$

and by the continuity of $(\hat{A} - \lambda)^{-1}$ that

$$\lim_{\nu \to \infty} S\xi_\nu = (\hat{A} - \lambda)^{-1}S(A - \lambda)\xi \ .$$

Since S is closed, it follows $\xi \in Y$ and

$$S\xi = (\hat{A} - \lambda)^{-1}S(A - \lambda)\xi$$

and, finally, $(\hat{A} - \lambda)S\xi = S(A - \lambda)\xi$. Hence (6.2.8) follows by Theorem 6.2.2 (ii).
'(ii)': Let S be bijective and $\lambda \in (-\infty, -\mu)$ be such that $(\hat{A} - \lambda)S\tilde{D}$ is dense in Z. Then $S^{-1} : (Z, \|\|_Z) \to (Y, \|\|_Y)$ is continuous by the inverse mapping theorem.[4] Hence it follows by the continuity of the inclusion $\iota_{Y \hookrightarrow X}$ that $\iota_{Y \hookrightarrow X}S^{-1}(\hat{A} - \lambda)^{-1}, (A - \lambda)^{-1}\iota_{Y \hookrightarrow X}S^{-1} \in L(Z, X)$. Further, it follows for every $\xi \in \tilde{D}$ that

$$\begin{aligned}\iota_{Y \hookrightarrow X}S^{-1}(\hat{A} - \lambda)^{-1}(\hat{A} - \lambda)S\xi &= S^{-1}S\xi = \xi \\ &= (A - \lambda)^{-1}\iota_{Y \hookrightarrow X}S^{-1}S(A - \lambda)\xi = (A - \lambda)^{-1}\iota_{Y \hookrightarrow X}S^{-1}(\hat{A} - \lambda)S\xi \ .\end{aligned}$$

Since $(\hat{A} - \lambda)S\tilde{D}$ is dense in Z, from this follows

$$\iota_{Y \hookrightarrow X}S^{-1}(\hat{A} - \lambda)^{-1} = (A - \lambda)^{-1}\iota_{Y \hookrightarrow X}S^{-1} \tag{6.2.10}$$

and hence

$$S(A - \lambda)^{-1} \supset (\hat{A} - \lambda)^{-1}S \ .$$

Therefore, (6.2.8) follows by Theorem 6.2.1. Further, (6.2.10) implies

[4] See, e.g., Theorem III.11 in the first volume of [179].

$$(\hat{A}-\lambda)^{-1} = S(A-\lambda)^{-1}S^{-1} .$$

Hence for every $\eta \in Z$

$$S(A-\lambda)S^{-1}(\hat{A}-\lambda)^{-1}\eta = S(A-\lambda)S^{-1}S(A-\lambda)^{-1}S^{-1}\eta = \eta$$

and for every $\xi \in D(\hat{A})$

$$(\hat{A}-\lambda)^{-1}S(A-\lambda)S^{-1}\xi = S(A-\lambda)^{-1}S^{-1}S(A-\lambda)S^{-1}\xi = \xi .$$

Hence it follows that $\hat{A}-\lambda = S(A-\lambda)S^{-1}$. As a consequence, it follows for every $\xi \in D(\hat{A})$ that

$$S^{-1}\xi \in D(A) , \ (A-\lambda)S^{-1}\xi = AS^{-1}\xi - \lambda S^{-1}\xi ,$$

$AS^{-1}\xi \in Y$ and

$$\hat{A}\xi = S(A-\lambda)S^{-1}\xi + \lambda\xi = SAS^{-1}\xi - \lambda\xi + \lambda\xi = SAS^{-1}\xi .$$

On the other hand, for every $\xi \in D(SAS^{-1})$ it follows

$$S^{-1}\xi \in D(A) \cap Y , \ AS^{-1}\xi \in Y , \ (A-\lambda)S^{-1}\xi \in Y$$

and hence $\xi \in D(\hat{A})$. Hence, finally, it follows that $\hat{A} = SAS^{-1}$. □

6.3 Nonexpansive Homomorphisms

Theorem 6.3.1. (Nonexpansive homomorphisms) Let $(X, \| \ \|_X)$, $(Z, \| \ \|_Z)$ be $\mathbb{K}$-Banach spaces. Further, let $\Phi : (L(X,X), +, ., \circ) \to (L(Z,Z), +, ., \circ)$ be a strongly sequentially continuous nonexpansive homomorphism, i.e., a homomorphism such that for every strongly convergent sequence $(A_\nu)_{\nu \in \mathbb{N}} \in (L(X,X))^{\mathbb{N}}$ it follows that

$$\text{s-}\lim_{\nu\to\infty} \Phi(A_\nu) = \Phi\left(\text{s-}\lim_{\nu\to\infty} A_\nu\right) ,$$

$\Phi(\text{id}_X) = \text{id}_Z$ and $\|\Phi(A)\|_{\text{Op},Z} \leqslant \|A\|_{\text{Op},X}$ for all $A \in L(X,X)$. Finally, let $T : [0,\infty) \to L(X,X)$ be a strongly continuous semigroup with corresponding infinitesimal generator $A : D(A) \to X$ and let $(c,\mu) \in [1,\infty) \times \mathbb{R}$ such that $\|T(t)\|_X \leqslant c\, e^{\mu t}$ for all $t \in [0,\infty)$. Then

(i) $\Phi T : [0,\infty) \to L(Z,Z)$ defined by $(\Phi T)(t) := \Phi(T(t))$ for all $t \in [0,\infty)$ is a strongly continuous semigroup and such that $\|(\Phi T)(t)\|_{\text{Op},Z} \leqslant c\, e^{\mu t}$ for all $t \in [0,\infty)$. Further, in case that A is continuous, the infinitesimal generator of that semigroup is given by $\Phi(A)$. Also in the remaining cases, we denote that infinitesimal generator by $\Phi(A)$.

(ii)

$$\Phi((A-\lambda)^{-1}) = (\Phi(A)-\lambda)^{-1}$$

for all $\lambda \in (-\infty, -\mu)$

(iii)

$$\Phi(A - \lambda) = \Phi(A) - \lambda \tag{6.3.1}$$

for every $\lambda \in \mathbb{C}$.

Proof. '(i)': For all $s, t \in [0, \infty)$, it follows

$$\begin{aligned}(\Phi T)(t+s) &= \Phi(T(t+s)) = \Phi(T(t)T(s)) = \Phi(T(t))\Phi(T(s)) \\ &= (\Phi T)(t)(\Phi T)(s)\ ,\ (\Phi T)(0) = \Phi(T(0)) = \Phi(\mathrm{id}_X) = \mathrm{id}_Z\ .\end{aligned}$$

Further, for every $t \in [0, \infty)$, it follows

$$\begin{aligned}\Phi(T(t_0)) &= \Phi\left(\mathrm{s}-\lim_{t\to t_0} T(t)\right) = \mathrm{s}-\lim_{t\to t_0} \Phi(T(t_0))\ , \\ \|(\Phi T)(t)\|_{\mathrm{Op,Z}} &= \|\Phi(T(t))\|_{\mathrm{Op,Z}} \leqslant \|T(t)\|_{\mathrm{Op,X}} \leqslant c\, e^{\mu t}\ .\end{aligned}$$

Hence ΦT is a strongly continuous semigroup and the spectrum of its generator $\bar{A}$ is contained in $(-\infty, -\mu) \times \mathbb{R}$. That $\bar{A} = \Phi(A)$, in case that A is continuous, can be concluded as follows. First, since A is densely-defined, continuous and closed, it follows by the linear extension theorem[5] that $D(A) = X$ and hence that $A \in L(X, X)$. Further, by Theorems 3.3.1, 4.1.1, it follows that $T(t) = \exp(-t.A)$ for all $t \in [0, \infty)$. From this, it follows by using the assumed properties of Φ that

$$\begin{aligned}(\Phi T)(t) = \Phi(\exp(-t.A)) &= \Phi\left(\lim_{\nu\to\infty} \sum_{k=0}^{\nu} \frac{(-t)^k}{k!} . A^k\right) \\ &= \lim_{\nu\to\infty} \sum_{k=0}^{\nu} \frac{(-t)^k}{k!} . (\Phi(A))^k = \exp(-t.\Phi(A))\end{aligned}$$

for every $t \in [0, \infty)$ and hence that $\Phi(A)$ is the infinitesimal generator of ΦT.
'(ii)': By Theorem 4.1.1, it follows for $\lambda \in (-\infty, -\mu)$

$$(\bar{A} - \lambda)^{-1} = \int_0^\infty \Phi(e^{\lambda t}\, T(t))\, dt$$

where integration is weak Lebesgue integration with respect to $L(Z, Z)$ and

$$(A - \lambda)^{-1} = \int_0^\infty e^{\lambda t}\, T(t)\, dt$$

where integration is weak Lebesgue integration with respect to $L(X, X)$. According to Corollary 4.2.2, there is a sequence $(A_\nu)_{\nu\in\mathbb{N}} \in (L(X, X))^{\mathbb{N}}$ such that

$$\|\exp(-t.A_\nu)\|_{\mathrm{Op,X}} \leqslant c\, e^{\mu t}$$

---
5 See, e.g., Theorem 4 in IV, § 3 [128].

and

$$\text{s}-\lim_{\nu\to\infty} \exp(-t.A_\nu) = T(t)$$

for all $t \in [0,\infty)$ and

$$\lim_{\nu\to\infty} A_\nu\xi = A\xi$$

for all $\xi \in D(A)$. Hence it follows for $\nu \in \mathbb{N}$ by Theorem 4.1.1 that $(-\infty,-\mu)$ is contained in the resolvent set of A_ν and that

$$(A_\nu - \lambda)^{-1} = \int_0^\infty f_{\lambda\nu}\, dt$$

where $f_{\lambda\nu} : (0,\infty) \to L(X,X)$ is defined by $f_{\lambda\nu}(t) := e^{\lambda t}.\exp(-t.A_\nu)$ for all $t > 0$ and integration is weak Lebesgue integration with respect to $L(X,X)$. By Theorem 3.3.1, $f_{\lambda\nu}$ is norm continuous and such that $\|f_{\lambda\nu}(t)\|_{\text{Op,X}} \leqslant c\, e^{-|\mu+\lambda|t}$ for all $t > 0$. Therefore, it follows by the norm continuity and nonexpansiveness of Φ the continuity of $\Phi \circ f_{\lambda\nu}$ and $\|(\Phi \circ f_{\lambda\nu})(t)\|_{\text{Op,Z}} \leqslant c\, e^{-|\mu+\lambda|t}$ for all $t > 0$. Hence by Theorem 3.2.5, there is a sequence $(s_n)_{n\in\mathbb{N}}$ of step functions with support contained in $(0,\infty)$ and range contained in $\text{Ran}(f_{\nu\lambda}) \cup 0_{L(X,X)}$ such that

$$\lim_{n\to\infty} \|s_n(t) - f_{\lambda\nu}(t)\|_{\text{Op,X}}$$

for almost all $t > 0$ and

$$\lim_{n\to\infty} \left\| \int_0^\infty s_n(t)\, dt - \int_0^\infty f_{\lambda\nu}(t)\, dt \right\|_{\text{Op,X}} = 0\,.$$

Hence it follows by the norm continuity of Φ that

$$\lim_{n\to\infty} \|\Phi(s_n(t)) - (\Phi \circ f_{\lambda\nu})(t)\|_{\text{Op,Z}} = 0 \tag{6.3.2}$$

for almost all $t > 0$ and

$$\lim_{n\to\infty} \left\| \int_0^\infty \Phi(s_n(t))\, dt - \Phi\left(\int_0^\infty f_{\lambda\nu}(t)\, dt \right) \right\|_{\text{Op,Z}} = 0\,.$$

Further, it follows by (6.3.2), Theorem 3.2.11 and Lebesgue's dominated convergence Theorem that

$$\lim_{n\to\infty} \left\| \int_0^\infty \Phi(s_n(t))\, dt - \int_0^\infty (\Phi \circ f_{\lambda\nu})(t)\, dt \right\|_{\text{Op,Z}} = 0$$

and hence, finally,

$$\Phi\left(\int_0^\infty f_{\lambda\nu}(t)\, dt \right) = \int_0^\infty (\Phi \circ f_{\lambda\nu})(t)\, dt\,. \tag{6.3.3}$$

Further, since Φ maps strongly continuous sequences in $L(X,X)$ to strongly continuous sequences in $L(Z,Z)$, it follows that

$$\text{s}-\lim_{\nu\to\infty}(\Phi\circ f_{\lambda\nu})(t)=e^{\lambda t}.\Phi(T(t))$$

for all $t>0$. Further, it follows for every $\xi\in Z$

$$\left(\int_0^\infty(\Phi\circ f_{\nu\lambda})(t)\,dt\right)\xi=\int_0^\infty(\Phi\circ f_{\nu\lambda})(t)\,\xi\,dt$$

where the integration on the right hand side of the equality is weak Lebesgue integration with respect to $L(Z,\mathbb{K})$. By Theorem 3.2.11, it follows

$$\left\|\int_0^\infty(\Phi\circ f_{\nu\lambda})(t)\,\xi\,dt-\int_0^\infty e^{\lambda t}.\Phi(T(t))\xi\,dt\right\|_Z$$
$$\leqslant\int_0^\infty\|(\Phi\circ f_{\nu\lambda})(t)\,\xi-e^{\lambda t}.\Phi(T(t))\xi\|_Z\,dt$$

and since $(\|(\Phi\circ f_{\nu\lambda})(t)\,\xi-e^{\lambda t}.\Phi(T(t))\xi\|_z)_{\nu\in\mathbb{N}}$ is everywhere pointwise on $(0,\infty)$ convergent to $0_{(0,\infty)\to\mathbb{R}}$ and is majorized by the summable function $2e^{-|\mu+\lambda|t}$ by Lebesgue's dominated convergence Theorem that

$$\lim_{\nu\to\infty}\int_0^\infty(\Phi\circ f_{\nu\lambda})(t)\,\xi\,dt=\int_0^\infty e^{\lambda t}.\Phi(T(t))\xi\,dt$$

and hence

$$\text{s}-\lim_{\nu\to\infty}\int_0^\infty(\Phi\circ f_{\nu\lambda})(t)\,dt=\int_0^\infty e^{\lambda t}.\Phi(T(t))\,dt\,. \tag{6.3.4}$$

Further, it follows for every $\xi\in X$

$$\left(\int_0^\infty f_{\nu\lambda}(t)\,dt\right)\xi=\int_0^\infty f_{\nu\lambda}(t)\,\xi\,dt$$

where the integration on the right hand side of the equality is weak Lebesgue integration with respect to $L(X,\mathbb{K})$. By Theorem 3.2.11, it follows

$$\left\|\int_0^\infty f_{\nu\lambda}(t)\,\xi\,dt-\int_0^\infty e^{\lambda t}.T(t)\xi\,dt\right\|_X\leqslant\int_0^\infty\|f_{\nu\lambda}(t)\,\xi-e^{\lambda t}.T(t)\xi\|_X\,dt$$

and, since $(\|f_{\nu\lambda}(t)\,\xi-e^{\lambda t}.T(t)\xi\|_X)_{\nu\in\mathbb{N}}$ is everywhere pointwise on $(0,\infty)$ convergent to $0_{(0,\infty)\to\mathbb{R}}$ as well as majorized by the summable function $2e^{-|\mu+\lambda|t}$, by Lebesgue's dominated convergence Theorem that

$$\lim_{\nu\to\infty}\int_0^\infty f_{\nu\lambda}(t)\,\xi\,dt=\int_0^\infty e^{\lambda t}.T(t)\xi\,dt$$

and hence

$$\text{s}-\lim_{\nu\to\infty}\int_0^\infty f_{\nu\lambda}(t)\,dt=\int_0^\infty e^{\lambda t}.T(t)\,dt\,.$$

Since Φ maps strongly continuous sequences in $L(X,X)$ to strongly continuous sequences in $L(Z,Z)$, this implies

$$\text{s-}\lim_{\nu\to\infty} \Phi\left(\int_0^\infty f_{\nu\lambda}(t)\,dt\right) = \Phi\left(\int_0^\infty e^{\lambda t}.T(t)\,dt\right) . \tag{6.3.5}$$

Finally, from (6.3.3), (6.3.4) and (6.3.5) it follows

$$(\bar{A}-\lambda)^{-1} = \int_0^\infty \Phi(e^{\lambda t}\,T(t))\,dt = \Phi\left(\int_0^\infty e^{\lambda t}\,T(t))\,dt\right) = \Phi\left((A-\lambda)^{-1}\right) .$$

'(iii)': First, obviously, for $\lambda \in \mathbb{C}$ by $T_\lambda(t) := e^{\lambda t}.T(t)$ for $t \in [0,\infty)$, there is defined a strongly continuous semigroup on X. We denote its infinitesimal generator by A_λ. Further, because of

$$\frac{1}{t}\,(T_\lambda(t)\xi - \xi) = e^{\lambda t}\cdot\frac{1}{t}\,(T(t)\xi - \xi) + \frac{e^{\lambda t}-1}{t}\,\xi$$

for every $\xi \in X$ and $t > 0$, it follows that $A_\lambda = A - \lambda$. Hence by (i) $\Phi T_\lambda : [0,\infty) \to L(Z,Z)$ defined by $\Phi T_\lambda(t) := \Phi(T_\lambda(t)) = e^{\lambda t}.\Phi(T(t))$, for all $t \in [0,\infty)$, is a strongly continuous semigroup with infinitesimal generator $\Phi(A-\lambda)$. Finally, applying the same reasoning to ΦT_λ, as has been applied to T_λ, leads to (6.3.1). □

7

Examples of Constrained Systems

This chapter gives examples of linearly constrained systems, i.e., systems of partial differential equations describing the evolution of a physical system whose states are required to belong to the kernel of a linear operator at every time t. A standard example are Maxwell's equations on flat space which are considered in Chapter 7.3. In all examples, it turns out that that linear operator 'is' an intertwining operator between the system of partial differential equations describing the evolution of the system and a system of partial differential equations describing the evolution of the constraints. For the motivation of the examples in Chapters 7.1, 7.2, see the beginning of Chapter 5.3. Theorem 7.2.1 below is a new result. For examples of applications of the results of the previous chapter in General Relativity, see [152, 181].

7.1 1-D Wave Equations with Sommerfeld Boundary Conditions

First, for motivation, we consider in a formal manner the $1 + 1$ wave equation

$$\frac{\partial^2 u}{\partial t^2} - \frac{\partial^2 u}{\partial x^2} + Vu = 0 \tag{7.1.1}$$

on the interval I with boundary conditions

$$\left(\frac{\partial u}{\partial t} - \frac{\partial u}{\partial x}\right)\bigg|_{x=0} = 0\,, \quad \left(\frac{\partial u}{\partial t} + \frac{\partial u}{\partial x}\right)\bigg|_{x=a} = 0$$

where V is some element of $C(\bar{I}, \mathbb{C})$.
Introduction of new dependent variables v_1, v_2 by

$$v_1 := \frac{\partial u}{\partial t} - \frac{\partial u}{\partial x}\,, \; v_2 := \frac{\partial u}{\partial t} + \frac{\partial u}{\partial x}$$

leads by (7.1.1) to the following first order system for $\mathbf{u} := {}^t(u, v_1, v_2)$

$$\frac{\partial \mathbf{u}}{\partial t} = -\mathbf{A} \cdot \mathbf{u} , \tag{7.1.2}$$

where

$$\mathbf{A} := \begin{pmatrix} 0 & -1/2 & -1/2 \\ V & \frac{\partial}{\partial x} & 0 \\ V & 0 & -\frac{\partial}{\partial x} \end{pmatrix}$$

with boundary conditions

$$v_1(0) = v_2(a) = 0 . \tag{7.1.3}$$

If in addition the constraint

$$\frac{\partial u}{\partial x} + \frac{1}{2}(v_1 - v_2) = 0$$

is satisfied for all t, the first component of $\mathbf{u}$ can be seen to satisfy (7.1.1) as a consequence of (7.1.2). Moreover, it follows by (7.1.2) that this constraint is 'propagated' in the sense that

$$\frac{\partial}{\partial t}\left[\frac{\partial u}{\partial x} + \frac{1}{2}(v_1 - v_2)\right] = 0 .$$

Theorem 7.1.1. Let $a > 0$, I the open interval of $\mathbb{R}$ defined by $I := (0, a)$ and $V \in C(\bar{I}, \mathbb{C})$. We define the densely-defined linear operator A_0 in $X := \left(L^2_{\mathbb{C}}(I)\right)^3$ by

$$A_0(u, v_1, v_2) := \left(-\frac{1}{2}(v_1 + v_2), v_1' + Vu, -v_2' + Vu\right)$$

for all $u \in C(\bar{I}, \mathbb{C})$ and $v_1, v_2 \in C^1(\bar{I}, \mathbb{C})$ such that

$$\lim_{x \to 0+} v_1(x) = \lim_{x \to a-} v_2(x) = 0 .$$

In addition, we define $\hat{A}$ as the trivial operator on $Z := L^2_{\mathbb{C}}(I)$, i.e., $\hat{A}f := 0$ for all $f \in L^2_{\mathbb{C}}(I)$. Finally, we define the dense subspace Y_0 of X by $Y_0 := C^1(\bar{I}, \mathbb{C}) \times (C(\bar{I}, \mathbb{C}))^2$, the Z-valued linear operator $S_0 : Y_0 \to Z$ in X by

$$S_0(u, v_1, v_2) := u' + \frac{1}{2}(v_1 - v_2)$$

for all $(u, v_1, v_2) \in Y_0$ and the subspace $\tilde{D}$ of $D(A_0) \cap Y_0$ by

$$\tilde{D} := \left(C^1(\bar{I}, \mathbb{C})\right)^3 \cap D(A_0) .$$

(i) A_0 is closable and $A := \bar{A}_0$ is the generator of a quasi-contractive strongly continuous semigroup.

(ii) S_0 is closable,

$$S_0 A_0(u, v_1, v_2) = \hat{A} S_0(u, v_1, v_2)$$

for all $(u, v_1, v_2) \in \tilde{D}$ and for all $\lambda \leqslant 0$ of large enough absolute value

$$(A_0 - \lambda)\tilde{D} = Y_0 . \tag{7.1.4}$$

(iii) $S := \bar{S}_0$ satisfies

$$S\,T(t) \supset \hat{T}(t) S$$

where $T : [0, \infty) \to L(X, X)$ is the strongly continuous semigroup on X generated by A and $\hat{T} = ([0, \infty) \to L(Z, Z), t \mapsto \mathrm{id}_Z)$ is the strongly continuous semigroup on Z generated by $\hat{A}$. In particular, $\ker S$ is left invariant by $T(t)$ for every $t \in [0, \infty)$.

Proof. '(i)': First, it follows that A_0 is the sum of the two linear operators $A_1 : D(A_0) \to X$ and $B : X \to X$ defined by

$$A_1(u, v_1, v_2) := (0, v_1', -v_2') = (0, A_r v_1, A_l v_2)$$

for all $(u, v_1, v_2) \in D(A_0)$ where A_r, A_l are defined according to Definition 5.1.1 and Definition 5.1.3, respectively, and

$$B(u, v_1, v_2) := \left(-\frac{1}{2}(v_1 + v_2), Vu, Vu\right)$$

for all $(u, v_1, v_2) \in X$. By Theorem 5.1.2, Corollary 5.1.4 and Theorem 4.2.6, it follows that A_1 is closable, that its closure is the generator of a contractive strongly continuous semigroup and that

$$\bar{A}_1 = 0 \times \bar{A}_r \times \bar{A}_l$$

where $0 \times \bar{A}_r \times \bar{A}_l : L^2_{\mathbb{C}}(I) \times D(\bar{A}_r) \times D(\bar{A}_l) \to X$ is defined by

$$0 \times \bar{A}_r \times \bar{A}_l\,(u, v_1, v_2) := (0, \bar{A}_r v_1, \bar{A}_l v_2)$$

for all $(u, v_1, v_2) \in L^2_{\mathbb{C}}(I) \times D(\bar{A}_r) \times D(\bar{A}_l)$. Further, it follows because of

$$\begin{aligned}
\|B(u, v_1, v_2)\|^2 &= \frac{1}{4}\|v_1 + v_2\|_2^2 + 2\|Vu\|_2^2 \\
&\leqslant \frac{1}{4}\left(\|v_1\|_2^2 + \|v_2\|_2^2 + 2\|v_1\|_2\|v_2\|_2\right) + 2\|V\|_\infty^2\|u\|_2^2 \\
&\leqslant \frac{1}{2}\left(\|v_1\|_2^2 + \|v_2\|_2^2\right) + 2\|V\|_\infty^2\|u\|_2^2 \\
&\leqslant \left(2\|V\|_\infty^2 + \frac{1}{2}\right)\|(u, v_1, v_2)\|^2
\end{aligned}$$

for every $(u, v_1, v_2) \in X$, that $B \in L(X, X)$ and

$$\|B\| \leqslant \frac{1}{\sqrt{2}} \left(1 + 4 \|V\|_\infty^2\right)^{1/2} .$$

Hence it follows by Theorems 4.4.3, 3.1.3 (vi) that A_0 is closable and that

$$A := \bar{A}_0 = 0 \times \bar{A}_r \times \bar{A}_l + B$$

is the generator of a quasi-contractive strongly continuous semigroup $T : [0, \infty) \to L(X, X)$ such that

$$\|T(t)\| \leqslant e^{\left(1+4\|V\|_\infty^2\right)^{1/2} t/\sqrt{2}}$$

for all $t \in [0, \infty)$.
'(ii)': Obviously, S_0^* is densely-defined and hence S_0 is closable. Also

$$\begin{aligned} S_0 A_0(u, v_1, v_2) &= S_0 \left(-\frac{1}{2} (v_1 + v_2), v_1' + Vu, -v_2' + Vu \right) \\ &= -\frac{1}{2} (v_1' + v_2') + \frac{1}{2} (v_1' + Vu + v_2' - Vu) = 0 = \hat{A} S_0(u, v_1, v_2) \end{aligned}$$

for all $(u, v_1, v_2) \in \tilde{D}$. In the following, let $\lambda < 0$, $(f, g_1, g_2) \in X$ and $(u, v_1, v_2) \in L^2_{\mathbb{C}}(I) \times D(\bar{A}_r) \times D(\bar{A}_l)$ such that

$$\begin{aligned} (A - \lambda)(u, v_1, v_2) &= \left(-\frac{1}{2} (v_1 + v_2) - \lambda u, \bar{A}_r v_1 + Vu - \lambda v_1, \bar{A}_l v_2 + Vu - \lambda v_2 \right) \\ &= (f, g_1, g_2) . \end{aligned} \tag{7.1.5}$$

(7.1.5) is equivalent to the system of equations

$$\begin{aligned} u &= -\frac{1}{2\lambda} (v_1 + v_2 + 2f) \\ (\bar{A}_r \times \bar{A}_l - \lambda + B')(v_1, v_2) &= \left(g_1 + \frac{1}{\lambda} V f , g_2 + \frac{1}{\lambda} V f \right) \end{aligned}$$

where

$$B'(w_1, w_2) := \left(-\frac{V}{2\lambda} (w_1 + w_2), -\frac{V}{2\lambda} (w_1 + w_2) \right)$$

for all $w_1, w_2 \in L^2_{\mathbb{C}}(I)$. Because of

$$\|B'(w_1, w_2)\|^2 \leqslant \frac{\|V\|_\infty^2}{|\lambda|^2} \cdot \|(w_1, w_2)\|^2$$

for all $w_1, w_2 \in L^2_{\mathbb{C}}(I)$, it follows that $B' \in L(L^2_{\mathbb{C}}(I) \times L^2_{\mathbb{C}}(I), L^2_{\mathbb{C}}(I) \times L^2_{\mathbb{C}}(I))$ and

$$\|B'\| \leqslant \frac{\|V\|_\infty}{|\lambda|} .$$

In addition,

$$\left\| B'\left(\bar{A}_r \times \bar{A}_l - \lambda\right)^{-1} \right\| \leqslant \frac{\|V\|_\infty}{|\lambda|^2} \, .$$

Hence it follows for $\lambda < -\|V\|_\infty^{1/2}$ the bijectivity of $\bar{A}_r \times \bar{A}_l - \lambda + B'$ and for $(f, g_1, g_2) \in Y_0$ by the proofs of Theorem 5.1.2, Corollary 5.1.4 the existence of $(u, v_1, v_2) \in \tilde{D}$ such that

$$(A_0 - \lambda)(u, v_1, v_2) = (f, g_1, g_2) \, .$$

and therefore finally (7.1.4).
'(iii)': The statement is an immediate consequence of parts (i), (ii) along with Theorem 6.2.2 (i). □

7.2 1-D Wave Equations with Engquist-Majda Boundary Conditions

The next example considers again the wave equation (7.1.1), but with Engquist-Majda boundary conditions [58]

$$\left(\frac{\partial}{\partial t} - \frac{\partial}{\partial x}\right)^2 u \Bigg|_{x=0} = 0 \, , \, \left(\frac{\partial}{\partial t} + \frac{\partial}{\partial x}\right)^2 u \Bigg|_{x=a} = 0$$

where V is some element of $C^1(\bar{I}, \mathbb{C})$. The first part of the discussion proceeds in formal manner. Introduction of new dependent variables v_1, v_2, v_3, v_4 by

$$v_1 := \left(\frac{\partial}{\partial t} - \frac{\partial}{\partial x}\right) u \, , \, v_2 := \left(\frac{\partial}{\partial t} + \frac{\partial}{\partial x}\right) u \, ,$$
$$v_3 := \left(\frac{\partial}{\partial t} - \frac{\partial}{\partial x}\right)^2 u \, , \, v_4 := \left(\frac{\partial}{\partial t} + \frac{\partial}{\partial x}\right)^2 u$$

leads by (5.4.10) to the following first order system for $\mathbf{u} := {}^t(u, v_1, v_2, v_3, v_4)$

$$\frac{\partial \mathbf{u}}{\partial t} = -\mathbf{A} \cdot \mathbf{u} \qquad (7.2.1)$$

where

$$\mathbf{A} := \begin{pmatrix} 0 & -1/2 & -1/2 & 0 & 0 \\ V/2 & 0 & 0 & -1/2 & 0 \\ V/2 & 0 & 0 & 0 & -1/2 \\ -V' & V & 0 & \frac{\partial}{\partial x} & 0 \\ V' & 0 & V & 0 & -\frac{\partial}{\partial x} \end{pmatrix}$$

with boundary conditions

$$v_3(0) = v_4(a) = 0 \, .$$

If in addition the constraints

$$C_1 := \frac{\partial u}{\partial x} + \frac{1}{2}(v_1 - v_2) = 0 \ , \ C_2 := Vu + 2\frac{\partial v_1}{\partial x} + v_3 = 0$$

$$C_3 := Vu - 2\frac{\partial v_2}{\partial x} + v_4 = 0$$

are satisfied for all t, the first component of $\mathbf{u}$ can be seen to satisfy (5.4.10) as a consequence of (7.2.1). Moreover, it follows by (7.2.1) that these constraints are 'propagated', in the sense that

$$\frac{\partial C_1}{\partial t} = \frac{1}{4}(C_2 - C_3) \ , \ \frac{\partial C_2}{\partial t} = -VC_1 \ , \ \frac{\partial C_3}{\partial t} = VC_1 \ .$$

Theorem 7.2.1. Let $a > 0$, I the open interval of $\mathbb{R}$ defined by $I := (0, a)$ and $V \in C^1(\bar{I}, \mathbb{C})$. We define the densely-defined linear operator A_0 in $X := \left(L^2_{\mathbb{C}}(I)\right)^5$ by

$$A_0(u, v_1, v_2, v_3, v_4) := \left(-\frac{1}{2}(v_1 + v_2), \frac{V}{2}u - \frac{1}{2}v_3, \frac{V}{2}u - \frac{1}{2}v_4, \right.$$
$$\left. -V'u + Vv_1 + v_3', V'u + Vv_2 - v_4' \right)$$

for all $u, v_1, v_2 \in C(\bar{I}, \mathbb{C})$ and $v_3, v_4 \in C^1(\bar{I}, \mathbb{C})$ satisfying

$$\lim_{x \to 0+} v_3(x) = \lim_{x \to a-} v_4(x) = 0 \ .$$

In addition, we define $\hat{A} : Z \to Z$, where $Z := (L^2_{\mathbb{C}}(I))^3$, by

$$\hat{A}(f_1, f_2, f_3) := \left(-\frac{1}{4}f_2 + \frac{1}{4}f_3, Vf_1, -Vf_1 \right)$$

for all $f_1, f_2, f_3 \in L^2_{\mathbb{C}}(I)$. Finally, we define the dense subspace Y_0 of X by $Y_0 := \left(C^1(\bar{I}, \mathbb{C})\right)^3 \times \left(C(\bar{I}, \mathbb{C})\right)^2$, the Z-valued linear operator $S_0 : Y_0 \to Z$ in X by

$$S_0(u, v_1, v_2, v_3, v_4) := \left(u' + \frac{1}{2}(v_1 - v_2), Vu + 2v_1' + v_3, Vu - 2v_2' + v_4 \right)$$

for all $(u, v_1, v_2, v_3, v_4) \in Y_0$ and the subspace $\tilde{D}$ of $D(A_0) \cap Y_0$ by

$$\tilde{D} := \left(C^1(\bar{I}, \mathbb{C})\right)^5 \cap D(A_0) \ .$$

(i) A_0 is closable and $A := \bar{A}_0$ is the generator of a quasi-contractive strongly continuous semigroup.
(ii) $\hat{A} \in L(Z, Z)$.
(iii) S_0 is closable,

$$S_0 A_0(u, v_1, v_2, v_3, v_4) = \hat{A} S_0(u, v_1, v_2, v_3, v_4)$$

for all $(u, v_1, v_2) \in \tilde{D}$ and for all $\lambda \leqslant 0$ of large enough absolute value

$$(A_0 - \lambda)\tilde{D} = Y_0 \ . \tag{7.2.2}$$

(iv) $S := \bar{S}_0$ satisfies

$$S\,T(t) \supset \hat{T}(t)S$$

where $T : [0,\infty \to L(X,X)$ is the strongly continuous semigroup on X generated by A and $\hat{T} : [0,\infty) \to Z$ is the strongly continuous semigroup on Z generated by $\hat{A}$. In particular, $\ker S$ is left invariant by $T(t)$ for every $t \in [0,\infty)$.

Proof. '(i)': First, it follows that A_0 is the sum of the two linear operators $A_1 : D(A_0) \to X$ and $B : X \to X$ defined by

$$A_1(u, v_1, v_2, v_3, v_4) := (0, 0, 0, v_3', -v_4') = (0, 0, 0, A_r v_3, A_l v_4)$$

for all $(u, v_1, v_2, v_3, v_4) \in D(A_0)$ where A_r, A_l are defined according to Definition 5.1.1 and Definition 5.1.3, respectively, and

$$B(u, v_1, v_2, v_3, v_4) := \left(-\frac{1}{2}(v_1 + v_2), \frac{V}{2}u - \frac{1}{2}v_3, \frac{V}{2}u - \frac{1}{2}v_4, -V'u \right.$$
$$\left. +Vv_1, V'u + Vv_2 \right)$$

for all $(u, v_1, v_2, v_3, v_4) \in X$. By Theorem 5.1.2, Corollary 5.1.4 and Theorem 4.2.6, it follows that A_1 is closable, that its closure is the generator of a contractive strongly continuous semigroup and that

$$\bar{A}_1 = 0 \times 0 \times 0 \times \bar{A}_r \times \bar{A}_l$$

where $0 \times 0 \times 0 \times \bar{A}_r \times \bar{A}_l : \left(L^2_{\mathbb{C}}(I)\right)^3 \times D(\bar{A}_r) \times D(\bar{A}_l) \to X$ is defined by

$$0 \times 0 \times 0 \times \bar{A}_r \times \bar{A}_l\,(u, v_1, v_2, v_3, v_4) := (0, 0, 0, \bar{A}_r v_1, \bar{A}_l v_2)$$

for all $(u, v_1, v_2, v_3, v_4) \in \left(L^2_{\mathbb{C}}(I)\right)^3 \times D(\bar{A}_r) \times D(\bar{A}_l)$. Further, it follows, because of

$$\begin{aligned}
\|B(u, v_1, v_2, v_3, v_4)\|^2 &= \frac{1}{4}\left(\|v_1 + v_2\|_2^2 + \|Vu - v_3\|_2^2 + \|Vu - v_4\|_2^2\right) \\
&\quad + \|V'u - Vv_1\|_2^2 + \|V'u + Vv_2\|_2^2 \\
&\leqslant \frac{1}{2}\|(u, v_1, v_2, v_3, v_4)\|_2^2 + \|Vu\|_2^2 + 4\|V'u\|_2^2 + 2\|Vv_1\|_2^2 \\
&\quad + 2\|Vv_2\|_2^2 \leqslant \left(2\|V\|_\infty^2 + 4\|V'\|_\infty^2 + \frac{1}{2}\right)\|(u, v_1, v_2, v_3, v_4)\|^2
\end{aligned}$$

for every $(u, v_1, v_2, v_3, v_4) \in X$, that $B \in L(X,X)$ and

$$\|B\| \leqslant \frac{1}{\sqrt{2}}\left(1 + 4\|V\|_\infty^2 + 8\|V'\|_\infty^2\right)^{1/2}.$$

Hence it follows by Theorems 4.4.3, 3.1.3 (vi) that A_0 is closable and that

$$A := \bar{A}_0 = 0 \times 0 \times 0 \times \bar{A}_r \times \bar{A}_l + B$$

is the generator of a quasi-contractive strongly continuous semigroup $T : [0, \infty) \to L(X, X)$ such that

$$\|T(t)\| \leqslant e^{\left(1+4\|V\|_\infty^2+8\|V'\|_\infty^2\right)^{1/2} t/\sqrt{2}}$$

for all $t \in [0, \infty)$.
'(ii)': It follows

$$\|\hat{A}(f_1, f_2, f_3)\|^2 = \frac{1}{16}\|f_2 - f_3\|^2 + 2\|Vf_1\|^2 \leqslant \frac{1}{8}\left(\|f_2\|^2 + \|f_3\|^2\right)$$
$$+ 2\|V\|_\infty^2 \|f_1\|^2 \leqslant \left(2\|V\|_\infty^2 + \frac{1}{8}\right)\|(f_1, f_2, f_3\|^2$$

for all $f_1, f_2, f_3 \in L^2_{\mathbb{C}}(I)$ and hence that $\hat{A} \in L(Z, Z)$ as well as

$$\|\hat{A}\| \leqslant \frac{1}{2\sqrt{2}}\left(1 + 16\|V\|_\infty^2\right)^{1/2} .$$

'(iii)': Obviously, S_0^* is densely-defined and hence S_0 is closable. Also

$$\begin{aligned}
S_0 A_0(u, v_1, v_2, v_3, v_4) &= S_0\Bigg(-\frac{1}{2}(v_1 + v_2), \frac{V}{2}u - \frac{1}{2}v_3, \frac{V}{2}u - \frac{1}{2}v_4, \\
&\qquad -V'u + Vv_1 + v_3', V'u + Vv_2 - v_4'\Bigg) \\
&= \Bigg(-\frac{1}{2}(v_1' + v_2') + \frac{1}{2}\left(\frac{V}{2}u - \frac{1}{2}v_3 - \frac{V}{2}u + \frac{1}{2}v_4\right), \\
&\qquad -\frac{V}{2}(v_1 + v_2) + (Vu - v_3)' - V'u + Vv_1 + v_3', \\
&\qquad -\frac{V}{2}(v_1 + v_2) - (Vu - v_4)' + V'u + Vv_2 - v_4'\Bigg) \\
&= \Bigg(-\frac{1}{2}(v_1' + v_2') + \frac{1}{4}(v_4 - v_3), \frac{V}{2}(v_1 - v_2) + Vu', \\
&\qquad -\frac{V}{2}(v_1 - v_2) - Vu'\Bigg) \\
&= \hat{A}\left(u' + \frac{1}{2}(v_1 - v_2), Vu + 2v_1' + v_3, Vu - 2v_2' + v_4\right) \\
&= \hat{A}S_0(u, v_1, v_2, v_3, v_4)
\end{aligned}$$

for all $(u, v_1, v_2, v_3, v_4) \in \tilde{D}$. In the following, let $\lambda < -(\|V\|_\infty/2)^{1/2}$, $(f, g_1, g_2, g_3, g_4) \in X$ and $(u, v_1, v_2, v_3, v_4) \in \left(L^2_{\mathbb{C}}(I)\right)^3 \times D(\bar{A}_r) \times D(\bar{A}_l)$ such that

$$(A-\lambda)(u,v_1,v_2,v_3,v_4) \tag{7.2.3}$$

$$= \left(-\frac{1}{2}(v_1+v_2)-\lambda u, \frac{V}{2}u-\frac{1}{2}v_3-\lambda v_1, \frac{V}{2}u-\frac{1}{2}v_4-\lambda v_2,\right.$$

$$\left.-V'u+Vv_1+(\bar{A}_r-\lambda)v_3, V'u+Vv_2+(\bar{A}_l-\lambda)v_4\right) \tag{7.2.4}$$

$$= (f,g_1,g_2,g_3,g_4)\,.$$

(7.2.3) is equivalent to the system of equations

$$u = \frac{1}{2\lambda^2+V}\left[g_1+g_2-2\lambda f+\frac{1}{2}(v_3+v_4)\right]$$

$$v_1 = \frac{1}{2\lambda^2+V}\left\{-\left(2\lambda+\frac{V}{2\lambda}\right)\left[\left(g_1+\frac{V}{2\lambda}f\right)+\frac{1}{2}v_3\right]\right.$$

$$\left.+\frac{V}{2\lambda}\left[\left(g_2+\frac{V}{2\lambda}f\right)+\frac{1}{2}v_4\right]\right\}$$

$$v_2 = \frac{1}{2\lambda^2+V}\left\{\frac{V}{2\lambda}\left[\left(g_1+\frac{V}{2\lambda}f\right)+\frac{1}{2}v_3\right]\right.$$

$$\left.-\left(2\lambda+\frac{V}{2\lambda}\right)\left[\left(g_2+\frac{V}{2\lambda}f\right)+\frac{1}{2}v_4\right]\right\}$$

$$(\bar{A}_r\times\bar{A}_l-\lambda+B')(v_3,v_4) = (h_1,h_2)$$

where

$$B'(w_1,w_2) := \left(-\frac{V^2+2\lambda V'+4\lambda^2 V}{4\lambda(2\lambda^2+V)}w_1+\frac{V^2-2\lambda V'}{4\lambda(2\lambda^2+V)}w_2,\right.$$

$$\left.\frac{V^2+2\lambda V'}{4\lambda(2\lambda^2+V)}w_1-\frac{V^2-2\lambda V'+4\lambda^2 V}{4\lambda(2\lambda^2+V)}w_2\right)$$

for all $w_1, w_2 \in L^2_{\mathbb{C}}(I)$ and

$$h_1 := g_3+\frac{V'}{2\lambda^2+V}(g_1+g_2-2\lambda f)$$

$$-\frac{V}{2\lambda^2+V}\left\{-\left(2\lambda+\frac{V}{2\lambda}\right)\left(g_1+\frac{V}{2\lambda}f\right)+\frac{V}{2\lambda}\left(g_2+\frac{V}{2\lambda}f\right)\right\}$$

$$h_2 := g_4-\frac{V'}{2\lambda^2+V}(g_1+g_2-2\lambda f)$$

$$-\frac{V}{2\lambda^2+V}\left\{\frac{V}{2\lambda}\left(g_1+\frac{V}{2\lambda}f\right)-\left(2\lambda+\frac{V}{2\lambda}\right)\left(g_2+\frac{V}{2\lambda}f\right)\right\}.$$

Obviously, it follows that $B' \in L(L^2_{\mathbb{C}}(I)\times L^2_{\mathbb{C}}(I), L^2_{\mathbb{C}}(I)\times L^2_{\mathbb{C}}(I))$ and the existence of $\lambda_0 < -(\|V\|_\infty/2)^{1/2}$, $C \geqslant 0$ such that

$$\|B'\| \leqslant \frac{C}{|\lambda|}$$

for all $\lambda \leqslant \lambda_0$. For such λ, it also follows

$$\left\| B' \, (\bar{A}_r \times \bar{A}_l - \lambda)^{-1} \right\| \leqslant \frac{C}{|\lambda|^2}$$

and hence the bijectivity of $\bar{A}_r \times \bar{A}_l - \lambda + B'$ for $\lambda < 0$ of large enough absolute value. In particular, it follows for such λ and $(f, g_1, g_2, g_3, g_4) \in Y_0$ by the proofs of Theorem 5.1.2, Corollary 5.1.4 the existence of $(u, v_1, v_2, v_3, v_4) \in \tilde{D}$ such that

$$(A_0 - \lambda)(u, v_1, v_2, v_3, v_4) = (f, g_1, g_2, g_3, g_4) \; .$$

and therefore finally (7.2.2).
'(iv)': The statement is an immediate consequence of parts (i)–(iii) along with Theorem 6.2.2 (i). □

7.3 Maxwell's Equations in Flat Space

The final example considers Maxwell's equations for the electromagnetic field E, B in Minkowski space and inertial coordinates t, x, y, z:

$$\frac{\partial E}{\partial t} = c \, \nabla \times B \; , \; \frac{\partial B}{\partial t} = -c \, \nabla \times E$$
$$\nabla \cdot E = \nabla \cdot B = 0$$

where c denotes the speed of light. The evolution equations form a symmetric hyperbolic system

$$\begin{aligned} \frac{\partial u}{\partial t} &= -c. \, (\partial^z u_5 - \partial^y u_6, \partial^x u_6 - \partial^z u_4, \partial^y u_4 - \partial^x u_5, \\ &\qquad \partial^y u_3 - \partial^z u_2, \partial^z u_1 - \partial^x u_3, \partial^x u_2 - \partial^y u_1) \\ &= -\left(A^x \cdot \frac{\partial u}{\partial x} + A^y \cdot \frac{\partial u}{\partial y} + A^z \cdot \frac{\partial u}{\partial x} \right) \end{aligned}$$

where $u := {}^t(E_1, E_2, E_3, B_1, B_2, B_3)$, the dot denotes matrix multiplication and

$$A^x := c \, . \begin{pmatrix} 0 & 0 & 0 & 0 & 0 & 0 \\ 0 & 0 & 0 & 0 & 0 & 1 \\ 0 & 0 & 0 & 0 & -1 & 0 \\ 0 & 0 & 0 & 0 & 0 & 0 \\ 0 & 0 & -1 & 0 & 0 & 0 \\ 0 & 1 & 0 & 0 & 0 & 0 \end{pmatrix} , \; A^y := c \, . \begin{pmatrix} 0 & 0 & 0 & 0 & 0 & -1 \\ 0 & 0 & 0 & 0 & 0 & 0 \\ 0 & 0 & 0 & 1 & 0 & 0 \\ 0 & 0 & 1 & 0 & 0 & 0 \\ 0 & 0 & 0 & 0 & 0 & 0 \\ -1 & 0 & 0 & 0 & 0 & 0 \end{pmatrix} ,$$

$$A^z := c \, . \begin{pmatrix} 0 & 0 & 0 & 0 & 1 & 0 \\ 0 & 0 & 0 & -1 & 0 & 0 \\ 0 & 0 & 0 & 0 & 0 & 0 \\ 0 & -1 & 0 & 0 & 0 & 0 \\ 1 & 0 & 0 & 0 & 0 & 0 \\ 0 & 0 & 0 & 0 & 0 & 0 \end{pmatrix} .$$

For the analysis of Maxwell's equations, we define for every $n \in \mathbb{N}^*$ and every multi-index $\alpha \in \mathbb{N}^n$ the densely-defined linear operator ∂^α in $L^2_{\mathbb{C}}(\mathbb{R}^n)$ by

$$\partial^\alpha := (-1)^{|\alpha|} \cdot \left(C_0^\infty(\mathbb{R}^n, \mathbb{C}) \to L^2_{\mathbb{C}}(\mathbb{R}^n), f \mapsto \frac{\partial^\alpha f}{\partial x^\alpha} \right)^*$$

where

$$|\alpha| := \sum_{j=1}^{n} \alpha_j .$$

Moreover, define as usual for any $k \in \mathbb{N}$

$$W^k_{\mathbb{C}}(\mathbb{R}^n) := \bigcap_{\alpha \in \mathbb{N}^n, |\alpha| \leqslant k} D(\partial^\alpha) .$$

Equipped with the scalar product

$$\langle , \rangle_k \ : (W^k_{\mathbb{C}}(\mathbb{R}^n))^2 \to \mathbb{C}$$

defined by

$$\langle f, g \rangle_k := \sum_{\alpha \in \mathbb{N}^n, |\alpha| \leqslant k} \langle \partial^\alpha f | \partial^\alpha g \rangle_2$$

for all $f, g \in W^k_{\mathbb{C}}(\mathbb{R}^n)$, $W^k_{\mathbb{C}}(\mathbb{R}^n)$ is a Hilbert space. In particular, we denote by $\|\ \|_k$ the norm on $W^k_{\mathbb{C}}(\mathbb{R}^n)$ which is induced by $\langle , \rangle_k$.

Theorem 7.3.1. We define the densely-defined linear operator $A_0 : \left(W^1_{\mathbb{C}}(\mathbb{R}^3)\right)^6 \to X$ in $X := \left(L^2_{\mathbb{C}}(\mathbb{R}^3)\right)^6$ by

$$(A_0 u)_i := \sum_{j=1}^{6} \left(A^x_{ij}\, \partial^x u_j + A^y_{ij}\, \partial^y u_j + A^z_{ij}\, \partial^z u_j \right) ,$$

$i \in \{1, \ldots, 6\}$, for all $u \in \left(W^1_{\mathbb{C}}(\mathbb{R}^3)\right)^6$. In addition, we define $\hat{A}$ as the trivial operator on $Z := \left(L^2_{\mathbb{C}}(\mathbb{R}^3)\right)^2$, i.e., $\hat{A} f := 0$ for all $f \in Z$. Finally, we define the dense subspace Y_0 of X by $Y_0 := \left(W^1_{\mathbb{C}}(\mathbb{R}^3)\right)^6$, the Z-valued linear operator $S_0 : Y_0 \to Z$ in X by

$$S_0 u := (\partial^x u_1 + \partial^y u_2 + \partial^z u_3, \partial^x u_4 + \partial^y u_5 + \partial^z u_6)$$

for all $u \in Y_0$ and the subspace $\tilde{D}$ of $D(A_0) \cap Y_0$ by

$$\tilde{D} := \left(W^2_{\mathbb{C}}(\mathbb{R}^3)\right)^6 .$$

(i) A_0 is closable and its closure A is skew-symmetric ($:\Leftrightarrow -iA$ is self-adjoint) and hence the infinitesimal generator of a unitary one-parameter group.

(ii) Define for every $\lambda \in \mathbb{C}^*$ the bounded linear operator $B_\lambda : \left(W^2_{\mathbb{C}}(\mathbb{R}^3)\right)^6 \to X$ by

$$B_\lambda u := \Big(\frac{c}{\lambda}\, \partial^x(\partial^x u_1 + \partial^y u_2 + \partial^z u_3) + \partial^y u_6 - \partial^z u_5 - \frac{\lambda}{c}\, u_1$$
$$\frac{c}{\lambda}\, \partial^y(\partial^x u_1 + \partial^y u_2 + \partial^z u_3) + \partial^z u_4 - \partial^x u_6 - \frac{\lambda}{c}\, u_2,$$
$$\frac{c}{\lambda}\, \partial^z(\partial^x u_1 + \partial^y u_2 + \partial^z u_3) + \partial^x u_5 - \partial^y u_4 - \frac{\lambda}{c}\, u_3,$$
$$\frac{c}{\lambda}\, \partial^x(\partial^x u_4 + \partial^y u_5 + \partial^z u_6) + \partial^z u_2 - \partial^y u_3 - \frac{\lambda}{c}\, u_4,$$
$$\frac{c}{\lambda}\, \partial^y(\partial^x u_4 + \partial^y u_5 + \partial^z u_6) + \partial^x u_3 - \partial^z u_1 - \frac{\lambda}{c}\, u_5,$$
$$\frac{c}{\lambda}\, \partial^z(\partial^x u_4 + \partial^y u_5 + \partial^z u_6) + \partial^y u_1 - \partial^x u_2 - \frac{\lambda}{c}\, u_6 \Big)$$

for all $u \in \left(W^2_{\mathbb{C}}(\mathbb{R}^3)\right)^6$. Then (using obvious notation)

$$(A - \lambda)^{-1} u = \frac{1}{c}\, B_\lambda \bigtimes_{i=1}^{6} [-\triangle + (\lambda/c)^2]^{-1} u \tag{7.3.1}$$

for all $u \in X$ and $\lambda \in \mathbb{C}\backslash i\mathbb{R}$ where $\triangle : W^2_{\mathbb{C}}(\mathbb{R}^3) \to L^2_{\mathbb{C}}(\mathbb{R}^3)$ is defined by $\triangle f := (\partial^x\partial^x + \partial^y\partial^y + \partial^z\partial^z) f$ for all $f \in W^2_{\mathbb{C}}(\mathbb{R}^3)$.

(iii) S_0 is closable,

$$S_0 A_0 u = \hat{A} S_0 u$$

for all $u \in \tilde{D}$ and for all $\lambda \in \mathbb{C}\backslash i\mathbb{R}$

$$(A_0 - \lambda)\tilde{D} = Y_0\,. \tag{7.3.2}$$

(iv) $S := \bar{S}_0$ satisfies

$$S\, T(t) \supset \hat{T}(t) S$$

where $\hat{T} = ([0, \infty) \to Z, t \mapsto \mathrm{id}_Z)$ is the strongly continuous semigroup on Z generated by $\hat{A}$. In particular, $\ker S$ is left invariant by $T(t)$ for every $t \in [0, \infty)$.

Proof. Part (i) is a consequence of Corollary 5.5.3.
'(ii)': First, it follows for every $\lambda \in \mathbb{C}\backslash i\mathbb{R}$ after some calculation that

$$(A_0 - \lambda) B_\lambda u = c \bigtimes_{i=1}^{6} [-\triangle + (\lambda/c)^2] u$$

for every $u \in \left(W^3_{\mathbb{C}}(\mathbb{R}^3)\right)^6$ and hence that

$$(A_0 - \lambda)\, \frac{1}{c}\, B_\lambda \bigtimes_{i=1}^{6} [-\triangle + (\lambda/c)^2]^{-1} u = u$$

for all $u \in \left(W^1_{\mathbb{C}}(\mathbb{R}^3)\right)^6$. Therefore, since $\left(W^1_{\mathbb{C}}(\mathbb{R}^3)\right)^6$ is dense in X and

$$\frac{1}{c}\,B_\lambda \bigtimes_{i=1}^{6} [-\triangle + (\lambda/c)^2]^{-1} \in L(X,X) \ ,$$

finally, it follows (7.3.1).
'(iii)': First, we notice that S_0 defines a continuous linear operator from $\left(W^1_{\mathbb{C}}(\mathbb{R}^3),\right.$ $\left.\|\,\|_1\right)^6$ to X. In addition, it follows

$$\begin{aligned}\langle (f_1,f_2)|S_0u\rangle &= \langle f_1|\partial^x u_1 + \partial^y u_2 + \partial^z u_3\rangle_2 + \langle f_2|\partial^x u_4 + \partial^y u_5 + \partial^z u_6\rangle_2 \\ &= -\langle (f_{1,x}, f_{1,y}, f_{1,z}, f_{2,x}, f_{2,y}, f_{2,z})|u\rangle\end{aligned}$$

for all $u \in Y_0$ and $f_1, f_2 \in C^\infty_0(\mathbb{R}^3, \mathbb{C})$. Hence S_0^* is densely-defined and therefore S_0 is closable. Further, it follows

$$S_0 A_0 u = (0,0) = \hat{A} S_0 u$$

for all $u \in \tilde{D}$. Finally, it follows by (ii) that

$$(A-\lambda)^{-1} Y_0 \subset \tilde{D}$$

and hence (7.3.2) for all $\lambda \in \mathbb{C}\backslash i\mathbb{R}$.
'(iv)': The statement is an immediate consequence of parts (i), (iii) along with Theorem 6.2.2 (i). □

8

Kernels, Chains, and Evolution Operators

This chapter lays the foundation for the treatment of non-autonomous linear equations of the form

$$u'(t) = -A(t)u(t) \tag{8.0.1}$$

where t is from some non-empty open subinterval I of $\mathbb{R}$. Here $'$ denotes the ordinary derivative of functions with values in a Banach space X, $(A(t))_{t\in I}$ is a family of infinitesimal generators of strongly continuous semigroups on X and $u : I \to X$ is such that $u(t)$ is contained in the domain of $A(t)$ for every $t \in I$. Such equations are called 'linear evolution equations'. There is a large literature on such systems. For instance, see [42, 46, 50, 57, 60, 90, 104, 107, 108, 111, 118, 120, 121, 127, 148, 149, 151, 160, 168, 199, 203, 204, 208, 210, 219, 220]. Since in general explicitly time-dependent, the solutions of a linear evolution equation corresponding to the same data given at different initial times are in general not related by a time translation. As a consequence, the concept of a semigroup is replaced by the concept of a 'chain' which includes two time parameters. Chains are considered in Chapter 8.2. In physics they are usually referred to as 'propagators'. The convolution calculus developed in Chapter 8.1 is used in the treatment of perturbations of chains.

The method used for solving linear evolution equations in Chapter 9 is analogous to Euler's method in the theory of ordinary differential equations. For this, it is necessary to join chains which is considered in Chapters 8.3, 8.4. In addition, the domains of the operators $A(t)$ should not 'vary too strongly' with t. This is made precise in the notion of 'stable families of generators' which is considered in Chapter 8.6. For an extension of Kato's stability condition, see [159]. See also [59, 103, 153, 165].

Evolution operators are considered in Chapter 8.5. These are special chains whose definition involves two Banach spaces. They are defined in view of Chapter 9. There it will be proved the well-posedness of the initial value problem for linear evolution equations under certain assumptions. The result will be strong enough to allow for variable domains of the operators occurring in the equations and at the same time suitable to conclude the well-posedness (local in time) of the initial value problem for quasi-linear evolution equations in Chapter 11 by iteration.

8.1 A Convolution Calculus with Operator-Valued Kernels

Lemma 8.1.1. Let $\mathbb{K} \in \{\mathbb{R}, \mathbb{C}\}$, $(X, \| \, \|)$ a $\mathbb{K}$-Banach space.

(i) In addition, let $(Y, \| \, \|_Y)$ be a $\mathbb{K}$-Banach space, K a non-empty compact subset of some normed vector space and I some non-empty closed interval of $\mathbb{R}$. We denote by $C_*(K, L(X,Y))$, $PC_*(I, L(X,Y))$ the vector spaces of strongly continuous and piecewise strongly continuous, respectively, functions on K and I, respectively, with values in $L(X,Y)$. Then by

$$\|U\|_\infty := \sup\{\|U(x)\|_{\mathrm{Op}} : x \in K\}$$

there is defined a norm on $C_*(K, L(X,Y))$. Further, $(C_*(K, L(X,Y)), \| \, \|_\infty)$ is a complex Banach space.

(ii) In addition, let $n \in \mathbb{N}^*$, S_1, S_2, S_3 be non-empty subsets of $\mathbb{R}^n$, $f_1 : S_1 \to L(X,X)$, $f_2 : S_2 \to L(X,X)$ strongly continuous and $f_3 : S_3 \to X$ continuous. Then $f_1 f_2 := (S_1 \cap S_2 \to L(X,X), x \mapsto f_1(x) \circ f_2(x))$ is strongly continuous and $f_1 f_3 := (S_1 \cap S_3 \to X, x \mapsto f_1(x) f_3(x))$ is continuous.

(iii) In addition, let $a, b \in \mathbb{R}$ be such that $a \leqslant b$, $I := [a,b]$,

$$\Delta(I) := \{(t,s) \in \mathbb{R}^2 : a \leqslant s \leqslant t \leqslant b\} ,$$

$U, V, W \in C_*(\Delta(I), L(X,X))$ and $B, C \in PC_*(I, L(X,X))$.

a) Then by

$$[UBV](t,r) := \int_r^t U(t,s)B(s)V(s,r)\,ds$$

for every $(t,r) \in \Delta(I)$, there is defined an element [UBV] of $C_*(\Delta(I), L(X,X))$. Here integration is weak Lebesgue integration with respect to $L(X,\mathbb{K})$. In particular,

$$\|[UBV]\|_\infty \leqslant (b-a)\,\|B\|_\infty\,\|U\|_\infty\,\|V\|_\infty \tag{8.1.1}$$

where $\|B\|_\infty := \sup\{\|B(x)\|_{\mathrm{Op}} : x \in I\}$.

b) $[[UBV]CW] = [UB[VCW]]$.

c) The sequence $(V_\nu)_{\nu\in\mathbb{N}} \in (C_*(\Delta(I), L(X,X)))^{\mathbb{N}}$, recursively defined by $V_0 := U$ and $V_{\nu+1} := [UBV_\nu] = [V_\nu BU]$ for every $\nu \in \mathbb{N}$, is absolutely summable in $(C_*(\Delta(I), L(X,X)), \| \, \|_\infty)$. We define:

$$\mathrm{volt}(U,B) := \sum_{\nu\in\mathbb{N}} V_\nu .$$

In particular, $\mathrm{volt}(U,B)$ is the uniquely determined solution of the equation $U + [UBW] = W$ in $C_*(\Delta(I), L(X,X))$ and also the uniquely determined solution of the equation $U + [WBU] = W$ in $C_*(\Delta(I), L(X,X))$.

d) $\mathrm{volt}(\mathrm{volt}(U,B),C) = \mathrm{volt}(U, B+C)$.

Proof. '(i)': First, it follows for every $U \in C_*(K, L(X,Y))$ and $\xi \in X$ that $U\xi := (K \to Y, x \mapsto U(x)\xi)$ is continuous and hence by the compactness of K that $\operatorname{Ran} U\xi$ is bounded. Therefore, it follows by the principle of uniform boundedness[1] that $\operatorname{Ran} U$ is bounded, too. Hence $\|U\|_\infty := \sup\{\|U(x)\|_{\mathrm{Op}} : x \in K\}$ exists and is $\geqslant 0$. Further, $\|0_{K\to L(X,X)}\|_\infty = 0$ and if $U \in C_*(K, L(X,Y))$ is such that $\|U\|_\infty = 0$, it follows that $U = 0_{K\to L(X,X)}$. For $U \in C_*(K, L(X,Y))$ and $\lambda \in \mathbb{C}$, it follows that

$$\{\|(\lambda.U)(x)\|_{\mathrm{Op}} : x \in K\} = |\lambda|.\{\|U(x)\|_{\mathrm{Op}} : x \in K\}$$

and hence $\|\lambda.U\|_\infty = |\lambda| \cdot \|U\|_\infty$. Finally, it follows for $U, V \in C_*(K, L(X,Y))$ and every $x \in K$ that $\|(U+V)(x)\|_{\mathrm{Op}} \leqslant \|U(x)\|_{\mathrm{Op}} + \|V(x)\|_{\mathrm{Op}}$ and hence that

$$\|(U+V)(x)\|_{\mathrm{Op}} \leqslant \|U\|_\infty + \|V\|_\infty$$

which implies that $\|U+V\|_\infty \leqslant \|U\|_\infty + \|V\|_\infty$. Note that for every $U \in C_*(K, L(X,Y))$, it follows because of $\|U\|_\infty \geqslant \|U(x)\|_{\mathrm{Op}}$ that

$$\|U\xi\|_{\max} = \max\{\|U(x)\xi\|_Y : x \in K\} \leqslant \|U\|_\infty \cdot \|\xi\| \tag{8.1.2}$$

for all $\xi \in X$. Further, let $(U_\nu)_{\nu\in\mathbb{N}}$ be a Cauchy sequence in $(C_*(K, L(X,Y)), \| \, \|_\infty)$. Since $\|U\|_\infty \geqslant \|U(x)\|_{\mathrm{Op}}$ and $(L(X,Y), \|\,\|_{\mathrm{Op}})$ is complete, note that for this to be true only the completeness $(Y, \|\,\|_Y)$ is needed, it follows for every $x \in K$ that $(U_\nu(x))_{\nu\in\mathbb{N}}$ is uniformly convergent to some $U(x) \in L(X,Y)$. Note that this implies also that $(U_\nu(x)\xi)_{\nu\in\mathbb{N}}$ is convergent to $U(x)\xi$ for every $\xi \in X$. Further, the map $U := (X \to Y, x \mapsto U(x))$ is strongly continuous. To prove this, we notice that, as a consequence of (8.1.2), $(U_\nu\xi)_{\nu\in\mathbb{N}}$ is also a Cauchy sequence in $(C(K,Y), \|\,\|_{\max})$ and hence, as a consequence of the completeness of this space, convergent to some $F_\xi \in C(K,Y)$. Because of

$$\|U_\nu(x)\xi - F_\xi(x)\|_Y \leqslant \|U_\nu\xi - F_\xi\|_\infty$$

for all $x \in K$, it follows that $U\xi = F_\xi \in C(K,Y)$ for all $\xi \in X$ and hence that $U \in C_*(K, L(X,Y))$. Finally, let $\varepsilon > 0$ and ν_0 such that $\|U_\mu - U_\nu\|_\infty < \varepsilon/2$ for $\mu, \nu \geqslant \nu_0$. Then for $x \in K$, there is $\mu_0 \geqslant \nu_0$ such that $\|U_{\mu_0}(x) - U(x)\|_{\mathrm{Op}} < \varepsilon/2$. Hence it follows for $\nu \in \mathbb{N}$ such that $\nu \geqslant \nu_0$

$$\begin{aligned}\|U_\nu(x) - U(x)\|_{\mathrm{Op}} &\leqslant \|U_\nu(x) - U_{\mu_0}(x)\|_{\mathrm{Op}} + \|U_{\mu_0}(x) - U(x)\|_{\mathrm{Op}} \\ &\leqslant \|U_\nu - U_{\mu_0}\|_\infty + \|U_{\mu_0}(x) - U(x)\|_{\mathrm{Op}} < \varepsilon\end{aligned}$$

and hence also $\|U_\nu - U\|_\infty < \varepsilon$. Hence it follows also that

$$\lim_{\nu\to\infty} \|U_\nu - U\|_\infty = 0.$$

'(ii)': For this, let $x \in S_1 \cap S_2$ and $(x_\nu)_{\nu\in\mathbb{N}}$ be a sequence of elements of $S_1 \cap S_2$ converging to x. Then

[1] See, e.g, Theorem III.9 in the first volume of [179].

$$\begin{aligned}
&\|(f_1 f_2)(x_\nu)\xi - (f_1 f_2)(x)\xi\| = \|f_1(x_\nu)f_2(x_\nu)\xi - f_1(x)f_2(x)\xi\| \\
&\quad = \|f_1(x_\nu)(f_2(x_\nu)\xi - f_2(x)\xi) + f_1(x_\nu)f_2(x)\xi - f_1(x)f_2(x)\xi\| \\
&\quad \leqslant \|f_1(x_\nu)\|_{\mathrm{Op}}\,\|f_2(x_\nu)\xi - f_2(x)\xi\| + \|f_1(x_\nu)f_2(x)\xi - f_1(x)f_2(x)\xi\|
\end{aligned}$$

for every $\xi \in X$. Since $\mathrm{s}-\lim_{\nu\to\infty} f_1(x_\nu) = f_1(x)$, it follows that $(\|f_1(x_\nu)\eta\|)_{\nu\in\mathbb{N}}$ is bounded for every $\eta \in X$. Hence it follows by the principle of uniform boundedness[2] the existence of some $C \geqslant 0$ such that $\|f_1(x_\nu)\|_{\mathrm{Op}} \leqslant C$ for all $\nu \in \mathbb{N}$. Hence

$$\begin{aligned}
\|(f_1 f_2)(x_\nu)\xi - (f_1 f_2)(x)\xi\| &\leqslant C\,\|f_2(x_\nu)\xi - f_2(x)\xi\| \\
&\quad + \|f_1(x_\nu)f_2(x)\xi - f_1(x)f_2(x)\xi\|
\end{aligned}$$

and $\lim_{\nu\to\infty}\|(f_1 f_2)(x_\nu)\xi - (f_1 f_2)(x)\xi\| = 0$. Further, if $x \in S_1 \cap S_3$ and $(x_\nu)_{\nu\in\mathbb{N}}$ is a sequence of elements of $S_1 \cap S_3$ converging to x, then

$$\begin{aligned}
&\|(f_1 f_3)(x_\nu) - (f_1 f_3)(x)\| = \|f_1(x_\nu)f_3(x_\nu) - f_1(x)f_3(x)\| \\
&\quad = \|f_1(x_\nu)(f_3(x_\nu) - f_3(x)) + f_1(x_\nu)f_3(x) - f_1(x)f_3(x)\| \\
&\quad \leqslant \|f_1(x_\nu)\|_{\mathrm{Op}}\,\|f_3(x_\nu) - f_3(x)\| + \|f_1(x_\nu)f_3(x) - f_1(x)f_3(x)\| \\
&\quad \leqslant C\,\|f_3(x_\nu) - f_3(x)\| + \|f_1(x_\nu)f_3(x) - f_1(x)f_3(x)\|
\end{aligned}$$

where $C \geqslant 0$ is such that $\|f_1(x_\nu)\|_{\mathrm{Op}} \leqslant C$ for all $\nu \in \mathbb{N}$, and hence

$$\lim_{\nu\to\infty}\|(f_1 f_3)(x_\nu)\xi - (f_1 f_3)(x)\xi\| = 0\ .$$

'(iii)a)': Let $(t,r) \in \Delta(I)$. Then it follows by (ii) that $G_{(t,r)} := U(t,\cdot)BV(\cdot,r)$ is an element of $PC_*([r,t], L(X,X))$ which is uniformly bounded by

$$C := \|B\|_\infty\,\|U\|_\infty\,\|V\|_\infty\ .$$

Hence by Theorem 3.2.11, $\hat{G}_{(t,r)}$, where $\hat{G}_{(t,r)} : \mathbb{R} \to L(X,X)$ is defined by $\hat{G}_{(t,r)}(s) := G_{(t,r)}(s)$ if $s \in (r,t)$ and by $\hat{G}_{(t,r)}(s) := 0_{L(X,X)}$ otherwise, is weakly Lebesgue summable with respect to $L(X,\mathbb{K})$. Further, for $(t,r) \in \Delta(I)$, $((t_\nu, r_\nu))_{\nu\in\mathbb{N}} \in \Delta(I)^{\mathbb{N}}$ such that $\lim_{\nu\to\infty}(t_\nu, r_\nu) = (t,r)$ and $\xi \in X$, it follows that by $(\hat{G}_{(t_\nu,r_\nu)}\xi - \hat{G}_{(t,r)}\xi)_{\nu\in\mathbb{N}}$, there is given a sequence of bounded, almost everywhere continuous functions whose norm is majorized by $2C\|\xi\|.\chi(I)$ where $\chi(I)$ denotes the characteristic function of I. Hence by Theorem 3.2.5

$$\|[UBV](t_\nu, r_\nu)\xi - [UBV](t,r)\xi\| \leqslant \int_{\mathbb{R}} \|\hat{G}_{(t_\nu,r_\nu)}\xi - \hat{G}_{(t,r)}\xi\|\,dv^1$$

for every $\nu \in \mathbb{N}$. Further, $(\|\hat{G}_{(t_\nu,r_\nu)}\xi - \hat{G}_{(t,r)}\xi\|)_{\nu\in\mathbb{N}}$ is almost everywhere pointwise convergent to $0_{\mathbb{R}\to\mathbb{R}}$ and majorized by the summable function $2C\|\xi\|.\chi(I)$. Hence it follows by Lebesgue's dominated convergence Theorem that

$$\lim_{\nu\to\infty}\|[UBV](t_\nu, r_\nu)\xi - [UBV](t,r)\xi\| = 0\ .$$

[2] See, e.g, Theorem III.9 in the first volume of [179].

Finally, it follows by Theorem 3.2.5

$$\|[UBV](t,r)\xi\| \leqslant \int_r^t \|U(t,\cdot)BV(\cdot,r)\xi\|\,ds \leqslant C\,(b-a)\,\|\xi\|\,.$$

'(iii)b)': Let $(t,r) \in \Delta(I)$, $\xi,\eta \in X$ and $\text{pr}_1 := (\mathbb{R}^2 \to \mathbb{R}, (s,s') \mapsto s)$ and $\text{pr}_2 := (\mathbb{R}^2 \to \mathbb{R}, (s,s') \mapsto s')$. Then it follows by (ii) and the principle of uniform boundedness[3] that $(U(t,\cdot)\circ \text{pr}_1)(B\circ \text{pr}_1)V(C\circ \text{pr}_2)(W(\cdot,r)\circ \text{pr}_2)$ is a norm bounded, almost everywhere strongly continuous function on $\Delta([r,t])$. Hence it follows by the compactness of $\Delta([r,t])$ and Theorem 3.2.11 the weak integrability of this function with respect to $L(X,\mathbb{K})$. Further, it follows by (iii)a), the Theorem of Fubini and the change of variable formula that

$$\begin{aligned}
[UB[VCW]](t,r)\xi &= \int_r^t \left[\int_r^s U(t,s)B(s)V(s,s')C(s')W(s',r)\xi\,ds'\right] ds \\
&= \int_{r\leqslant s'\leqslant s\leqslant t} U(t,s)B(s)V(s,s')C(s')W(s',r)\xi\,ds\,ds' \\
&= \int_{r\leqslant s\leqslant s'\leqslant t} U(t,s')B(s')V(s',s)C(s)W(s,r)\xi\,ds\,ds' \\
&= \int_r^t \left[\int_s^t U(t,s')B(s')V(s',s)C(s)W(s,r)\xi ds'\right] ds \\
&= \int_r^t [UBV](t,s)C(s)W(s,r)\xi\,ds = [[UBV]CW](t,r)\xi\,.
\end{aligned}$$

'(iii)c)': First, using (iii)b), it follows by induction that $V_{\nu+1} = [V_\nu BU]$ for every $\nu \in \mathbb{N}$ because $V_1 = [UBV_0] = [UBU] = [V_0BU]$ and if $V_{\nu+1} = [V_\nu BU]$ for some $\nu \in \mathbb{N}$, then $V_{\nu+2} = [UBV_{\nu+1}] = [UB[V_\nu BU]] = [[UBV_\nu]BU] = [V_{\nu+1}BU]$. For $\beta \geqslant 0$, it follows that $e^{-\beta(\text{pr}_1-\text{pr}_2)}.U \in C_*(\Delta, L(X,X))$ and hence

$$\|U(t,r)\|_{\text{Op}} \leqslant Me^{\beta(t-r)} \tag{8.1.3}$$

for all $(t,r) \in \Delta(I)$ where $M := \|e^{-\beta(\text{pr}_1-\text{pr}_2)}.U\|_\infty$. Further, if $K > 0$ is such that $\|B(t)\|_{\text{Op}} \leqslant K$ for all $t \in I$, it follows by induction that

$$\|V_\nu(t,r)\|_{\text{Op}} \leqslant Me^{\beta(t-r)} \cdot \frac{(MK(t-r))^\nu}{\nu!} \tag{8.1.4}$$

for all $(t,r) \in \Delta(I)$. For $\nu = 0$ inequality (8.1.4) coincides with (8.1.3). If (8.1.4) is true for some $\nu \in \mathbb{N}$, then it follows by Theorem 3.2.5 for $(t,r) \in \Delta(I)$ and $\xi \in X$:

$$\begin{aligned}
\|V_{\nu+1}(t,r)\xi\| &\leqslant \int_r^t \|U(t,s)B(s)V_\nu(s,r)\xi\|\,ds \\
&\leqslant Me^{\beta(t-r)} \cdot \frac{(MK)^{\nu+1}}{\nu!} \cdot \|\xi\| \int_r^t (t-s)^\nu\,ds \\
&= Me^{\beta(t-r)} \cdot \frac{(MK(t-r))^{\nu+1}}{(\nu+1)!} \cdot \|\xi\|
\end{aligned}$$

[3] See, e.g, Theorem III.9 in the first volume of [179].

and hence also (8.1.4) where ν is replaced by $\nu+1$. Hence it follows for every $\mu \in \mathbb{N}$:

$$\sum_{\nu=0}^{\mu} \|V_\nu\|_\infty \leqslant Me^{\beta(b-a)} \cdot \sum_{\nu=0}^{\mu} \frac{(MK(b-a))^\nu}{\nu!} \leqslant Me^{(\beta+MK)(b-a)} . \tag{8.1.5}$$

Further, it follows by (iii)a) that $[UB\,\cdot]$ and $[\cdot\, BU]$ define bounded linear operators on $(C_*(\triangle(I), L(X,X)), \|\,\|_\infty)$. Hence it follows from

$$\left[UB\sum_{\nu=0}^{\mu} V_\nu\right] = \sum_{\nu=0}^{\mu}[UBV_\nu] = \sum_{\nu=0}^{\mu} V_{\nu+1} = \sum_{\nu=0}^{\mu+1} V_\nu - U$$

$$\left[\left(\sum_{\nu=0}^{\mu} V_\nu\right)BU\right] = \sum_{\nu=0}^{\mu}[V_\nu BU] = \sum_{\nu=0}^{\mu} V_{\nu+1} = \sum_{\nu=0}^{\mu+1} V_\nu - U$$

for every $\mu \in \mathbb{N}$ that

$$U + [UB\,\mathrm{volt}(U,B)] = \mathrm{volt}(U,B) \;\text{ and }\; U + [\mathrm{volt}(U,B)BU] = \mathrm{volt}(U,B) .$$

Finally, let $W \in C_*(\triangle(I), L(X,X))$ be a fixed point of $[UB\cdot]$ ($[\cdot BU]$). Then we define the sequence $(W_\nu)_{\nu\in\mathbb{N}} \in (C_*(\triangle, L(X,X)))^{\mathbb{N}}$ recursively by $W_0 := W$ and $W_{\nu+1} := [UBW_\nu]$ ($[W_\nu BU]$) for every $\nu \in \mathbb{N}$. Then it follows inductively that $W_\nu = W$ and

$$\|W_\nu(t,s)\|_{\mathrm{Op}} \leqslant M'e^{\beta(t-s)} \cdot \frac{(MK(t-s))^\nu}{\nu!}$$

for all $(t,s) \in \triangle(I)$, $\nu \in \mathbb{N}$ where $M' := \|e^{-\beta(\mathrm{pr}_1-\mathrm{pr}_2)}.W\|_\infty$. This implies

$$\begin{aligned}(\mu+1)\cdot\|W_\nu\|_\infty &\leqslant M'e^{\beta(b-a)} \cdot \sum_{\nu=0}^{\mu} \frac{(MK(b-a))^\nu}{\nu!}\\ &\leqslant M'e^{(\beta+MK)(b-a)}\end{aligned}$$

for every $\mu \in \mathbb{N}$ and hence $W(t,r) = 0_{L(X,X)}$ for all $(t,r) \in \triangle(I)$.
'(iii)d)': Let $V := \mathrm{volt}(U,B)$ and $W := \mathrm{volt}(V,C)$. Hence according to (iii)c), the following equations hold

$$U + [UBV] = V \,,\; V + [VCW] = W .$$

From this follows by (iii)b)

$$\begin{aligned}[U(B+C)W] &= [UBW] + [UCW]\\ &= [UB(V + [VCW])] + [(V - [UBV])CW]\\ &= [UBV] + [UB[VCW]] + [VCW] - [[UBV]CW]\\ &= V - U + W - V = W - U\end{aligned}$$

and hence it follows by (iii)c) that $W = \mathrm{volt}(U, B+C)$. □

Lemma 8.1.2. (**Distance of Volterra kernels**) Let $\mathbb{K} \in \{\mathbb{R}, \mathbb{C}\}$, $(X, \| \ \|)$ a $\mathbb{K}$-Banach space, $a, b \in \mathbb{R}$ such that $a \leqslant b$, $I := [a, b]$, $U, \bar{U} \in C_*(\triangle(I), L(X, X))$ and $B, \bar{B} \in PC_*(I, L(X, X))$. Then

(i) $\text{id} - [UB\cdot]$, where 'id' denotes the identity operator on $C_*(\triangle(I), L(X, X))$, defines a bijective bounded linear operator on $C_*(\triangle(I), L(X, X))$. In particular,

$$\| (\text{id} - [UB\cdot])^{-1} \|_{\text{Op}} \leqslant e^{\|U\|_\infty \|B\|_\infty (b-a)} \tag{8.1.6}$$

and

$$\left\| \left((\text{id} - [UB\cdot])^{-1} V \right) (t, r)\, \xi \right\| \leqslant M_\mu \, \|V\xi\|_\infty \, e^{(\mu + c\|B\|_\infty)(t-r)} \tag{8.1.7}$$

for every $V \in C_*(\triangle(I), L(X, X))$, $(t, r) \in \triangle(I)$ and $\xi \in X$ where $\mu \in \mathbb{R}, c \geqslant 0$ are such that $\|U(t, r)\|_{\text{Op}} \leqslant c\, e^{\mu(t-r)}$ for all $(t, r) \in \triangle(I)$, $M_\mu := 1, e^{|\mu|(b-a)}$ for $\mu \geqslant 0$ and $\mu < 0$, respectively, and $V\xi := (\triangle(I) \to X, (t, r) \mapsto V(t, r)\xi)$.

(ii) For every $V \in C_*(\triangle(I), L(X, X))$

$$(\text{id} - [UB\cdot])^{-1} V = V + [\text{volt}(U, B)BV] \,. \tag{8.1.8}$$

(iii) The following identity holds

$$\begin{aligned} \text{volt}(\bar{U}, \bar{B}) - \text{volt}(U, B) &= (\text{id} - [\bar{U}\bar{B}\cdot])^{-1} \{\bar{U} - U \\ &+ [(\bar{U} - U)B\, \text{volt}(U, B)] + [\bar{U}(\bar{B} - B)\, \text{volt}(U, B)]\} \,. \end{aligned} \tag{8.1.9}$$

Proof. '(i)': By Lemma 8.1.1 (iii)a) and c) (compare also the proof of (iii)c)), it follows that $\text{id} - [UB\cdot]$ defines an injective bounded linear operator on $C_*(\triangle(I), L(X, X))$. Let $V \in C_*(\triangle(I), L(X, X))$. Then we define the sequence $(W_\nu)_{\nu \in \mathbb{N}} \in (C_*(\triangle(I), L(X, X)))^{\mathbb{N}}$ recursively by $W_0 := V$ and $W_{\nu+1} := [UBW_\nu]$ for every $\nu \in \mathbb{N}$. Then it follows by induction that

$$\|W_\nu(t, r)\|_{\text{Op}} \leqslant M' e^{\beta(t-r)} \cdot \frac{(MK(t-r))^\nu}{\nu!} \tag{8.1.10}$$

for all $(t, r) \in \triangle(I)$ where $\beta \geqslant 0$, $M := \|e^{-\beta(\text{pr}_1 - \text{pr}_2)}.U\|_\infty$ which implies that

$$\|U(t, r)\|_{\text{Op}} \leqslant M e^{\beta(t-r)} \tag{8.1.11}$$

for all $(t, r) \in \triangle(I)$, $K > 0$ is such that $\|B(t)\|_{\text{Op}} \leqslant K$ for all $t \in I$ and $M' := \|e^{-\beta(\text{pr}_1 - \text{pr}_2)}.V\|_\infty$. For $\nu = 0$, inequality (8.1.10) is trivially satisfied. If (8.1.10) is true for some $\nu \in \mathbb{N}$, it follows by Theorem 3.2.5 for $(t, r) \in \triangle(I)$ and $\xi \in X$:

$$\begin{aligned} \|W_{\nu+1}(t, r)\xi\| &\leqslant \int_r^t \|U(t, s)B(s)W_\nu(s, r)\xi\|\, ds \\ &\leqslant M' e^{\beta(t-r)} \cdot \frac{(MK)^{\nu+1}}{\nu!} \cdot \|\xi\| \int_r^t (t-s)^\nu\, ds \\ &= M' e^{\beta(t-r)} \cdot \frac{(MK(t-r))^{\nu+1}}{(\nu+1)!} \cdot \|\xi\| \end{aligned}$$

and hence also (8.1.10) where ν is replaced by $\nu+1$. Hence it follows for every $\mu \in \mathbb{N}$:

$$\sum_{\nu=0}^{\mu} \|W_\nu\|_\infty \leqslant M' e^{\beta(b-a)} \cdot \sum_{\nu=0}^{\mu} \frac{(MK(b-a))^\nu}{\nu!}$$
$$\leqslant M' e^{(\beta+MK)(b-a)} \leqslant e^{(2\beta+MK)(b-a)} \cdot \|V\|_\infty$$

and therefore the absolutely summability of the sequence $(W_\nu)_{\nu\in\mathbb{N}}$ in $(C_*(\triangle(I), L(X,X)), \|\|_\infty)$ and

$$\left\| \sum_{\nu=0}^{\infty} W_\nu \right\|_\infty \leqslant e^{(2\beta+MK)(b-a)} \cdot \|V\|_\infty \, .$$

In addition, it follows

$$\left[UB\left(\sum_{\nu=0}^{\mu} W_\nu \right) \right] = \sum_{\nu=0}^{\mu} [UBW_\nu] = \sum_{\nu=0}^{\mu} W_{\nu+1} = \sum_{\nu=0}^{\mu+1} W_\nu - V$$

for every $\mu \in \mathbb{N}$ and hence

$$(\mathrm{id} - [UB\,\cdot]) \sum_{\nu=0}^{\infty} W_\nu = V$$

and the bijectivity of $\mathrm{id} - [UB\,\cdot]$ and by choosing $\beta = 0$ the validness of (8.1.6). Analogously to (8.1.10), it follows for every $\xi \in X$ that

$$\|W_\nu(t,r)\xi\| \leqslant M'_\xi e^{\mu(t-r)} \cdot \frac{(c\,K(t-r))^\nu}{\nu!} \tag{8.1.12}$$

for all $(t,r) \in \triangle(I)$ where $M'_\xi := \|(\triangle(I) \to X, (t,r) \mapsto e^{-\mu(t-r)}.V(t,r)\xi)\|_\infty$ and $\mu \in \mathbb{R}, c \geqslant 0$ are such that $\|U(t,r)\|_{\mathrm{Op}} \leqslant c\,e^{\mu(t-r)}$ for all $(t,r) \in \triangle(I)$. Hence it follows for every $(t,r) \in \triangle(I)$ that

$$\left\| \left((\mathrm{id} - [UB\,\cdot])^{-1}\, V \right)(t,r)\,\xi \right\| = \left\| \left(\sum_{\nu=0}^{\infty} W_\nu(t,r) \right) \xi \right\| = \left\| \sum_{\nu=0}^{\infty} W_\nu(t,r)\xi \right\|$$
$$\leqslant \sum_{\nu=0}^{\infty} \|W_\nu(t,r)\xi\| \leqslant M'_\xi e^{(\mu+c\,K)(t-r)}$$

and hence, finally, (8.1.7).
'(ii)': By Lemma 8.1.1 (iii) b), c), it follows for every $V \in C_*(\triangle(I), L(X,X))$ that

$$\begin{aligned}
&(\mathrm{id} - [UB\,\cdot])\,(V + [\mathrm{volt}(U,B)BV]) \\
&= V - [UBV] + [\mathrm{volt}(U,B)BV] - [UB\,[\mathrm{volt}(U,B)BV]] \\
&= V - [UBV] + [\mathrm{volt}(U,B)BV] - [\,[UB\,\mathrm{volt}(U,B)]BV] \\
&= V - [UBV] + [\mathrm{volt}(U,B)BV] - [(\mathrm{volt}(U,B) - U)BV] = V
\end{aligned}$$

and hence (8.1.8).

'(iii)': By Lemma 8.1.1 (iii)c), it follows

$$
\begin{aligned}
&[(\bar{U}-U)B\,\mathrm{volt}(U,B)] + [\bar{U}(\bar{B}-B)\,\mathrm{volt}(U,B)] \\
&= [\bar{U}B\,\mathrm{volt}(U,B)] - [UB\,\mathrm{volt}(U,B)] + [\bar{U}\bar{B}\,\mathrm{volt}(U,B)] - [\bar{U}B\,\mathrm{volt}(U,B)] \\
&= [\bar{U}\bar{B}\,\mathrm{volt}(U,B)] - [UB\,\mathrm{volt}(U,B)] = [\bar{U}\bar{B}\,\mathrm{volt}(U,B)] + U - \mathrm{volt}(U,B) \\
&= [\bar{U}\bar{B}\,(\mathrm{volt}(U,B) - \mathrm{volt}(\bar{U},\bar{B}))] + [\bar{U}\bar{B}\,\mathrm{volt}(\bar{U},\bar{B})] + U - \mathrm{volt}(U,B) \\
&= [\bar{U}\bar{B}\,(\mathrm{volt}(U,B) - \mathrm{volt}(\bar{U},\bar{B}))] + \mathrm{volt}(\bar{U},\bar{B}) - \bar{U} + U - \mathrm{volt}(U,B)\ .
\end{aligned}
$$

Hence it follows

$$
\begin{aligned}
&(\mathrm{id} - [\bar{U}\bar{B}\cdot])\,(\mathrm{volt}(\bar{U},\bar{B}) - \mathrm{volt}(U,B)) \\
&\quad = \bar{U} - U + [(\bar{U}-U)B\,\mathrm{volt}(U,B)] + [\bar{U}(\bar{B}-B)\,\mathrm{volt}(U,B)]
\end{aligned}
$$

and, finally, by (i) (8.1.9). □

Lemma 8.1.3. **(Approximation of Volterra kernels)** Let $\mathbb{K} \in \{\mathbb{R}, \mathbb{C}\}$, $(X, \|\ \|)$ a $\mathbb{K}$-Banach space, $a, b \in \mathbb{R}$ such that $a \leqslant b$, $I := [a, b]$, $U \in C_*(\triangle(I), L(X, X))$, $B \in PC_*(I, L(X, X))$, $(U_\nu)_{\nu\in\mathbb{N}} \in (C_*(\triangle(I), L(X, X)))^{\mathbb{N}}$, $(B_\nu)_{\nu\in\mathbb{N}} \in (PC_*(I, L(X, X)))^{\mathbb{N}}$ be bounded sequences and such that

$$\mathrm{s-}\lim_{\nu\to\infty} U_\nu(t,r) = U(t,r)$$

for all $(t,r) \in \triangle(I)$ and

$$\mathrm{s-}\lim_{\nu\to\infty} B_\nu(s) = B(s)$$

for almost all $s \in I$. Then

$$\mathrm{s-}\lim_{\nu\to\infty} \mathrm{volt}(U_\nu, B_\nu)(t,r) = \mathrm{volt}(U,B)(t,r) \tag{8.1.13}$$

for all $(t,r) \in \triangle(I)$.

Proof. For this, let $c, K \geqslant 0$ be such that $\|U\|_\infty \leqslant c$, $\|B\|_\infty \leqslant K$ and $\|U_\nu\|_\infty \leqslant c$, $\|B_\nu\|_\infty \leqslant K$ for all $\nu \in \mathbb{N}$. Then it follows by Lemma 8.1.1 (8.1.5) where $\beta = 0$ that

$$\|\mathrm{volt}(U_\nu, B_\nu)\|_\infty \leqslant c\, e^{cK(b-a)}\ ,\ \|\mathrm{volt}(U,B)\|_\infty \leqslant c\, e^{cK(b-a)}\ .$$

By Lemma 8.1.2 (ii), (iii), it follows for every $\nu \in \mathbb{N}$

$$
\begin{aligned}
&\mathrm{volt}(U_\nu, B_\nu) - \mathrm{volt}(U,B) \\
&= (\mathrm{id} - [U_\nu B_\nu\cdot])^{-1}\,\{U_\nu - U + [(U_\nu - U)B\,\mathrm{volt}(U,B)] \\
&\quad + [U_\nu(B_\nu - B)\,\mathrm{volt}(U,B)]\} \\
&= U_\nu - U + [(U_\nu - U)B\,\mathrm{volt}(U,B)] + [U_\nu(B_\nu - B)\,\mathrm{volt}(U,B)] \\
&\quad + [\mathrm{volt}(U_\nu, B_\nu)B_\nu\,\{U_\nu - U + [(U_\nu - U)B\,\mathrm{volt}(U,B)] \\
&\quad + [U_\nu(B_\nu - B)\,\mathrm{volt}(U,B)]\}]
\end{aligned}
$$

$$= U_\nu - U + [(U_\nu - U)B \operatorname{volt}(U,B)] + [U_\nu(B_\nu - B) \operatorname{volt}(U,B)]$$
$$+ [\operatorname{volt}(U_\nu, B_\nu)B_\nu(U_\nu - U)] + [\operatorname{volt}(U_\nu, B_\nu)B_\nu[(U_\nu - U)B \operatorname{volt}(U,B)]]$$
$$+ [\operatorname{volt}(U_\nu, B_\nu)B_\nu[U_\nu(B_\nu - B) \operatorname{volt}(U,B)]] .$$

Hence, obviously, it follows by the proof of Lemma 8.1.2 (iii)b), Theorem 3.2.5 and Lebesgue's dominated convergence Theorem the validity of (8.1.13) for every $(t,r) \in \triangle(I)$. □

8.2 Chains

Lemma 8.2.1. Let $\mathbb{K} \in \{\mathbb{R}, \mathbb{C}\}$, $(X, \| \|)$ a $\mathbb{K}$-Banach space. In addition, let $a, b \in \mathbb{R}$ be such that $a \leqslant b$ and $I := [a,b]$. Finally, let $U \in C_*(\triangle(I), L(X,X))$ be a chain, i.e., such that $U(s,s) = \mathrm{id}_X$ for all $s \in I$ and

$$U(t,r) = U(t,s)\, U(s,r)$$

for all $(t,r) \in \triangle(I)$ and $s \in [r,t]$, and $B \in PC_*(I, L(X,X))$. Then $\operatorname{volt}(U,B)$ is a chain, too.

Proof. Let $V := \operatorname{volt}(U,B)$. By Lemma 8.1.1, V satisfies $U + [UBV] = V$. Hence for all $(t,r) \in \triangle(I)$, $s \in [r,t]$ and $\xi \in X$:

$$\begin{aligned}
V(t,s)V(s,r)\xi &= (U + [UBV])(t,s)V(s,r)\xi \\
&= U(t,s)V(s,r)\xi + [UBV](t,s)V(s,r)\xi \\
&= U(t,s)(U + [UBV])(s,r)\xi + [UBV](t,s)V(s,r)\xi \\
&= U(t,s)U(s,r)\xi + U(t,s)[UBV](s,r)\xi + [UBV](t,s)V(s,r)\xi \\
&= U(t,r)\xi + U(t,s)[UBV](s,r)\xi + [UBV](t,s)V(s,r)\xi .
\end{aligned}$$

By Theorem 3.2.4 (ii)

$$\begin{aligned}
U(t,s)[UBV](s,r)\xi &= U(t,s)\int_r^s U(s,s')B(s')V(s',r)\xi\, ds' \\
&= \int_r^s U(t,s)U(s,s')B(s')V(s',r)\xi\, ds' \\
&= \int_r^s U(t,s')B(s')V(s',r)\xi\, ds' \\
&= [UBV](t,r)\xi - \int_s^t U(t,s')B(s')V(s',r)\xi\, ds' ,
\end{aligned}$$

$$[UBV](t,s)V(s,r)\xi = \int_s^t U(t,s')B(s')V(s',s)V(s,r)\xi\, ds'$$

and hence

$$V(t,s)V(s,r)\xi = V(t,r)\xi + \int_s^t U(t,s')B(s')\,(V(s',s)V(s,r) - V(s',r))\,\xi ds'$$

and

$$W_r(t,s)\xi = \int_s^t U(t,s')B(s')W_r(s',s)\xi\, ds' \tag{8.2.1}$$

where

$$W_r := V(V(\cdot,r)\circ \mathrm{pr}_2) - V(\cdot,r)\circ \mathrm{pr}_1$$

and $\mathrm{pr}_1 := (\mathbb{R}^2 \to \mathbb{R}, (t,s) \mapsto t), \mathrm{pr}_2 := (\mathbb{R}^2 \to \mathbb{R}, (t,s) \mapsto s)$. Note that it follows by Lemma 8.1.1 (ii) that $W_r \in C_*(\Delta([r,b]), L(X,X))$. Hence it follows from (8.2.1) by the proof of Lemma 8.1.1 (iii)c) that $W_r = 0_{\Delta([r,b])\to L(X,X)}$. Finally, for every $s \in I$,

$$V(s,s) = U(s,s) + \int_s^s U(s,s')B(s')V(s',s)\, ds' = \mathrm{id}_X\,.$$

□

8.3 Juxtaposition of Chains

Lemma 8.3.1. Let $\mathbb{K} \in \{\mathbb{R},\mathbb{C}\}$ and $(X, \|\ \|)$ a $\mathbb{K}$-Banach space. In addition, let $a', b', b'' \in \mathbb{R}$ be such that $a' \leqslant b' \leqslant b''$, $I' := [a',b']$ and $I'' := [b',b'']$ be the two corresponding adjoining intervals and $I := I' \cup I''$. Finally, let $U' \in C_*(\Delta(I'), L(X,X))$ and $U'' \in C_*(\Delta(I''), L(X,X))$ be chains.

(i) There is a uniquely determined chain $U' \cup U'' \in C_*(\Delta(I), L(X,X))$ such that $U' \cup U''|_{\Delta(I')} = U'$ and $U' \cup U''|_{\Delta(I'')} = U''$.
(ii) In addition, let $B \in PC_*(I, L(X,X))$. Then

$$\mathrm{volt}(U' \cup U'', B) = \mathrm{volt}(U', B|_{I'}) \cup \mathrm{volt}(U'', B|_{I''})\,.$$

Proof. '(i)': First, it follows that $\Delta(I) \setminus (\Delta(I') \cup \Delta(I'')) = I'' \times I'$. We define $U : \Delta(I) \to L(X,X)$ by $U(t,r) := U'(t,r)$ for $(t,r) \in \Delta(I')$, $U(t,r) := U''(t,r)$ for $(t,r) \in \Delta(I'')$, note that $U'(b',b') = \mathrm{id}_X = U''(b',b')$, and

$$U(t,r) := U''(t,b')\,U'(b',r)$$

for $(t,r) \in I'' \times I'$. Note that $U''(b',b')\,U'(b',r) = U'(b',r)$ for all $r \in I'$ and $U''(t,b')\,U'(b',b') = U''(t,b')$ for all $t \in I''$. Then $U \in C_*(\Delta(I), L(X,X))$ because of the strong continuity of U', U'' and since $U|_{I''\times I'}$ is equal to $[U''(\cdot,b')\circ \mathrm{pr}_1][U'(b',\cdot)\circ \mathrm{pr}_2]$ which is strongly continuous by Lemma 8.1.1 (ii). Here $\mathrm{pr}_1 := (\mathbb{R}^2 \to \mathbb{R}, (t,r) \mapsto t)$ and $\mathrm{pr}_2 := (\mathbb{R}^2 \to \mathbb{R}, (t,r) \mapsto r)$. If $(t,r) \in \Delta(I'), \Delta(I'')$, then $[r,t] \subset I', I''$ and hence for every $s \in [r,t]$

$$U(t,r) = U'(t,r) = U'(t,s)\,U'(s,r) = U(t,s)\,U(s,r)$$

and

$$U(t,r) = U''(t,r) = U''(t,s)\,U''(s,r) = U(t,s)\,U(s,r)$$

respectively. If $(t,r) \in I'' \times I'$ and $s \in [r,t]$, we consider the cases $s > b'$ and $s \leqslant b'$. In the first case,

$$U(t,s)\,U(s,r) = U''(t,s)\,U''(s,b')\,U'(b',r) = U''(t,b')\,U'(b',r) = U(t,r)\,,$$

and in the second case,

$$U(t,s)\,U(s,r) = U''(t,b')\,U'(b',s)\,U'(s,r) = U''(t,b')\,U'(b',r) = U(t,r)\,.$$

Hence U is a chain. Finally, let $\hat{U} \in C_*(\triangle(I), L(X,X))$ be a chain and such that $\hat{U}|_{\triangle(I')} = U'$ and $\hat{U}|_{\triangle(I'')} = U''$. Then it follows for every $(t,r) \in I'' \times I'$

$$\hat{U}(t,r) = \hat{U}(t,b')\,\hat{U}(b',r) = U''(t,b')\,U'(b',r)$$

and hence $\hat{U} = U$.
'(ii)': Let $V := \text{volt}(U,B)$. Then $U + [UBV] = V$ and hence for every $(t,r) \in \triangle(I')$

$$\begin{aligned} V|_{\triangle(I')}(t,r) &= U(t,r) + [UBV](t,r) \\ &= U'(t,r) + \int_r^t U(t,s)B(s)V(s,r)\,ds \\ &= U'(t,r) + \int_r^t U'(t,s)B|_{I'}(s)V|_{\triangle(I')}(s,r)\,ds\,. \end{aligned}$$

Hence it follows by Lemma 8.1.1 (iii) that $V|_{\triangle(I')} = \text{volt}(U', B|_{I'})$. Analogously, it follows for every $(t,r) \in \triangle(I'')$ that

$$\begin{aligned} V|_{\triangle(I'')}(t,r) &= U(t,r) + [UBV](t,r) = U''(t,r) + \int_r^t U(t,s)B(s)V(s,r)\,ds \\ &= U''(t,r) + \int_r^t U''(t,s)B|_{I''}(s)V|_{\triangle(I'')}(s,r)\,ds \end{aligned}$$

and hence by Lemma 8.1.1 (iii) that $V|_{\triangle(I'')} = \text{volt}(U'', B|_{I''})$. Finally, by (i) it follows that $V = \text{volt}(U', B|_{I'}) \cup \text{volt}(U'', B|_{I''})$. □

8.4 Finitely Generated Chains

Lemma 8.4.1. Let $\mathbb{K} \in \{\mathbb{R}, \mathbb{C}\}$, $(X, \|\ \|)$ a $\mathbb{K}$-Banach space, $a, b \in \mathbb{R}$ be such that $a \leqslant b$ and $I := [a,b]$.

(i) In addition, let $A : D(A) \to X$ be the infinitesimal generator of a strongly continuous semigroup $T : [0,\infty) \to L(X,X)$. Then $U : \triangle(I) \to L(X,X)$ defined by $U(t,r) := T(t-r)$ for all $(t,r) \in \triangle(I)$ is a strongly continuous chain.
(ii) In addition, let $n \in \mathbb{N}^*$, $(a_0,\dots,a_n)$ be a partition of I, $I_j := [a_j, a_{j+1}]$ for $j = 0,\dots,n-1$ and $A_j : D(A_j) \to X$ the infinitesimal generator of a strongly continuous semigroup $T_j : [0,\infty) \to L(X,X)$ for every $j \in \{0,\dots,n-1\}$.
  a) Then there is uniquely determined chain $U \in C_*(\triangle(I), L(X,X))$ such that $U(t,r) = T_j(t-r)$ for all $(t,r) \in \triangle(I_j)$ and $j \in \{0,\dots,n-1\}$. We say that U is generated by the family $(I_0,A_0),\dots,(I_{n-1},A_{n-1})$.

b) In addition, let $U \in C_*(\Delta(I), L(X,X))$ as in a) and $B_0, \ldots, B_{n-1} \in L(X,X)$. Then the family $(I_0, A_0 + B_0), \ldots, (I_{n-1}, A_{n-1} + B_{n-1})$ generates $\mathrm{volt}(U, -B)$ where $B \in PC_*(I, L(X,X))$ is defined by $B(t) := B_j$ for all $t \in [a_j, a_{j+1})$ and $j \in \{0, \ldots, n-1\}$ and $B(b) := B_{n-1}$.

Proof. '(i)': U is strongly continuous as composition of the strongly continuous map T and the continuous map $(\mathrm{pr}_1 - \mathrm{pr}_2)|_{\Delta(I) \to \mathbb{R}}$ where $\mathrm{pr}_1 := (\mathbb{R}^2 \to \mathbb{R}, (t,r) \mapsto t)$ and $\mathrm{pr}_2 := (\mathbb{R}^2 \to \mathbb{R}, (t,r) \mapsto r)$. Further,

$$U(t,r) = T(t-r) = T(t-s+s-r) = T(t-s)\,T(s-r) = U(t,s)\,U(s,r)$$

for all $(t,r) \in \Delta(I)$, $s \in [t,r]$ and $U(s,s) = T(0) = \mathrm{id}_X$ for all $s \in I$.
'(ii)a)' The statement is a simple consequence of Lemma 8.3.1 (i).
'(ii)b)': By Theorem 4.4.3, for every $j \in \{0, \ldots, n-1\}$ the corresponding $A_j + B_j$ generates a strongly continuous semigroup $\hat{T}_j : [0, \infty) \to L(X,X)$. Let $\hat{U} \in C_*(\Delta(I), L(X,X))$ be the corresponding chain such that $\hat{U}(t,r) = \hat{T}_j(t-r)$ for all $(t,r) \in \Delta(I_j)$ and $j \in \{0, \ldots, n-1\}$. In order to prove the statement, according to Lemma 8.3.1, it is sufficient to show for each $j \in \{0, \ldots, n-1\}$ that $\mathrm{volt}(U|_{\Delta(I_j)}, -B|_{I_j}) = \hat{U}|_{\Delta(I_j)}$. For this, let $(t,r) \in \Delta(I_j)$, $\xi \in D(A_j)$ and $u, f_j : [r, \infty) \to X$ be defined by $u(s) := \hat{T}_j(s-r)\xi$ and $f_j(s) := -B_j u(s)$ for all $s \geqslant r$. Then u, f_j are continuous, $\mathrm{Ran}(u) \subset D(A_j)$ and u is differentiable on (r, ∞) such that

$$u'(s) + A_j u(s) = f_j(s)$$

for all $s > r$. Hence it follows by Lemma 4.6.1

$$\begin{aligned}\hat{T}_j(t-r)\xi = u(t) &= T_j(t-r)\xi - \int_r^t T_j(t-s) B_j u(s)\, ds \\ &= T_j(t-r)\xi - \int_r^t T_j(t-s) B_j \hat{T}_j(s-r)\xi\, ds\,.\end{aligned}$$

Since $D(A)$ is dense in X, it follows

$$U|_{\Delta(I_j)} + [U|_{\Delta(I_j)}(-B|_{I_j})\hat{U}|_{\Delta(I_j)}] = \hat{U}|_{\Delta(I_j)}$$

and hence finally by Lemma 8.1.1 (iii)c) that

$$\hat{U}|_{\Delta(I_j)} = \mathrm{volt}(U|_{\Delta(I_j)}, -B|_{I_j})\,.$$

□

8.5 Evolution Operators

Definition 8.5.1. (Evolution operators) Let $\mathbb{K} \in \{\mathbb{R}, \mathbb{C}\}$, $a, b \in \mathbb{R}$ such that $a \leqslant b$ and $I := [a,b]$. Further, let $(X, \|\ \|)$ be a $\mathbb{K}$-Banach space, Y a dense subspace of X and $\|\ \|_Y$ a norm on Y such that $(Y, \|\ \|_Y)$ is a Banach space and such that the inclusion $\iota_{Y \hookrightarrow X}$ of $(Y, \|\ \|_Y)$ into X is continuous. Finally, let $A \in PC_*(I, L((Y, \|\ \|_Y), X))$. Then $U \in C_*(\Delta(I), L(X,X))$ is called a 'Y/X-evolution operator for A' if the following conditions are satisfied

(i) U is a chain.
(ii) For all $(t,r) \in \triangle(I)$ the corresponding $U(t,r)$ leaves Y invariant. Moreover, the map U_Y associating to every $(t,r) \in \triangle(I)$ the part $U_Y(t,r)$ of $U(t,r)$ in Y is an element of $C_*(\triangle(I), L(Y,Y))$.
(iii) For all $(t,r) \in \triangle(I)$ and $\xi \in Y$

$$\int_r^t A(s)U(s,r)\xi\,ds = \int_r^t U(t,s)A(s)\xi\,ds = \xi - U(t,r)\xi$$

where integration denotes weak Lebesgue integration with respect to $L(X,\mathbb{K})$.

Lemma 8.5.2. (Associated differential equations, Uniqueness) Let $\mathbb{K} \in \{\mathbb{R},\mathbb{C}\}$ and $a,b \in \mathbb{R}$ be such that $a \leqslant b$ and $I := [a,b]$. Further, let $(X, \| \ \|)$ a $\mathbb{K}$-Banach space, Y a dense subspace of X and $\| \ \|_Y$ a norm on Y such that $(Y, \| \ \|_Y)$ is a Banach space and such that the inclusion $\iota_{Y \hookrightarrow X}$ of $(Y, \| \ \|_Y)$ into X is continuous. Finally, let $A \in PC_*(I, L((Y, \| \ \|_Y), X))$.

(i) In addition, let $U \in C_*(\triangle(I), L(X,X))$ be a 'Y/X-evolution operator for A' and $r,t \in I$, $\xi \in Y$. Then $U(\cdot,r)\xi : [r,b] \to X$, $U(t,\cdot)\xi : [a,t] \to X$ are differentiable with derivatives

$$(U(\cdot,r)\xi)'(s) = -A(s)U(s,r)\xi$$

in every strong continuity point s of A in (r,b) and

$$(U(t,\cdot)\xi)'(s) = U(t,s)A(s)\xi$$

in every strong continuity point s of A in (a,t), respectively. Here $(U(\cdot,r)\xi)(s) := U(s,r)\xi$ for every $s \in [r,b]$ and $(U(t,\cdot)\xi)(s) := U(t,s)\xi$ for every $s \in [a,t]$.
(ii) In addition, let $U,V \in C_*(\triangle(I), L(X,X))$ be 'Y/X-evolution operators for A'. Then $V = U$.

Proof. '(i)': It follows for $\xi \in Y$ and every $r \in I$, $s,s' \in [r,b]$ because of

$$\begin{aligned}&\|A(s')U(s',r)\xi - A(s)U(s,r)\xi\| \\ &\quad\leqslant \|A\|_{\infty,\mathrm{Y,X}} \cdot \|U_{\mathrm{Y}}(s',r)\xi - U_{\mathrm{Y}}(s,r)\xi\|_{\mathrm{Y}} + \|A(s')U(s,r)\xi - A(s)U(s,r)\xi\|\end{aligned}$$

that $f_1 := ([r,b] \to X, s \mapsto A(s)U(s,r)\xi)$ is bounded by $\|A\|_{\infty,\mathrm{Y,X}}\, \|U_Y\|_{\infty,\mathrm{Y,Y}}\, \|\xi\|_{\mathrm{Y}}$ as well as continuous in the points of strong continuity of A in $[r,b]$ and hence also almost everywhere continuous, and for $t \in I$, $s,s' \in [a,t]$

$$\begin{aligned}&\|U(t,s')A(s')\xi - U(t,s)A(s)\xi\| \\ &\quad\leqslant \|U\|_{\infty,\mathrm{X,X}} \cdot \|A(s')\xi - A(s)\xi\| + \|U(t,s')A(s)\xi - U(t,s)A(s)\xi\|\end{aligned}$$

that $f_2 := ([a,t] \to X, s \mapsto U(t,s)A(s)\xi)$ is bounded by $\|U_Y\|_{\infty,\mathrm{X,X}}\, \|A\|_{\infty,\mathrm{Y,X}}\, \|\xi\|_{\mathrm{Y}}$, continuous in the points of strong continuity of A in $[a,t]$ and hence also almost everywhere continuous. Hence, obviously, the statement of this lemma follows by Theorem 3.2.9 from property (iii) of Definition 8.5.1.

'(ii)': For this, let $(r,t) \in \triangle(I)$ and $\xi \in Y$. Because of

$$\begin{aligned}
&\|V(t,s')U(s',r)\xi - V(t,s)U(s,r)\xi\| \\
&\leqslant \|\,(V(t,s') - V(t,s))\,(U(s',r) - U(s,r))\,\xi\| \\
&\quad + \|\,(V(t,s') - V(t,s))\,U(s,r)\xi\| + \|V(t,s)\,(U(s',r)\xi - U(s,r)\xi)\,\| \\
&\leqslant 3\|V\|_{\infty,X,X} \cdot \|U(s',r)\xi - U(s,r)\xi\| + \|V(t,s')U(s,r)\xi - V(t,s)U(s,r)\xi\|
\end{aligned}$$

for every $s, s' \in [r,t]$, it follows that

$$F := ([r,t] \to X, s \mapsto V(t,s)U(s,r)\xi)$$

is continuous. Further, for any strong continuity point s of A in (r,t) and $h \in (r-s, t-s)\backslash\{0\}$, it follows

$$\begin{aligned}
&\left\|\frac{1}{h}.\,[V(t,s+h)U(s+h,r)\xi - V(t,s)U(s,r)\xi]\right\| \\
&= \left\|\frac{1}{h}.(V(t,s+h) - V(t,s))(U(s+h,r) - U(s,r))\xi\right. \\
&\quad \left. + \frac{1}{h}.(V(t,s+h) - V(t,s))U(s,r)\xi + V(t,s)\frac{1}{h}(U(s+h,r) - U(s,r))\xi\right\| \\
&\leqslant 2\|V\|_{\infty,X,X} \cdot \left\|\left[\frac{1}{h}.(U(s+h,r) - U(s,r))\xi + A(s)U(s,r)\xi\right]\right\| \\
&\quad + \|V(t,s+h)A(s)U(s,r)\xi - V(t,s)A(s)U(s,r)\xi\| \\
&\quad + \left\|\frac{1}{h}.(V(t,s+h) - V(t,s))U(s,r)\xi + V(t,s)\frac{1}{h}(U(s+h,r) - U(s,r))\xi\right\|
\end{aligned}$$

and hence by (i) along with $V \in C_*(\triangle(I), L(X,X))$ that F is differentiable in s with derivative 0_X. Hence it follows by the fundamental theorem of calculus that F is constant and hence that

$$V(t,r)\xi = V(t,r)U(r,r)\xi = F(r)\xi = F(t)\xi = V(t,t)U(t,s)\xi = U(t,s)\xi\ .$$

□

Lemma 8.5.3. Let $\mathbb{K} \in \{\mathbb{R}, \mathbb{C}\}$, $a, b \in \mathbb{R}$ such that $a \leqslant b$ and $I := [a,b]$. In addition, let $n \in \mathbb{N}^*$, $(a_0, \ldots, a_n)$ be a partition of I, $I_j := [a_j, a_{j+1}]$ for $j = 0, \ldots, n-1$ and $A_j : D(A_j) \to X$ be the infinitesimal generator of a strongly continuous semigroup on X for every $j \in \{0, \ldots, n-1\}$ and $U \in C_*(\triangle(I), L(X,X))$ be the chain generated by the family $(I_0, A_0), \ldots, (I_{n-1}, A_{n-1})$. Further, let Y be a dense subspace of X and $\|\ \|_Y$ be a norm on Y such that $(Y, \|\ \|_Y)$ is a Banach space and such that the inclusion $\iota_{Y\hookrightarrow X}$ of $(Y, \|\ \|_Y)$ into X is continuous. Finally, let Y be contained in $D(A_j)$ as well as A_j-admissible for every $j \in \{0, \ldots, n-1\}$, $A_j|_Y \in L(Y,X)$ for every $j \in \{0, \ldots, n-1\}$ and $A \in PC_*(I, L((Y, \|\ \|_Y), X))$ be defined by $A(t) := A_j|_Y$ for all $t \in [a_j, a_{j+1})$ and $j \in \{0, \ldots, n-1\}$ and $A(b) := A_{n-1}|_Y$. Then U is a Y/X-evolution operator for A.

Proof. For every $j \in \{0, \ldots, n-1\}$, let $T_j : [0, \infty) \to L(X, X)$ be the strongly continuous semigroup generated by $A_j : D(A_j) \to X$. Then $U(t, r) = T_j(t - r)$ for all $(t, r) \in \triangle(I_j)$ and $j \in \{0, \ldots, n-1\}$. Further, it follows for every $j \in \{0, \ldots, n-1\}$ since Y is A_j-admissible that Y is left invariant by $T_j(t)$ for all $t \in [0, \infty)$ and that the family $T_{jY}(t)$ of restrictions of $T_j(t)$ to Y in domain and in image, $t \in [0, \infty)$, defines a strongly continuous semigroup $T_{jY} : [0, \infty) \to L(Y, Y)$ on $(Y, \| \ \|_Y)$ with infinitesimal generator given by the part A_{jY} of A_j in Y. Therefore, every $U(t, r)$ with (t, r) contained in $\triangle(I_0) \cup \ldots \triangle(I_{n-1})$ leaves Y invariant. Hence it follows by the proof of Lemma 8.3.1 (i) and Lemma 8.4.1 (i) that $U(t, r)$ leaves Y invariant for all $(t, r) \in \triangle(I)$. Let in addition $U_Y \in C_*(\triangle(I), L(Y, Y))$ be the chain generated by the family $(I_0, A_{0Y}), \ldots, (I_{n-1}, A_{jY})$. Then $U_Y(t, r)$ is identical to the the restriction of $U(t, r)$ to Y in domain and in range for every $(t, r) \in \triangle(I_0) \cup \ldots \triangle(I_{n-1})$. Hence it follows by the proof of Lemma 8.3.1 (i) and Lemma 8.4.1 (i) that $U_Y(t, r)$ is identical to the restriction of $U(t, r)$, to Y in domain and in range for every $(t, r) \in \triangle(I)$. Further, let $\xi \in Y$, $j \in \{0, \ldots, n-1\}$, $r \in I_j$, $F_r := ([r, a_{j+1}] \to X, s \mapsto U(s, r)\xi = T_{jY}(s - r)\xi)$ and $f_r := ([r, a_{j+1}] \to X, s \mapsto -A_j U(s, r)\xi = -U(s, r)A_j\xi)$. Then f_r and F_r are continuous. Further, $F_r|_{(r, a_{j+1})}$ is differentiable with derivative $f_r|_{(r, a_{j+1})}$. Hence it follows the weak Lebesgue integrability of f_r with respect to $L(X, \mathbb{K})$ and by the fundamental theorem of calculus

$$
\begin{aligned}
U(t, r)\xi - \xi &= T_{jY}(t - r)\xi - T_{jY}(0)\xi = F_r(t) - F_r(r) = \int_r^t f_r(s)\, ds \\
&= -\int_r^t A(s)U(s, r)\xi\, ds
\end{aligned}
$$

for every $t \in [r, a_{j+1}]$. If $t > a_{j+1}$ and $j_t \in \{0, \ldots, n-1\}$ are such that $t \in [a_{j_t}, a_{j_t+1}]$, then it follows

$$
\begin{aligned}
&- \int_r^t A(s)U(s, r)\xi\, ds = -\int_r^{a_{j+1}} A(s)U(s, r)\xi\, ds - \sum_{k=j+1}^{j_t-1} \int_{a_k}^{a_{k+1}} A(s)U(s, r)\xi\, ds \\
&- \int_{a_{j_t}}^t A(s)U(s, r)\xi\, ds = -\int_r^{a_{j+1}} A(s)U(s, r)\xi\, ds \\
&- \sum_{k=j+1}^{j_t-1} \int_{a_k}^{a_{k+1}} A(s)U(s, a_k)U(a_k, r)\xi\, ds - \int_{a_{j_t}}^t A(s)U(s, a_{j_t})U(a_{j_t}, r)\xi\, ds \\
&= U(a_{j+1}, r)\xi - \xi + \sum_{k=j+1}^{j_t-1} [U(a_{k+1}, a_k)U(a_k, r)\xi - U(a_k, r)\xi] \\
&\quad + U(t, a_{j_t})U(a_{j_t}, r)\xi - U(a_{j_t}, r)\xi \\
&= U(a_{j+1}, r)\xi - \xi + \sum_{k=j+1}^{j_t-1} [U(a_{k+1}, r)\xi - U(a_k, r)\xi] + U(t, r)\xi - U(a_{j_t}, r)\xi \\
&= U(t, r)\xi - \xi\,.
\end{aligned}
$$

Finally, let $\xi \in Y$, $j \in \{0, \ldots, n-1\}$, $t \in I_j$,

$$F_t := \left([a_j, t] \to X, s \mapsto U(t,s)\xi = T_{jY}(t-s)\xi \right)$$

and

$$f_t := \left([a_j, t] \to X, s \mapsto A_j U(t,s)\xi = U(t,s)A_j\xi \right) .$$

Then f_t and F_t are both continuous. Further, $F_t|_{(a_j,t)}$ is differentiable with derivative $f_t|_{(a_j,t)}$. Hence it follows the weak Lebesgue integrability of f_t with respect to $L(X, \mathbb{K})$ and by the fundamental theorem of calculus that

$$\begin{aligned}
\xi - U(t,r)\xi = T_{jY}(0)\xi - T_{jY}(t-r)\xi = F_t(t) - F_t(r) = \int_r^t f_t(s)\,ds \\
= \int_r^t U(t,s)A(s)\xi\,ds
\end{aligned}$$

for every $r \in [a_j, t]$. If $r < a_j$ and $j_r \in \{1, \ldots, n\}$ are such that $r \in [a_{j_r-1}, a_{j_r}]$, then it follows

$$\begin{aligned}
&\int_r^t U(t,s)A(s)\xi\,ds = \int_r^{a_{j_r}} U(t,s)A(s)\xi\,ds + \sum_{k=j_r}^{j-1} \int_{a_k}^{a_{k+1}} U(t,s)A(s)\xi\,ds \\
&+ \int_{a_j}^t U(t,s)A(s)\xi\,ds = \int_r^{a_{j_r}} U(t,a_{j_r})U(a_{j_r},s)A(s)\xi\,ds \\
&+ \sum_{k=j_r}^{j-1} \int_{a_k}^{a_{k+1}} U(t,a_{k+1})U(a_{k+1},s)A(s)\xi\,ds + \int_{a_j}^t U(t,s)A(s)\xi\,ds \\
&= U(t,a_{j_r}) \int_r^{a_{j_r}} U(a_{j_r},s)A(s)\xi\,ds \\
&+ \sum_{k=j_r}^{j-1} U(t,a_{k+1}) \int_{a_k}^{a_{k+1}} U(a_{k+1},s)A(s)\xi\,ds + \int_{a_j}^t U(t,s)A(s)\xi\,ds \\
&= U(t,a_{j_r})[\xi - U(a_{j_r},r)\xi] + \sum_{k=j_r}^{j-1} U(t,a_{k+1})[\xi - U(a_{k+1},a_k)\xi] + \xi - U(t,a_j)\xi \\
&= U(t,a_{j_r})\xi - U(t,r)\xi + \sum_{k=j_r}^{j-1} [U(t,a_{k+1})\xi - U(t,a_k)\xi] + \xi - U(t,a_j)\xi \\
&= \xi - U(t,r)\xi .
\end{aligned}$$

□

A reinspection of the proof of Lemma 8.5.3 suggests the following

Lemma 8.5.4. Let $\mathbb{K} \in \{\mathbb{R}, \mathbb{C}\}$, $(X, \| \, \|)$ a $\mathbb{K}$-Banach space, $a, b \in \mathbb{R}$ such that $a \leqslant b$ and $I := [a, b]$. In addition, let $n \in \mathbb{N}^*$, $(a_0, \ldots, a_n)$ be a partition of I, $I_j := [a_j, a_{j+1}]$ for $j = 0, \ldots, n-1$ and $A_j : D(A_j) \to X$ the infinitesimal generator of a strongly continuous semigroup on X for every $j \in \{0,\ldots,n-1\}$ and $U \in C_*(\triangle(I), L(X,X))$ be the chain generated by the family $(I_0, A_0), \ldots, (I_{n-1}, A_{n-1})$. Further, let Y be subspace of X contained in $D(A_j)$ for every $j \in \{0,\ldots,n-1\}$ and A be defined by $A(t) := A_j|_Y$ for all $t \in [a_j, a_{j+1})$ and $j \in \{0,\ldots,n-1\}$ and $A(b) := A_{n-1}|_Y$. Then

$$\int_r^t U(t,s)A(s)\xi \, ds = \xi - U(t,r)\xi$$

for all $(t,r) \in \triangle(I)$ and $\xi \in Y$ where integration denotes weak Lebesgue integration with respect to $L(X,\mathbb{K})$.

Proof. For every $j \in \{0,\ldots,n-1\}$ let $T_j : [0,\infty) \to L(X,X)$ be the strongly continuous semigroup generated by $A_j : D(A_j) \to X$. Then $U(t,r) = T_j(t-r)$ for all $(t,r) \in \triangle(I_j)$ and $j \in \{0,\ldots,n-1\}$. Further, let $\xi \in Y$, $j \in \{0,\ldots,n-1\}$, $t \in I_j$, $F_t := ([a_j,t] \to X, s \mapsto U(t,s)\xi = T_j(t-s)\xi)$ and $f_t := ([a_j,t] \to X, s \mapsto A_jU(t,s)\xi = U(t,s)A_j\xi)$. Then f_t and F_t are both continuous. Further, $F_t|_{(a_j,t)}$ is differentiable with derivative $f_t|_{(a_j,t)}$. Hence it follows the weak Lebesgue integrability of f_t with respect to $L(X,\mathbb{K})$ and by the fundamental theorem of calculus that

$$\begin{aligned}\xi - U(t,r)\xi = T_j(0)\xi - T_j(t-r)\xi = F_t(t) - F_t(r) = \int_r^t f_t(s)\,ds \\ = \int_r^t U(t,s)A(s)\xi\,ds\end{aligned}$$

for every $r \in [a_j,t]$. If $r < a_j$ and $j_r \in \{1,\ldots,n\}$ such that $r \in [a_{j_r-1}, a_{j_r}]$, then it follows

$$\begin{aligned}\int_r^t U(t,s)A(s)\xi\,ds &= \int_r^{a_{j_r}} U(t,s)A(s)\xi\,ds + \sum_{k=j_r}^{j-1}\int_{a_k}^{a_{k+1}} U(t,s)A(s)\xi\,ds \\ &\quad + \int_{a_j}^t U(t,s)A(s)\xi\,ds = \int_r^{a_{j_r}} U(t,a_{j_r})U(a_{j_r},s)A(s)\xi\,ds \\ &\quad + \sum_{k=j_r}^{j-1}\int_{a_k}^{a_{k+1}} U(t,a_{k+1})U(a_{k+1},s)A(s)\xi\,ds + \int_{a_j}^t U(t,s)A(s)\xi\,ds \\ &= U(t,a_{j_r})\int_r^{a_{j_r}} U(a_{j_r},s)A(s)\xi\,ds \\ &\quad + \sum_{k=j_r}^{j-1} U(t,a_{k+1})\int_{a_k}^{a_{k+1}} U(a_{k+1},s)A(s)\xi\,ds + \int_{a_j}^t U(t,s)A(s)\xi\,ds \\ &= U(t,a_{j_r})[\xi - U(a_{j_r},r)\xi] + \sum_{k=j_r}^{j-1} U(t,a_{k+1})[\xi - U(a_{k+1},a_k)\xi] + \xi - U(t,a_j)\xi\end{aligned}$$

$$= U(t,a_{j_r})\xi - U(t,r)\xi + \sum_{k=j_r}^{j-1}\left[U(t,a_{k+1})\xi - U(t,a_k)\xi\right] + \xi - U(t,a_j)\xi$$
$$= \xi - U(t,r)\xi \, .$$

□

8.6 Stable Families of Generators

Definition 8.6.1. Let $\mathbb{K} \in \{\mathbb{R},\mathbb{C}\}$, $(X, \| \ \|)$ a $\mathbb{K}$-Banach space and I some closed interval of $\mathbb{R}$. A family $(A(t))_{t\in I}$ of infinitesimal generators of strongly continuous semigroups $(T(t))_{t\in I}$ on X is said to be stable if there are $\mu \in \mathbb{R}$ and $c \in [1,\infty)$ such that for all $n \in \mathbb{N}^*$ and all $t_1,\dots,t_n \in I$ satisfying $t_1 \leqslant \dots \leqslant t_n$

$$\left\|[T(t_n)](s_n)\dots[T(t_1)](s_1)\right\|_{\mathrm{Op}} \leqslant c\, e^{\mu\sum_{k=1}^{n} s_k} \, , \quad s_1,\dots,s_n \in [0,\infty) \, .$$

Lemma 8.6.2. (Exponential formula) Let $\mathbb{K} \in \{\mathbb{R},\mathbb{C}\}$, $(X, \| \ \|)$ a $\mathbb{K}$-Banach space and $A : D(A) \to X$ the infinitesimal generator of a strongly continuous semigroup $T : [0,\infty) \to L(X,X)$. Then

$$T(t) = \mathrm{s-}\lim_{n\to\infty}\left[\left(\mathrm{id}_X + \frac{t}{n}A\right)^{-1}\right]^n \tag{8.6.1}$$

for every $t \in [0,\infty)$. Also, this limit is uniform for every $\xi \in X$ on compact subsets of $[0,\infty)$.

Proof. By Theorem 4.1.1 there are $(c,\mu) \in [1,\infty) \times \mathbb{R}$ such that $\|T(t)\|_{\mathrm{Op}} \leqslant c\, e^{\mu t}$ for all $t \in [0,\infty)$. Further, by the same theorem, it follows that $(-\infty,-\mu) \times \mathbb{R}$ is contained in the resolvent set of A and in particular that

$$[(A-\lambda)^{-1}]^{(n+1)}\xi = \frac{1}{n!}\int_0^\infty s^n e^{\lambda s}.T(s)\xi\, ds$$

for every $n \in \mathbb{N}$, $\lambda \in (-\infty,-\mu) \times \mathbb{R}$ and $\xi \in X$. Here integration denotes weak Lebesgue integration with respect to $L(X,\mathbb{K})$. In the case $t = 0$, equation (8.6.1) is obviously true. In the following, let $t > 0$. Then

$$\left[\left(\mathrm{id}_X + \frac{t}{n}A\right)^{-1}\right]^{(n+1)}\xi = \left[\frac{n}{t}\left(A + \frac{n}{t}\right)^{-1}\right]^{(n+1)}\xi$$
$$= \frac{n^{n+1}}{n!}\int_0^\infty u^n e^{-nu}.T(tu)\xi\, du$$

for $n \in \mathbb{N}$ such that $n > \mu t$. Note that this is also true for the simple case that $A = 0_{\mathbb{C}\to\mathbb{C}}$ which implies

$$\frac{n^{n+1}}{n!}\int_0^\infty u^n e^{-nu}\, du = 1 \, .$$

Hence it follows

$$\left[\left(\mathrm{id}_X + \frac{t}{n}A\right)^{-1}\right]^{(n+1)} \xi - T(t)\xi = \frac{n^{n+1}}{n!}\int_0^\infty (ue^{-u})^n \, . \, (T(tu) - T(t))\,\xi\,du$$

and by Theorem 3.2.5 that

$$\left\| \left[\left(\mathrm{id}_X + \frac{t}{n}A\right)^{-1}\right]^{(n+1)} \xi - T(t)\xi \right\| \leqslant \frac{n^{n+1}}{n!}\int_0^\infty (ue^{-u})^n \cdot \|T(tu)\xi - T(t)\xi\|\,du\,.$$

Note that this is also true for the case $t = 0$. By Lemma 8.1.1 (ii), it follows that $h_\xi := \|T \circ ([0,\infty)^2 \to [0,\infty), (t,u) \mapsto tu) - T \circ pr_1\|$ is continuous where $pr_1 := (\mathbb{R}^2 \to \mathbb{R}, (t,u) \mapsto t)$. Hence for given $t_0 \geqslant 0, b > 1$, we assume that $n > |\mu|\, t_0$ in the following, also $h_\xi|_{[0,t_0]\times[0,b]}$ is continuous and for given $\varepsilon > 0$ and $t \in [0,t_0]$, since $h_\xi(t,1) = 0$, there exists $\delta_t > 0$ such that $h_\xi(t',u) < \varepsilon/6$ for all $(t',u) \in [0,t_0]\times[0,b]$ such that $|t'-t| < \delta_t$ and $|u-1| < \delta_t$. Since $[0,t_0]\times\{1\}$ is compact, it follows the existence of $\delta > 0$ such that $h_\xi(t,u) < \varepsilon/6$ for all $(t,u) \in [0,t_0]\times[1-\delta,1+\delta]$. Using that $g_n := (\mathbb{R} \to \mathbb{R}, u \mapsto (ue^{-u})^n)$ is strictly increasing on $[0,1]$, it follows that

$$\frac{n^{n+1}}{n!}\int_0^{1-\delta} (ue^{-u})^n \cdot h_\xi(t,u)\,du \leqslant M\,\frac{n^{n+1}}{n!}\int_0^{1-\delta} (ue^{-u})^n\,du$$

$$\leqslant M\,\frac{n^{n+1}}{n!}\,(1-\delta)^n e^{-n(1-\delta)} = M\,\frac{n^{n+1}}{n!}\,e^{-n(1+\delta')} \leqslant \frac{M}{e}\cdot n\,e^{-n\delta'}$$

where $M := \max\{h_\xi(t,u) : (t,u)\times[0,t_0]\times[0,1-\delta]\}$ and $\delta' := |\ln(1-\delta)| - \delta$. Note that because $g_n := (\mathbb{R} \to \mathbb{R}u \mapsto (ue^{-u})^n)$ is strictly increasing on $[0,1]$, it follows that $(1-\delta)^n e^{-n(1-\delta)} < e^{-n}$ and hence that $\delta' > 0$. Also it has been used that

$$\ln(n!) = \sum_{k=1}^{n} \ln(k) = \sum_{k=1}^{n-1} \ln(k+1) = \int_{\mathbb{R}} \sum_{k=1}^{n-1} \ln(k+1)\,\chi_{[k,k+1]}\,dv^1$$

$$\geqslant \int_1^n \ln(x)\,dx = \ln\left(e\,n^n e^{-n}\right)$$

and hence that

$$\frac{n^n \cdot e^{-n}}{n!} \leqslant \frac{1}{e}\,.$$

Further,

$$\frac{n^{n+1}}{n!}\int_{1-\delta}^{1+\delta} (ue^{-u})^n \cdot h_\xi(t,u)\,du \leqslant \frac{\varepsilon}{6}\,\frac{n^{n+1}}{n!}\int_{1-\delta}^{1+\delta} (ue^{-u})^n\,du \leqslant \frac{\varepsilon}{6}\,.$$

Finally, for $n > [1 + (1/\delta)]|\mu|\,t_0$

$$\frac{n^{n+1}}{n!}\int_{1+\delta}^\infty (ue^{-u})^n \cdot h_\xi(t,u)\,du \leqslant c\,\|\xi\|\frac{n^{n+1}}{n!}\int_{1+\delta}^\infty u^n e^{-nu}\left(e^{\mu tu} + e^{\mu t}\right)du$$

$$
\begin{aligned}
&\leqslant 2c\,\|\xi\|\,\frac{n^{n+1}}{n!}\int_{1+\delta}^{\infty} u^n e^{-(n-|\mu| t_0)\,u}\,du \\
&\leqslant 2c\,\|\xi\|\,\frac{n^{n+1}}{n!}\,e^{-n}e^{n\ln(1+\delta)}\int_{1+\delta}^{\infty} e^{-\left(\frac{n\delta}{1+\delta}-|\mu| t_0\right)u}\,du \\
&\leqslant \frac{2\,c\,(1+\delta)\,e^{|\mu|\,(1+\delta)\,t_0}}{e\,[n\delta-|\mu|\,(1+\delta)\,t_0]}\,\|\xi\|\,n\,(1+\delta)^n e^{-n\delta} \leqslant \frac{2\,c\,K\,(1+\delta)\,e^{|\mu|\,(1+\delta)\,t_0}}{e\,[n\delta-|\mu|\,(1+\delta)\,t_0]}\,\|\xi\|
\end{aligned}
$$

for some $K > 0$ where it has been used that because of the concavity of the natural logarithm function

$$
\ln(u) \leqslant \frac{1}{1+\delta}\,u + \ln(1+\delta) - 1\ ,\ \ln(u) < u - 1
$$

for all $u \in (1, \infty)$. Altogether, it follows the existence of a $n_0 \in \mathbb{N}$, not depending on t, such that $n_0 > [1 + (1/\delta)]|\mu|\, t_0$ and

$$
\left\| \left[\left(\mathrm{id}_X + \frac{t}{n} A \right)^{-1} \right]^{(n+1)} \xi - T(t)\xi \right\| \leqslant \frac{\varepsilon}{2}\ .
$$

for all $n \geqslant n_0$. Further, it follows by Theorem 4.1.1 that for every such n

$$
\begin{aligned}
&\left\| \left[\left(\mathrm{id}_X + \frac{t}{n} A \right)^{-1} \right]^{n} \xi - T(t)\xi \right\| \\
&\leqslant \left\| \left[\left(\mathrm{id}_X + \frac{t}{n} A \right)^{-1} \right]^{n} \right\|_{\mathrm{Op}} \cdot \left\| \xi - \left(\mathrm{id}_X + \frac{t}{n} A \right)^{-1} \xi \right\| + \frac{\varepsilon}{2} \\
&\leqslant c \left(1 - \frac{\mu t}{n} \right)^{-n} \cdot \left\| \xi - \left(\mathrm{id}_X + \frac{t}{n} A \right)^{-1} \xi \right\| + \frac{\varepsilon}{2}
\end{aligned}
$$

Using that $\ln(x) \geqslant h(x)$ for every $x \in [1/2, 3/2]$ where $h : [1/2, 3/2] \to \mathbb{R}$ is defined by $h(x) := (x-1)/2$ if $x \in [1, 3/2]$ and $h(x) := 2(x-1)$ if $x \in [1/2, 1)$, it follows for $n > 2|\mu| t_0$

$$
\left(1 - \frac{\mu t}{n} \right)^{-n} = e^{-n \ln\left(1 - \frac{\mu t}{n}\right)} \leqslant e^{2|\mu| t_0}\ .
$$

Further, it follows

$$
\left\| \eta - \left(\mathrm{id}_X + \frac{t}{n} A \right)^{-1} \eta \right\| \leqslant \frac{t_0}{n} \cdot \left\| \left(\mathrm{id}_X + \frac{t}{n} A \right)^{-1} A\eta \right\| \leqslant \frac{c\, t_0}{n - |\mu| t_0}\, \|A\eta\|
$$

for all $\eta \in D(A)$,

$$
\left\| \mathrm{id}_X - \left(\mathrm{id}_X + \frac{t}{n} A \right)^{-1} \right\|_{\mathrm{Op}} \leqslant C := 1 + \frac{c}{1 - \frac{|\mu| t_0}{n_0}}
$$

and

$$
c \left(1 - \frac{\mu t}{n} \right)^{-n} \cdot \left\| \xi - \left(\mathrm{id}_X + \frac{t}{n} A \right)^{-1} \xi \right\| \leqslant c\, e^{2|\mu| t_0} \left[C \|\xi - \eta\| + \frac{c\, t_0}{n - |\mu| t_0}\, \|A\eta\| \right]\ .
$$

Since $D(A)$ is dense in X, it follows the existence of an $n_1 \geqslant n_0$, not depending on t, such that

$$\left\| \left[\left(\mathrm{id}_X + \frac{t}{n} A \right)^{-1} \right]^n \xi - T(t)\xi \right\| \leqslant \varepsilon$$

for all $n \in \mathbb{N}$ such that $n \geqslant n_1$. □

Lemma 8.6.3. Let $\mathbb{K} \in \{\mathbb{R}, \mathbb{C}\}$, $(X, \| \ \|)$ a $\mathbb{K}$-Banach space, I some closed interval of $\mathbb{R}$ and $(A(t))_{t \in I}$ a family of infinitesimal generators of strongly continuous semigroups $(T(t))_{t \in I}$ on X. Then the following statements are equivalent.

(i) $(A(t))_{t \in I}$ is stable.

(ii) There are $\mu \in \mathbb{R}$ and $c \in [1, \infty)$ such that $(-\infty, -\mu)$ is contained in the resolvent set of every $A(t)$, $t \in I$ and such that for all $n \in \mathbb{N}^*$, $t_1, \dots, t_n \in I$ satisfying $t_1 \leqslant \dots \leqslant t_n$

$$\| (A(t_n) - \lambda_n)^{-1} \dots (A(t_1) - \lambda_1)^{-1} \|_{\mathrm{Op}} \leqslant c \prod_{k=1}^{n} |\lambda_k + \mu|^{-1} ,$$

$$\lambda_1, \cdots, \lambda_n \in (-\infty, -\mu) .$$

(iii) There are $\mu \in \mathbb{R}$ and $c \in [1, \infty)$ such that $(-\infty, -\mu)$ is contained in the resolvent set of every $A(t)$, $t \in I$ and such that for all $n \in \mathbb{N}^*$, $t_1, \dots, t_n \in I$ satisfying $t_1 \leqslant \dots \leqslant t_n$

$$\| (A(t_n) - \lambda)^{-1} \dots (A(t_1) - \lambda)^{-1} \|_{\mathrm{Op}} \leqslant c \, |\lambda + \mu|^{-n} , \ \lambda \in (-\infty, -\mu) .$$

Proof. '(i) ⇒ (ii)': Since $(A(t))_{t \in I}$ is stable, there are $\mu \in \mathbb{R}$ and $c \in [1, \infty)$ such that for all $n \in \mathbb{N}^*$ and all $t_1, \dots, t_n \in I$ satisfying $t_1 \leqslant \dots \leqslant t_n$

$$\| [T(t_n)](s_n) \dots [T(t_1)](s_1) \|_{\mathrm{Op}} \leqslant c \, e^{\mu \sum_{k=1}^{n} s_k} , \ s_1, \dots, s_n \in [0, \infty) .$$

This implies for every $t \in I$ that

$$\| [T(t)](s) \|_{\mathrm{Op}} \leqslant c \, e^{\mu s} , \ s \in [0, \infty)$$

and hence according to Theorem 4.1.1 that $(-\infty, -\mu)$ is contained in the resolvent set of $A(t)$. In a second step, it will now be proved that

$$\begin{aligned} &(A(t_n) - \lambda_n)^{-1} \dots (A(t_1) - \lambda_1)^{-1} \\ &= \int_{(0,\infty)^n} e^{\lambda_n s_n} \dots e^{\lambda_1 s_1} . [T(t_n)](s_n) \dots [T(t_1)](s_1) \, dv^n , \end{aligned} \tag{8.6.2}$$

for every $\mathbf{t} = (t_n, \dots, t_1) \in [0, \infty)^n$, $\boldsymbol{\lambda} = (\lambda_1, \dots, \lambda_n) \in (-\infty, -\mu)^n$ where integration is weak Lebesgue integration with respect to $L(X, \mathbb{K})$. First, it follows by Lemma 8.1.1 (ii) that

$$\begin{aligned} f_{(\mathbf{t}, \boldsymbol{\lambda})} := ((0, \infty)^n \to L(X, X), (s_n, \dots, s_1) \mapsto e^{\lambda_n s_n} \dots e^{\lambda_1 s_1} \\ [T(t_n)](s_n) \dots [T(t_1)](s_1)) \end{aligned}$$

is strongly continuous. In addition, $\|f_{(t,\lambda)}\|_{\text{Op}}$ is bounded by the summable function

$$h_{(t,\lambda)} := \left((0,\infty)^n \to \mathbb{R}, (s_n,\dots,s_1) \mapsto c^n e^{-|\lambda_n+\mu|s_n} \dots e^{-|\lambda_1+\mu|s_1}\right).$$

Hence it follows by Theorem 3.2.11 that $f_{(t,\lambda)}$ is weakly Lebesgue integrable with respect to $L(X,\mathbb{K})$. Further, it follows by Fubini's theorem and Theorem 4.1.1

$$\begin{aligned}
&\int_{(0,\infty)^n} e^{\lambda_n s_n} \dots e^{\lambda_1 s_1} . [T(t_n)](s_n) \dots [T(t_1)](s_1)\, dv^n\, \eta \\
&= \int_{(0,\infty)^n} e^{\lambda_n s_n} \dots e^{\lambda_1 s_1} [T(t_n)](s_n) \dots [T(t_1)](s_1)\, \eta\, dv^n \\
&= \int_{(0,\infty)^{n-1}} e^{\lambda_{n-1} s_{n-1}} \dots e^{\lambda_1 s_1} \left(\int_0^\infty e^{\lambda_n s_n} [T(t_n)](s_n) \dots [T(t_1)](s_1)\, \eta\, ds_n \right) dv^{n-1} \\
&= \int_{(0,\infty)^{n-1}} e^{\lambda_{n-1} s_{n-1}} \dots e^{\lambda_1 s_1} \cdot (A(t_n) - \lambda_n)^{-1} \\
&\quad [T(t_{n-1})](s_{n-1}) \dots [T(t_1)](s_1)\, \eta\, dv^{n-1} \\
&= (A(t_n) - \lambda_n)^{-1} \int_{(0,\infty)^{n-1}} e^{\lambda_{n-1} s_{n-1}} \dots e^{\lambda_1 s_1} . \\
&\quad [T(t_{n-1})](s_{n-1}) \dots [T(t_1)](s_1)\, dv^{n-1}\, \eta
\end{aligned}$$

for $\eta \in X$. In this way, recursively, it follows

$$\begin{aligned}
&\int_{(0,\infty)^n} e^{\lambda_n s_n} \dots e^{\lambda_1 s_1} . [T(t_n)](s_n) \dots [T(t_1)](s_1)\, dv^n\, \eta \\
&= (A(t_n) - \lambda_n)^{-1} \dots (A(t_1) - \lambda_1)^{-1} \eta
\end{aligned}$$

and hence finally (8.6.2). For the particular case that $t_1 \leqslant \cdots \leqslant t_n$, it follows by Theorem 3.2.5 that

$$\begin{aligned}
&\|(A(t_n) - \lambda_n)^{-1} \dots (A(t_1) - \lambda_1)^{-1} \xi\| \\
&\leqslant \int_{(0,\infty)^n} e^{\lambda_n s_n} \dots e^{\lambda_1 s_1} \|[T(t_n)](s_n) \dots [T(t_1)](s_1)\, \xi\|\, dv^n \\
&\leqslant c \left(\int_{(0,\infty)^n} e^{-|\lambda_n+\mu|s_n} \dots e^{-|\lambda_1+\mu|s_1}\, dv^n \right) \|\xi\| \\
&= c \left(\prod_{k=1}^n |\lambda_k + \mu|^{-1} \right) \|\xi\|
\end{aligned}$$

for all $\xi \in X$. The direction '(ii) $\Rightarrow$ (iii)' is obvious.
'(iii) $\Rightarrow$ (i)': For this, let $n \in \mathbb{N}^*$ and $t_1, \dots, t_n \in I$ such that $t_1 \leqslant \cdots \leqslant t_n$. Further, let $s_k = p_k/q_k$ where $p_k \in \mathbb{N}$ and $q_k \in \mathbb{N}^*$, for $k \in \{1, \cdots, n\}$. Then it follows for every $k \in \{1, \cdots, n\}$ and $N_k \in \mathbb{N}^*$

$$[T(t_k)]\left(\frac{p_k}{q_k}\right) = [T(t_k)]\left(\frac{p_k N_k}{q_k N_k}\right) = \left\{[T(t_k)]\left(\frac{1}{q_k N_k}\right)\right\}^{p_k N_k}$$

In particular for $N_k := \prod_{j=1, j\neq k}^{n} q_j$,

$$[T(t_k)]\left(\frac{p_k}{q_k}\right) = \left\{[T(t_k)]\left(\frac{1}{N}\right)\right\}^{p_k N_k},$$

where $N := \prod_{j=1}^{n} q_j$, and hence

$$[T(t_n)]\left(\frac{p_n}{q_n}\right)\dots[T(t_1)]\left(\frac{p_1}{q_1}\right) = \left\{[T(t_n)]\left(\frac{1}{N}\right)\right\}^{p_n N_n}\dots\left\{[T(t_1)]\left(\frac{1}{N}\right)\right\}^{p_1 N_1}$$

Further, by Lemmas 8.6.2, 8.1.1 (ii)

$$\begin{aligned}&[T(t_n)]\left(\frac{p_n}{q_n}\right)\dots[T(t_1)]\left(\frac{p_1}{q_1}\right)\\ &= \text{s}-\lim_{\nu\to\infty}\left[\left(\text{id}_X + \frac{1/N}{\nu}A(t_n)\right)^{-1}\right]^{p_n N_n \nu}\dots\left[\left(\text{id}_X + \frac{1/N}{\nu}A(t_1)\right)^{-1}\right]^{p_1 N_1 \nu}.\end{aligned}$$

Now,

$$\begin{aligned}&\left\|\left[\left(\text{id}_X + \frac{1/N}{\nu}A(t_n)\right)^{-1}\right]^{p_n N_n \nu}\dots\left[\left(\text{id}_X + \frac{1/N}{\nu}A(t_1)\right)^{-1}\right]^{p_1 N_1 \nu}\right\|_{\text{Op}}\\ &\leqslant c\left[\left(1-\frac{\mu}{\nu N}\right)^{-\nu N}\right]^{\sum_{k=1}^{n}\frac{p_k}{q_k}}\end{aligned}$$

and hence

$$\left\|[T(t_n)]\left(\frac{p_n}{q_n}\right)\dots[T(t_1)]\left(\frac{p_1}{q_1}\right)\right\|_{\text{Op}} \leqslant c\lim_{\nu\to\infty}\left[\left(1-\frac{\mu}{\nu N}\right)^{-\nu N}\right]^{\sum_{k=1}^{n}\frac{p_k}{q_k}} = c\,e^{\mu\sum_{k=1}^{n}\frac{p_k}{q_k}}.$$

Since by Lemma 8.1.1 (ii)

$$([0,\infty)^n \to L(X,X), (s_1,\dots,s_n)\mapsto [T(t_n)](s_n)\dots[T(t_1)](s_1))$$

is strongly continuous and the set consisting of the positive rational numbers is dense in $[0,\infty)$, from this follows

$$\|[T(t_n)]\,(s_n)\dots[T(t_1)]\,(s_1)\|_{\text{Op}} \leqslant c\,e^{\mu\sum_{k=1}^{n}s_k}.$$

for all $(s_1,\dots,s_n)\in[0,\infty)^n$ and hence that $(A(t))_{t\in I}$ is stable. □

Lemma 8.6.4. Let $\mathbb{K} \in \{\mathbb{R}, \mathbb{C}\}$, X be a $\mathbb{K}$-vector space, I some closed interval of $\mathbb{R}$ of length $l(I)$, $c \geqslant 0$, $\mu \in \mathbb{R}$, $(\| \; \|_t)_{t \in I}$ a family of norms on X such that $(X, \| \; \|_t)$ is a $\mathbb{K}$-Banach space for every $t \in I$ and such that

$$\|\xi\|_t \leqslant e^{c\,|t-s|}\,\|\xi\|_s$$

for all $\xi \in X, t, s \in I$. Finally, for every $t \in I$, let $A(t)$ be the infinitesimal generator of a strongly continuous semigroup $T(t)$ on $(X, \| \; \|_t)$ such that $\|[T(t)](s)\|_{\text{Op},t} \leqslant e^{\mu s}$ for all $s \in [0, \infty)$. Then $(A(t))_{t \in I}$ is a stable family of infinitesimal generators of strongly continuous semigroups on $(X, \| \|_\tau)$ with constants μ, $e^{2c\,l(I)}$, for every $\tau \in I$.

Proof. For this, let $\tau \in I$. Then it follows for every $t \in I$, because of the equivalence of the norms $\| \; \|_t$ and $\| \; \|_\tau$ that $A(t)$ is a densely-defined linear and closed operator in $(X, \| \|_\tau)$. Further, it follows by Theorem 4.1.1 that $(-\infty, -\mu) \times \mathbb{R}$ is contained in the resolvent set of $A(t)$ and in particular that for every $\lambda \in (-\infty, -\mu)$

$$\| \, (A(t) - \lambda)^{-1} \, \|_{\text{Op},t} \leqslant |\lambda + \mu|^{-1} \, .$$

Then it follows for $n \in \mathbb{N}^*$, $t_1, \ldots, t_n \in I$ such that $t_1 \leqslant \cdots \leqslant t_n$ that

$$\begin{aligned}
&\| \, (A(t_n) - \lambda)^{-1} \ldots (A(t_1) - \lambda)^{-1} \, \xi \|_{t_n} \\
&\leqslant \| \, (A(t_{n-1}) - \lambda)^{-1} \ldots (A(t_1) - \lambda)^{-1} \, \xi \|_{t_n} \, |\lambda + \mu|^{-1} \\
&\leqslant e^{c \cdot (t_n - t_{n-1})} \, \| \, (A(t_{n-1}) - \lambda)^{-1} \ldots (A(t_1) - \lambda)^{-1} \, \xi \|_{t_{n-1}} \, |\lambda + \mu|^{-1} \\
&\leqslant e^{c \cdot (t_n - t_{n-1})} \ldots e^{c \cdot (t_2 - t_1)} \, |\lambda + \mu|^{-n} \, \|\xi\|_{t_1} = e^{c \cdot (t_n - t_1)} \, |\lambda + \mu|^{-n} \, \|\xi\|_{t_1}
\end{aligned}$$

and hence

$$\begin{aligned}
&\| \, (A(t_n) - \lambda)^{-1} \ldots (A(t_1) - \lambda)^{-1} \, \xi \|_\tau \leqslant e^{c\,|\tau - t_n|} \, e^{c \cdot (t_n - t_1)} \, |\lambda + \mu|^{-n} \, e^{c\,|\tau - t_1|} \, \|\xi\|_\tau \\
&\quad e^{c\,[\,|\tau - t_n| + t_n - t_1 + |\tau - t_1|\,]} \, |\lambda + \mu|^{-n} \, \|\xi\|_\tau \, .
\end{aligned}$$

We consider three cases $\tau \geqslant t_n$, $t_n \geqslant \tau \geqslant t_1$ and $t_1 \geqslant \tau$. In the first case,

$$|\tau - t_n| + t_n - t_1 + |\tau - t_1| = \tau - t_n + t_n - t_1 + \tau - t_1 = 2\,(\tau - t_1) \leqslant 2\,l(I) \, ,$$

in the second case,

$$|\tau - t_n| + t_n - t_1 + |\tau - t_1| = t_n - \tau + t_n - t_1 + \tau - t_1 = 2\,(t_n - t_1) \leqslant 2\,l(I) \, ,$$

in the third case,

$$|\tau - t_n| + t_n - t_1 + |\tau - t_1| = t_n - \tau + t_n - t_1 + t_1 - \tau = 2\,(t_n - \tau) \leqslant 2\,l(I)$$

and hence

$$\| \, (A(t_n) - \lambda)^{-1} \ldots (A(t_1) - \lambda)^{-1} \, \|_{\text{Op},\tau} \leqslant e^{2c\,l(I)} \, |\lambda + \mu|^{-n} \, .$$

Hence, finally, it follows by Theorem 4.2.1 and Lemma 8.6.3 (iii) that $(A(t))_{t \in I}$ is a stable family of infinitesimal generators of strongly continuous semigroups on $(X, \| \|_\tau)$ with constants μ, $e^{2c\,l(I)}$. □

Theorem 8.6.5. Let $\mathbb{K} \in \{\mathbb{R}, \mathbb{C}\}$, $(X, \| \ \|)$ a $\mathbb{K}$-Banach space, I some closed interval of $\mathbb{R}$, $(A(t))_{t\in I}$ a stable family of infinitesimal generators of strongly continuous semigroups on X with constants $\mu \in \mathbb{R}$, $c \in [1, \infty)$ and $(B(t))_{t\in I}$ a bounded family of bounded linear operators on X with bound $K \in [0, \infty)$. Then $(A(t) + B(t))_{t\in I}$ is a stable family of infinitesimal generators of strongly continuous semigroups on X with constants $\mu + cK$ and c.

Proof. First, it follows by Theorem 4.4.3 and Lemma 8.6.3 that $A(t) + B(t)$ is the infinitesimal generator of a strongly continuous semigroups on X with constants $\mu + cK$ and c, for every $t \in I$. Further, it follows by the proof of Theorem 4.4.3 for every $t \in I$, $\lambda \in (-\infty, -(\mu + cK))$ the bijectivity $A(t) + B(t) - \lambda$ and

$$(A(t) + B(t) - \lambda)^{-1} = \sum_{k\in\mathbb{N}} (A(t) - \lambda)^{-1} \left[(-B(t))\,(A(t) - \lambda)^{-1}\right]^k$$

where the involved family of elements of $L(X, X)$ is absolutely summable. Hence it follows also for any $n \in \mathbb{N}^*$ and $t_1, \dots, t_n \in I$ satisfying $t_1 \leqslant \dots \leqslant t_n$ the absolute summability of

$$\Big((A(t_n) - \lambda)^{-1} \left[(-B(t_n))\,(A(t_n) - \lambda)^{-1}\right]^{k_1} \\ \dots (A(t_1) - \lambda)^{-1} \left[(-B(t_1))\,(A(t_1) - \lambda)^{-1}\right]^{k_n}\Big)_{k\in\mathbb{N}^n}$$

and

$$\begin{aligned}(A(t_n) &+ B(t_n) - \lambda)^{-1} \dots (A(t_1) + B(t_1) - \lambda)^{-1} \\ &= \sum_{k\in\mathbb{N}^n} (A(t_n) - \lambda)^{-1} \left[(-B(t_n))\,(A(t_n) - \lambda)^{-1}\right]^{k_1} \\ &\quad \dots (A(t_1) - \lambda)^{-1} \left[(-B(t_1))\,(A(t_1) - \lambda)^{-1}\right]^{k_n} .\end{aligned}$$

Hence it follows the absolute summability of

$$\Big(\Big\|(A(t_n) - \lambda)^{-1} \left[(-B(t_n))\,(A(t_n) - \lambda)^{-1}\right]^{k_1} \\ \dots (A(t_1) - \lambda)^{-1} \left[(-B(t_1))\,(A(t_1) - \lambda)^{-1}\right]^{k_n}\Big\|_{\text{Op}}\Big)_{k\in\mathbb{N}^n}$$

and

$$\begin{aligned}&\|(A(t_n) + B(t_n) - \lambda)^{-1} \dots (A(t_1) + B(t_1) - \lambda)^{-1}\|_{\text{Op}} \\ &\leqslant \sum_{k\in\mathbb{N}^n} \Big\|(A(t_n) - \lambda)^{-1} \left[(-B(t_n))\,(A(t_n) - \lambda)^{-1}\right]^{k_1} \\ &\quad \dots (A(t_1) - \lambda)^{-1} \left[(-B(t_1))\,(A(t_1) - \lambda)^{-1}\right]^{k_n}\Big\|_{\text{Op}} .\end{aligned} \tag{8.6.3}$$

Further, it follows by Lemma 8.6.3 including its proof that for every $k \in \mathbb{N}^n$

$$\Big\|(A(t_n) - \lambda)^{-1} \left[(-B(t_n))\,(A(t_n) - \lambda)^{-1}\right]^{k_1} \qquad (8.6.4)$$
$$\ldots (A(t_1) - \lambda)^{-1} \left[(-B(t_1))\,(A(t_1) - \lambda)^{-1}\right]^{k_n}\Big\|_{\mathrm{Op}}$$
$$\leqslant \frac{c^{|k|+1}}{|\lambda + \mu|^{|k|+n}} \cdot K^{|k|} = \frac{c}{|\lambda + \mu|^n} \cdot q^{|k|}$$

where $|k| := k_1 + \cdots + k_n$ and $q := c\,K/\,|\lambda + \mu|$. Because of $q < 1$, it follows the absolute summability of $(q^k)_{k\in\mathbb{N}}$ with sum $1/(1-q)$ and hence also for any $n \in \mathbb{N}^*$ the absolute summability of $(q^{|k|})_{k\in\mathbb{N}^n}$ with sum $1/(1-q)^n$. Hence it follows from (8.6.3), (8.6.4) that

$$\|(A(t_n) + B(t_n) - \lambda)^{-1} \ldots (A(t_1) + B(t_1) - \lambda)^{-1}\|_{\mathrm{Op}} \leqslant \frac{1}{(1-q)^n} \cdot \frac{c}{|\lambda + \mu|^n}$$
$$= \frac{c}{|\lambda + \mu + c\,K|^n}\,.$$

Hence, finally, it follows by Lemma 8.6.3 that $(A(t) + B(t))_{t\in I}$ is a stable family of infinitesimal generators of strongly continuous semigroups on X with constants $\mu + cK$ and c. □

9

The Linear Evolution Equation

In this chapter, it will be proved the well-posedness of the initial value problem for differential equations of the form

$$u'(t) = -A(t)u(t)$$

under certain assumptions.[1] Here t is from some non-empty open subinterval I of $\mathbb{R}$, $'$ denotes the ordinary derivative for functions assuming values in a Banach space X, $(A(t))_{t\in I}$ is a family of infinitesimal generators of strongly continuous semigroups on X and $u : I \to X$ is such that $u(t)$ is contained in the domain of $A(t)$ for every $t \in I$. Such equations are called 'linear evolution equations'. There is a large literature on such systems. For instance, see [42, 46, 50, 57, 59, 60, 90, 103, 104, 107, 108, 111, 118, 120, 121, 127, 148, 149, 151, 153, 159, 160, 165, 168, 199, 203, 204, 208, 210, 219, 220]. The result here will be strong enough to allow for variable domains of the operators occurring in the equations and at the same time suitable to conclude the well-posedness (local in time) for quasi-linear evolution equations in Chapter 11 by iteration.

Theorem 9.0.6. Let $\mathbb{K} \in \{\mathbb{R}, \mathbb{C}\}$, $(X, \| \ \|_X)$, $(Z, \| \ \|_Z)$ be $\mathbb{K}$-Banach spaces, $a, b \in \mathbb{R}$ such that $a \leqslant b$ and $I := [a, b]$, $(A(t))_{t\in I}$ a family of infinitesimal generators of strongly continuous semigroups $(T(t))_{t\in I}$ on X, $\Phi : (L(X, X), +, ., \circ) \to (L(Z, Z), +, ., \circ)$ a strongly sequentially continuous nonexpansive homomorphism, S a closed linear map from some dense subspace Y of X into Z and $B \in PC_*(I, L(Z, Z))$. Then $(Y, \| \ \|_Y)$, where $\| \ \|_y := \| \ \|_S$, is a $\mathbb{K}$-Banach space and the inclusion $\iota_{Y\hookrightarrow X}$ of $(Y, \| \ \|_Y)$ into X is continuous. Finally, let

(i) (**Stability**) $(A(t))_{t\in I}$ be stable with constants $\mu \in \mathbb{R}$, $c \in [1, \infty)$,
(ii) (**Continuity**) $Y \subset D(A(t))$, $A(t)|_Y \in L((Y, \| \ \|_Y), X)$ for every $t \in I$ and $A := (I \to L((Y, \| \ \|_Y), X), t \mapsto A(t)|_Y)$ be piecewise norm-continuous,

[1] Note that some of the assumptions in Theorem 9.0.6 can be weakened. See [104, 118, 206, 210, 219].

(iii) (**Intertwining relation**) for every $t \in I$

$$S \circ [T(t)](s) \supset [\hat{T}(t)](s) \circ S \tag{9.0.1}$$

for all $s \in [0, \infty)$ where $\hat{T}(t) : [0, \infty) \to L(Z, Z)$ is the strongly continuous semigroup generated by $\Phi(A(t)) + B(t)$.

Then there is a unique Y/X-evolution operator $U \in C_*(\triangle(I), L(X, X))$ for A. In particular, $\|U(t,r)\|_{\mathrm{Op,X}} \leqslant c\, e^{\mu(t-r)}$ and $\|U_Y(t,r)\|_{\mathrm{Op,Y}} \leqslant c\, e^{(\mu + c\|B\|_\infty)(t-r)}$ where $U_Y(t,r)$ denotes the part of $U(t,r)$ in Y for all $(t,r) \in \triangle(I)$.

Proof. First, it follows by Theorem 6.1.7 and as a consequence of (*iii*) for every $t \in I$ that the family $[T_Y(t)](s)$ of restrictions of $[T(t)](s)$ in domain and in image to Y, $s \in [0, \infty)$, defines a strongly continuous semigroup $T_Y(t) : [0, \infty) \to L(Y, Y)$ on $(Y, \| \ \|_Y)$ with corresponding infinitesimal generator given by the part $A_Y(t) := (\{\xi \in Y : A(t)\xi \in Y\} \to Y, \xi \mapsto A(t)\xi)$ of $A(t)$ in Y and that Y is a core for $A(t)$. In a first step, we construct a sequence of step functions $(A_\nu)_{\nu \in \mathbb{N}^*} \in (PC(I, L((Y, \| \ \|_Y), X)))_{\nu \in \mathbb{N}^*}$ that is uniformly convergent to A on I. Here $\| \ \|_{\mathrm{Op,Y,X}}$ denotes the operator norm on $L((Y, \| \ \|_Y), X)$ and $\| \ \|_{\infty,\mathrm{Y,X}}$ denotes the corresponding supremums norm on $PC(I, L((Y, \| \ \|_Y), X))$. Since A is piecewise norm-continuous and B is piecewise strongly continuous, there is some $n \in \mathbb{N}^*$ along with a partition $(a_0, \dots, a_n)$ of $[a, b]$ such that $A|_{(a_i, a_{i+1})}$ and $B|_{(a_i, a_{i+1})}$ both have an extension to a continuous function and a strongly continuous function, respectively, on $[a_i, a_{i+1}]$ for every $i \in \{0, \dots, n-1\}$. Note that for this reason, $A|_{(a_i, a_{i+1})}$ is uniformly continuous for every $i \in \{0, \dots, n-1\}$. We define for every $\nu \in \mathbb{N}^*$ the corresponding step function $A_\nu : I \to L((Y, \| \ \|_Y), X)$ by defining $A_\nu(a_i) := A(a_i)$ for all $i \in \{0, \dots, n\}$ and for all $i \in \{0, \dots, n-1\}$

$$A_\nu(t) := A(a_{i\nu k}),\ a_{i\nu k} := a_i + \frac{k}{\nu}(a_{i+1} - a_i)$$

for all $t \in I_{i\nu k} := [a_i + (k/\nu)(a_{i+1} - a_i),\ a_i + ((k+1)/\nu)(a_{i+1} - a_i)),\ k \in \{1, \dots, \nu - 1\}$,

$$A_\nu(t) := A(a_{i\nu 0})\ ,\ a_{i\nu 0} := a_i + \frac{1}{\nu}(a_{i+1} - a_i)$$

for all $t \in I_{i\nu 0} := (a_i, a_i + (1/\nu)(a_{i+1} - a_i))$. For given $\varepsilon > 0$, there is $\delta > 0$ such that $\|A(t) - A(s)\|_{\mathrm{Op,Y,X}} < \varepsilon$ whenever $t, s \in (a_i, a_{i+1})$ for some $i \in \{0, \dots, n-1\}$ are such that $|t - s| < \delta$. Hence it follows for $\nu > \max\{a_{i+1} - a_i : i \in \{0, \dots, n-1\}\}/\delta$ that $\|A_\nu(t) - A(t)\|_{\mathrm{Op,X,Y}} < \varepsilon$ for all $t \in I$. Let $U_\nu \in C_*(\triangle(I), L(X, X))$ be the chain generated by the family

$$(\bar{I}_{0\nu 0}, A(a_{0\nu 0})), \dots, (\bar{I}_{(n-1)\nu(\nu-1)}, A(a_{(n-1)\nu(\nu-1)}))\ .$$

U_ν is by Lemma 8.5.3 a Y/X-evolution operator for A_ν. Further, let $\hat{U}_\nu \in C_*(\triangle(I), L(Z, Z))$ be the chain generated by the family $(\bar{I}_{0\nu 0}, \Phi(A(a_{0\nu 0})) + B(a_{0\nu 0})), \dots, (\bar{I}_{(n-1)\nu(\nu-1)}, \Phi(A(a_{(n-1)\nu(\nu-1)})) + B(a_{(n-1)\nu(\nu-1)}))$. By Lemma 8.4.1

$$\hat{U}_\nu = \mathrm{volt}(\bar{U}_\nu, -B_\nu)$$

where $\bar{U}_\nu \in C_*(\Delta(I), L(Z,Z))$ is the chain generated by the family $(\bar{I}_{0\nu 0}, \Phi(A(a_{0\nu 0})))$, $\dots, (\bar{I}_{(n-1)\nu(\nu-1)}, \Phi(A(a_{(n-1)\nu(\nu-1)})))$ and the piecewise norm-continuous $B_\nu : I \to L(Z,Z))$ is defined by $B_\nu(a_i) := B(a_i)$ for all $i \in \{0,\dots,n\}$ and for all $i \in \{0,\dots,n-1\}$, $B_\nu(t) := B(a_{i\nu k})$ for all $t \in I_{i\nu k}$, $k \in \{1,\dots,\nu-1\}$ and $B_\nu(t) := B(a_{i\nu 0})$ for all $t \in I_{i\nu 0}$. Since Φ is a strongly sequentially continuous nonexpansive homomorphism, it follows that $\Phi \circ U_\nu \in C_*(\Delta(I), L(Z,Z))$ and further that $\Phi \circ U_\nu$ is a chain. In addition, it follows for every $i \in \{0,\dots,n-1\}, k \in \{0,\dots,\nu-1\}$, and $(t,r) \in \Delta(\bar{I}_{i\nu k})$ that

$$(\Phi \circ U_\nu)(t,r) = \Phi(T(a_{i\nu k})(t-r)) = (\Phi T(a_{i\nu k}))(t-r) = \bar{U}_\nu(t,r)$$

and hence by Lemma 8.4.1 that

$$\bar{U}_\nu = \Phi \circ U_\nu \, .$$

Finally, it follows from (9.0.1) and the proof of Lemma 8.3.1 (i) that

$$S \circ U_\nu(t,r) \supset \hat{U}_\nu(t,r) \circ S \tag{9.0.2}$$

for all $(t,s) \in \Delta(I)$. For the second step, we need some auxiliary estimates on U_ν and $\hat{U}_\nu$. By (i) and since U_ν is a chain, it follows for every $i \in \{0,\dots,n-1\}$, $k \in \{0,\dots,\nu-1\}$, $t \in \mathring{I}_{i\nu k}$ and $r \in [a,t]$ that

$$\|U_\nu(t,r)\|_{\mathrm{Op,X}} \leqslant c\, e^{\mu(t-r)} \, . \tag{9.0.3}$$

Further, it follows from (i), the nonexpansiveness of Φ along with Theorem 6.3.1 that $(\Phi(A(t)))_{t\in I}$ is a stable family of infinitesimal generators of continuous semigroups on Z with constants μ and c and hence by Theorem 8.6.5 that $(\Phi(A(t)) + B(t))_{t\in I}$ is a stable family of infinitesimal generators of continuous semigroups on Z with constants $\mu + c\|B\|_\infty$ and c. Since $\hat{U}_\nu$ is a chain, this implies

$$\|\hat{U}_\nu(t,r)\|_{\mathrm{Op,Z}} \leqslant c\, e^{(\mu+c\|B\|_\infty)(t-r)} \, . \tag{9.0.4}$$

Finally, by (9.0.2), (9.0.3), (9.0.4)

$$\begin{aligned}\|U_\nu(t,r)\,\xi\|_Y^2 &= \|U_\nu(t,r)\,\xi\|_X^2 + \|S\,U_\nu(t,r)\,\xi\|_Z^2 = \|U_\nu(t,r)\,\xi\|_X^2 + \|\hat{U}_\nu(t,r)S\,\xi\|_Z^2 \\ &\leqslant c^2 e^{2\mu(t-r)}\|\xi\|_X^2 + c^2 e^{2(\mu+c\|B\|_\infty)(t-r)}\|S\,\xi\|_Z^2 \leqslant c^2 e^{2(\mu+c\|B\|_\infty)(t-r)}\|\xi\|_Y^2\end{aligned}$$

for all $\xi \in Y$ and hence

$$\|U_\nu(t,r)|_Y\|_{\mathrm{Op,Y}} \leqslant c\, e^{(\mu+c\|B\|_\infty)(t-r)} \, . \tag{9.0.5}$$

Second, it follows for every $i \in \{0,\dots,n-1\}$, $k \in \{0,\dots,\nu-1\}$, $s \in \mathring{I}_{i\nu k}$ the differentiability of $U_\nu(\cdot,r)\xi : [r,b] \to X$ in s for every $r \in [a,s)$ and in particular

$$\Big(U_\nu(\cdot,r)\xi\Big)'(s) = -A(a_{i\nu k})U_\nu(s,r)\xi = -A_\nu(s)U_\nu(s,r)\xi$$

as well as the differentiability of $U_\nu(t,\cdot)\xi : [a,t] \to X$ in s for every $t \in (s,b]$ and in particular

$$\left(U_\nu(t,\cdot)\xi\right)'(s) = U_\nu(t,s)A(a_{i\nu k})\xi = U_\nu(t,s)A_\nu(s)\xi$$

for every $\xi \in Y$. Hence it follows also for every $\nu, \nu' \in \mathbb{N}^*$, $i \in \{0,\dots,n-1\}$, $k \in \{0,\dots,\nu-1\}$, $k' \in \{0,\dots,\nu'-1\}$, $s \in \mathring{I}_{i\nu k} \cap \mathring{I}_{i\nu' k'}$ the differentiability of $U_\nu(t,\cdot)U_{\nu'}(\cdot,r)\xi : [r,t] \to X$ in s for every $(t,r) \in \triangle(I)$ such that $t > s > r$ and in particular

$$\begin{aligned}\left(U_\nu(t,\cdot)U_{\nu'}(\cdot,r)\xi\right)'(s) &= U_\nu(t,s)A_\nu(s)U_{\nu'}(s,r)\xi - U_\nu(t,s)A_{\nu'}(s)U_{\nu'}(s,r)\xi \\ &= U_\nu(t,s)\left(A_\nu(s) - A_{\nu'}(s)\right)U_{\nu'}(s,r)\xi\end{aligned}$$

for every $\xi \in Y$. Note in particular that $U_\nu(t,\cdot)U_{\nu'}(\cdot,r)\xi$ is continuous and that $U_\nu(t,\cdot)\left(A_\nu - A_{\nu'}\right)U_{\nu'}(\cdot,r)\xi$ is piecewise continuous. Hence it follows by the fundamental theorem of calculus for every $(t,r) \in \triangle(I)$ that

$$U_{\nu'}(t,r)\xi - U_\nu(t,r)\xi = \int_r^t U_\nu(t,s)\left(A_\nu(s) - A_{\nu'}(s)\right)U_{\nu'}(s,r)\xi\, ds$$

where integration is weak Lebesgue integration with respect to $L(X,\mathbb{K})$. Hence by Theorem 3.2.5,

$$\begin{aligned}\|U_{\nu'}(t,r)\xi - U_\nu(t,r)\xi\|_X &\leqslant \int_r^t \|U_\nu(t,s)\left(A_\nu(s) - A_{\nu'}(s)\right)U_{\nu'}(s,r)\xi\|_X\, ds \\ &\leqslant c^2 e^{(\mu+c\|B\|_\infty)(b-a)} \cdot \|\xi\|_Y \cdot \int_a^b \|A_\nu(s) - A_{\nu'}(s)\|_{\mathrm{Op,Y,X}}\, ds\end{aligned}$$

and for every $\eta \in X$

$$\begin{aligned}\|U_{\nu'}(t,r)\eta - U_\nu(t,r)\eta\|_X &\leqslant \|U_{\nu'}(t,r)(\eta-\xi) - U_\nu(t,r)(\eta-\xi)\|_X \\ &+ \|U_{\nu'}(t,r)\xi - U_\nu(t,r)\xi\|_X \leqslant 2c\, e^{\mu(b-a)}\|\eta - \xi\| \\ &+ c^2 e^{(\mu+c\|B\|_\infty)(b-a)} \cdot \|\xi\|_Y \cdot \int_a^b \|A_\nu(s) - A_{\nu'}(s)\|_{\mathrm{Op,Y,X}}\, ds\,.\end{aligned}$$

Since

$$\lim_{\nu\to\infty} \|A_\nu - A\|_{\infty,\mathrm{Y,X}} = 0 \tag{9.0.6}$$

and Y is dense in X, it follows that $((\triangle(I) \to X, (t,r) \mapsto U_\nu(t,r)\eta))_{\nu\in\mathbb{N}^*}$ is a Cauchy sequence in $C(\triangle(I),X)$ and hence convergent to some $(\triangle(I) \to X, (t,r) \mapsto U(t,r)\eta) \in C(\triangle(I),X)$. Further, it follows by the linearity of $U_\nu(t,r)$ and the chain property of U_ν along with (9.0.3) for every $\nu \in \mathbb{N}^*$, the linearity of $U(t,r)$ and the chain property of U and by (9.0.3)

$$\|U(t,r)\eta\|_X \leqslant c\, e^{\mu(t-r)}\|\eta\|_X$$

for every $\eta \in X$ and hence $U(t,r) \in L(X,X)$ and

$$\|U(t,r)\|_{\text{Op,X}} \leqslant c\, e^{\mu(t-r)} \tag{9.0.7}$$

for all $(t,r) \in \triangle(I)$. Further, since Φ is a strongly sequentially continuous nonexpansive homomorphism, it follows that

$$\text{s}-\lim_{\nu\to\infty} \bar{U}_\nu(t,r) = \bar{U}(t,r)$$

where $\bar{U} := \Phi \circ U$, $\bar{U} \in C_*(\triangle(I), L(Z,Z))$ and

$$\|\bar{U}_\nu(t,r)\|_{\text{Op,Z}} \leqslant c\, e^{\mu(t-r)}$$

for every $\nu \in \mathbb{N}$ and $(t,r) \in \triangle(I)$. In addition, obviously, it also follows

$$\lim_{\nu\to\infty} \|B_\nu\xi - B\xi\|_\infty = 0$$

for every $\xi \in Z$ and $\|B_\nu\|_\infty \leqslant \|B\|_\infty$. Hence it follows by Lemma 8.1.3 that

$$\text{s}-\lim_{\nu\to\infty} \hat{U}_\nu(t,r) = \hat{U}(t,r) := \text{volt}(\bar{U}, -B)(t,r) \tag{9.0.8}$$

and employing (9.0.4)

$$\|\hat{U}(t,r)\|_{\text{Op,Z}} \leqslant c\, e^{(\mu+c\|B\|_\infty)(t-r)} \tag{9.0.9}$$

for all $(t,r) \in \triangle(I)$ and in particular that $\hat{U} \in C_*(\triangle(I), L(Z,Z))$. Therefore, it follows from (9.0.2) for every $(t,r) \in \triangle(I)$ and $\eta \in Y$ that

$$\lim_{\nu\to\infty} S\, U_\nu(t,r)\eta = \lim_{\nu\to\infty} \hat{U}_\nu(t,r) S\eta = \hat{U}(t,r) S\eta$$

and since

$$\text{s}-\lim_{\nu\to\infty} U_\nu(t,r) = U(t,r) \tag{9.0.10}$$

and S is in particular closed that $U(t,r)\eta \in Y$ and $S\,U(t,r)\eta = \hat{U}(t,r)S\eta$ or, since this is true for every $\eta \in Y$, that

$$S \circ U(t,r) \supset \hat{U}(t,r) \circ S\ .$$

Since for all $\eta \in Y$

$$\begin{aligned}\|U_\nu(t,r)\eta - U(t,r)\eta\|_Y^2 &= \|U_\nu(t,r)\eta - U(t,r)\eta\|_X^2 + \|S\,U_\nu(t,r)\eta - S\,U(t,r)\eta\|_Z^2 \\ &= \|U_\nu(t,r)\eta - U(t,r)\eta\|_X^2 + \|\hat{U}_\nu(t,r)S\eta - \hat{U}(t,r)\eta\|_Z^2\ ,\end{aligned}$$

it follows by (9.0.10), (9.0.8) that

$$\text{s}_Y-\lim_{\nu\to\infty} U_\nu(t,r) = U(t,r) \tag{9.0.11}$$

for all $(t,r) \in \triangle(I)$. Further,

$$\begin{aligned}
&\|U(t',r')\eta - U(t,r)\eta\|_Y^2 \\
&\quad = \|U(t',r')\eta - U(t,r)\eta\|_X^2 + \|S\,U(t',r')\eta - S\,U(t,r)\eta\|_Z^2 \\
&\quad = \|U(t',r')\eta - U(t,r)\eta\|_X^2 + \|\hat{U}(t',r')S\eta - \hat{U}(t,r)S\eta\|_Z^2
\end{aligned}$$

which is true for every $\eta \in Y$, $(t,r), (t',r') \in \triangle(I)$, along with

$$U \in C_*(\triangle(I), L(X,X))$$

and

$$\hat{U} \in C_*(\triangle(I), L(Z,Z))$$

implies that $U_Y \in C_*(\triangle(I), L(Y,Y))$. Also, (9.0.7) and (9.0.9) imply that

$$\|U_Y(t,r)\|_{\text{Op,Y}} \leqslant c\, e^{(\mu + c\|B\|_\infty)(t-r)} \tag{9.0.12}$$

for all $(t,r) \in \triangle(I)$. Further, since for every $\nu \in \mathbb{N}$ the corresponding U_ν is a Y/X-evolution operator for A_ν, it follows for all $(t,r) \in \triangle(I)$ and $\xi \in Y$ that

$$\int_r^t A_\nu(s)U_\nu(s,r)\xi\,ds = \int_r^t U_\nu(t,s)A_\nu(s)\xi\,ds = \xi - U_\nu(t,r)\xi \tag{9.0.13}$$

where integration denotes weak Lebesgue integration with respect to $L(X,\mathbb{K})$. In addition, $A_\nu \circ U_{\nu Y}(\cdot,r)\xi, U_\nu(t,\cdot)\circ A_\nu\xi, A\circ U_Y(\cdot,r)\xi, U(t,\cdot)\circ A\xi$ are almost everywhere continuous. Hence it follows by Theorem 3.2.5 that

$$\begin{aligned}
&\left\| \int_r^t A_\nu(s)U_{\nu Y}(s,r)\xi\,ds - \int_r^t A(s)U_Y(s,r)\xi\,ds \right\| \\
&\quad \leqslant \int_r^t \|A_\nu(s)U_{\nu Y}(s,r)\xi - A(s)U_Y(s,r)\xi\|\,ds\,, \\
&\left\| \int_r^t U_\nu(t,s)A_\nu(s)\xi\,ds - \int_r^t U(t,s)A(s)\xi\,ds \right\| \\
&\quad \leqslant \int_r^t \|U_\nu(t,s)A_\nu(s)\xi - U(t,s)A(s)\xi\|\,ds\,.
\end{aligned}$$

Because of

$$\begin{aligned}
&\|A_\nu(s)U_{\nu Y}(s,r)\xi - A(s)U_Y(s,r)\xi\| \leqslant \|A\|_{\infty,\text{Y,X}} \cdot \|U_{\nu Y}(s,r)\xi - U_Y(s,r)\xi\|_{\text{Y}} \\
&\quad + c\,e^{(|\mu| + c\|B\|_\infty)(b-a)}\,\|\xi\|_Y \cdot \|A_\nu - A\|_{\infty,\text{Y,X}} \\
&\quad \leqslant 4c\,e^{(|\mu| + c\|B\|_\infty)(b-a)}\,\|\xi\|_Y \cdot \|A\|_{\infty,\text{Y,X}}\,, \\
&\|U_\nu(t,s)A_\nu(s)\xi - U(t,s)A(s)\xi\| \leqslant c\,e^{|\mu|(b-a)}\,\|\xi\|_Y \cdot \|A_\nu - A\|_{\infty,\text{Y,X}} \\
&\quad + \|U_\nu(t,s)A(s)\xi - U(t,s)A(s)\xi\| \leqslant 4c\,e^{|\mu|(b-a)}\,\|\xi\|_Y \cdot \|A\|_{\infty,\text{Y,X}}
\end{aligned}$$

(9.0.7), (9.0.12), (9.0.3), (9.0.5), (9.0.6), (9.0.10), (9.0.11) and Lebesgue's dominated convergence Theorem, this implies that

$$\lim_{\nu\to\infty} \int_r^t A_\nu(s)U_{\nu Y}(s,r)\xi\, ds = \int_r^t A(s)U_Y(s,r)\xi\, ds$$

$$\lim_{\nu\to\infty} \int_r^t U_\nu(t,s)A_\nu(s)\xi\, ds = \int_r^t U(t,s)A(s)\xi\, ds$$

and hence, finally, by (9.0.13), (9.0.10) that

$$\int_r^t A(s)U(s,r)\xi\, ds = \int_r^t U(t,s)A(s)\xi\, ds = \xi - U(t,r)\xi\ . \tag{9.0.14}$$

Finally, the uniqueness of U follows by Lemma 8.5.2. □

Corollary 9.0.7. In addition to the assumptions of Theorem 9.0.6, let $U \in C_*(\triangle(I), L(X,X))$ be the Y/X-evolution operator for A. Further, let $t_0 \in I$, $t_1 \in (t_0, b]$, $f \in C([t_0,t_1),(Y,\|\ \|_Y))$, $\xi \in Y$, $u : [t_0,t_1) \to X$ be defined by

$$u(t) := U(t,t_0)\xi + \int_{t_0}^t U(t,s)f(s)\, ds$$

for all $t \in [t_0,t_1)$ where integration denotes weak Lebesgue integration with respect to $L(X,\mathbb{K})$.

(i) Then $u \in C([t_0,t_1),(Y,\|\ \|_Y))$, $u(t_0) = \xi$, $u|_{(t_0,t_1)}$ is differentiable and in particular

$$u'(t) = -A(t)u(t) + f(t) \tag{9.0.15}$$

for all continuity points t of A in (t_0,t_1).

(ii) If $v \in C([t_0,t_1),X)$ is such that $v(t_0) = \xi$, $\operatorname{Ran} v \subset Y$, $v|_{(t_0,t_1)}$ is differentiable and in particular $v'(t) = -A(t)v(t) + f(t)$ for all continuity points t of A in (t_0,t_1), then $v = u$.

Proof. (i) First, it follows trivially that $u(t_0) = \xi$. Further, it follows for every $t \in (t_0,t_1)$, $s \in [t_0,t]$ and $h \in [t_0 - s, t - s]$:

$$\begin{aligned}&\|U(t,s+h)f(s+h) - U(t,s)f(s)\|_Y\\ &= \|U(t,s+h)f(s+h) - U(t,s+h)f(s) + U(t,s+h)f(s) - U(t,s)f(s)\|_Y\\ &\leqslant \|U_Y\|_{\infty,Y,Y} \cdot \|f(s+h) - f(s)\|_Y + \|U(t,s+h)f(s) - U(t,s)f(s)\|_Y\end{aligned}$$

and hence because of $U_Y \in C_*(\triangle(I), L((Y,\|\ \|_Y),(Y,\|\ \|_Y))$, $f \in C([t_0,t_1),(Y,\|\ \|_Y))$ the continuity of $([t_0,t] \to (Y,\|\ \|_Y), s \mapsto U(t,s)f(s))$ and therefore also the continuity of $([t_0,t] \to X, s \mapsto U(t,s)f(s))$ since the inclusion $\iota_{Y\hookrightarrow X}$ of $(Y,\|\ \|_Y)$ into X is continuous. Note that the last also implies that $f \in C([t_0,t_1),X)$. As a consequence, by Theorem 3.2.4 (ii), it follows that $u(t) \in Y$ and

$$\int_{t_0}^t U(t,s)f(s)\, ds = \int_{t_0,Y}^t U(t,s)f(s)\, ds\ ,$$

for all $t \in [t_0, t_1)$ where the index Y denotes weak Lebesgue integration with respect to $L((Y, \| \ \|_Y), \mathbb{K})$. In particular, it follows by Theorem 3.2.4 (ii) and Theorem 3.2.5 for every $t \in [t_0, t_1)$ and $h \geqslant 0$ such that $t + h \in [t_0, t_1)$

$$\left\| \int_{t_0,\mathrm{Y}}^{t+h} U(t+h,s)f(s)\,ds - \int_{t_0,\mathrm{Y}}^{t} U(t,s)f(s)\,ds \right\|_{\mathrm{Y}}$$
$$\leqslant \left\| \int_{t_0,\mathrm{Y}}^{t} (U(t+h,s)f(s) - U(t,s)f(s))\,ds \right\|_{\mathrm{Y}} + \left\| \int_{t,\mathrm{Y}}^{t+h} U(t+h,s)f(s)\,ds \right\|_{\mathrm{Y}}$$
$$\leqslant \left\| (U(t+h,t) - U(t,t)) \int_{t_0,\mathrm{Y}}^{t} U(t,s)f(s)\,ds \right\|_{\mathrm{Y}} + \|U\|_{\infty,\mathrm{Y},\mathrm{Y}} \cdot \int_{t,\mathrm{Y}}^{t+h} \|f(s)\|_Y\,ds\,,$$

for $h \leqslant 0$ such that $t + h \in [t_0, t_1)$

$$\left\| \int_{t_0,\mathrm{Y}}^{t+h} U(t+h,s)f(s)\,ds - \int_{t_0,\mathrm{Y}}^{t} U(t,s)f(s)\,ds \right\|_{\mathrm{Y}}$$
$$\leqslant \left\| \int_{t_0,\mathrm{Y}}^{t+h} (U(t,s)f(s) - U(t+h,s)f(s))\,ds \right\|_{\mathrm{Y}} + \left\| \int_{t+h,\mathrm{Y}}^{t} U(t,s)f(s)\,ds \right\|_{\mathrm{Y}}$$
$$\leqslant \int_{t_0,\mathrm{Y}}^{t+h} \|U(t,s)f(s) - U(t+h,s)f(s)\|_{\mathrm{Y}}\,ds + \|U\|_{\infty,\mathrm{Y},\mathrm{Y}} \cdot \int_{t+h,\mathrm{Y}}^{t} \|f(s)\|_Y\,ds$$

and hence by $U_Y \in C_*(\triangle(I), L((Y, \| \ \|_Y), (Y, \| \ \|_Y))$ and Lebesgue's dominated convergence theorem that $u \in C([t_0, t_1), (Y, \| \ \|_Y))$. Since for every $s \in [t_0, t_1)$ and $h \in \mathbb{R}$ such that $s + h \in [t_0, t_1)$

$$\|A(s+h)u(s+h) - A(s)u(s)\| \leqslant \|\,(A(s+h) - A(s))\,u(s)\|$$
$$+ \|\,(A(s+h) - A(s))\,(u(s+h) - u(s))\,\| + \|A(s)\,(u(s+h) - u(s))\,\|$$
$$\leqslant \|\,(A(s+h) - A(s))\,u(s)\| + 3\|A\|_{\infty,\mathrm{Y},\mathrm{X}} \cdot \|u(s+h) - u(s)\|_Y\,,$$

this implies that $[t_0, t_1) \to X, s \mapsto A(s)u(s)$ is continuous in all points of continuity of A and hence also almost everywhere continuous on $[t_0, t_1)$. Finally, it follows for every

$$(t, r) \in \mathrm{D} := \{(t, r) : r \in [t_0, t_1) \wedge t \in [r, b]\}$$

and all $h, h' \in \mathbb{R}$ such that $(t, r) + (h, h') \in \mathrm{D}$

$$\|A(t+h')U(t+h', r+h)f(r+h) - A(t)U(t,r)f(r)\|$$
$$= \|A(t+h')\,[U(t+h', r+h)f(r+h) - U(t,r)f(r)]$$
$$+ (A(t+h') - A(t))\,U(t,r)f(r)\|$$
$$\leqslant \|A\|_{\infty,\mathrm{Y},\mathrm{X}} \cdot \|U(t+h', r+h)f(r+h) - U(t,r)f(r)\|_{\mathrm{Y}}$$
$$+ \|\,(A(t+h') - A(t))\,U(t,r)f(r)\|$$

$$
\begin{aligned}
&= \|A\|_{\infty,\mathrm{Y},\mathrm{X}} \|U(t+h',r+h)\,(f(r+h)-f(r)) \\
&\quad + (U(t+h',r+h)-U(t,r))f(r)\|_{\mathrm{Y}} \\
&\quad + \|\,(A(t+h')-A(t))\,U(t,r)f(r)\| \\
&\leqslant \|A\|_{\infty,\mathrm{Y},\mathrm{X}} \cdot \|U_{\mathrm{Y}}\|_{\infty,\mathrm{Y},\mathrm{Y}} \cdot \|f(r+h)-f(r)\|_{\mathrm{Y}} \\
&\quad + \|A\|_{\infty,\mathrm{Y},\mathrm{X}} \cdot \|U(t+h',r+h)-U(t,r))f(r)\|_{\mathrm{Y}} \\
&\quad + \|\,(A(t+h')-A(t))\,U(t,r)f(r)\|
\end{aligned}
$$

and hence that $(D \to X, (t,r) \mapsto A(t)U(t,r)f(r))$ is almost everywhere continuous on D. As consequence, it follows by Theorem 3.2.4 (ii), Theorem 3.2.5, the Theorem of Fubini, and (9.0.14),

$$
\begin{aligned}
&\xi + \int_{t_0}^{t} (f(s) - A(s)u(s))\, ds \\
&= \xi + \int_{t_0}^{t} \left[f(s) - A(s) \left(U(s,t_0)\xi + \int_{t_0,\mathrm{Y}}^{s} U(s,s')f(s')\, ds' \right) \right] ds \\
&= \xi + \int_{t_0}^{t} \left[f(s) - A(s)U(s,t_0)\xi - \int_{t_0}^{s} A(s)U(s,s')f(s')\, ds' \right] ds \\
&= \xi + \int_{t_0}^{t} f(s)\, ds - \int_{t_0}^{t} A(s)U(s,t_0)\xi\, ds - \int_{t_0}^{t} \left(\int_{t_0}^{s} A(s)U(s,s')f(s')\, ds' \right) ds \\
&= \xi + \int_{t_0}^{t} f(s)\, ds - (\xi - U(t,t_0)\xi) - \int_{t_0}^{t} \left(\int_{s'}^{t} A(s)U(s,s')f(s')\, ds \right) ds' \\
&= U(t,t_0)\xi + \int_{t_0}^{t} f(s)\, ds - \int_{t_0}^{t} (f(s') - U(t,s')f(s'))\, ds' \\
&= U(t,t_0)\xi + \int_{t_0}^{t} U(t,s)f(s)\, ds = u(t)
\end{aligned}
$$

for all $t \in [t_0, t_1)$. Hence, finally, it follows by the auxiliary result in the proof of Lemma 8.5.2 the relation (9.0.15) for all continuity points t of A in (t_0, t_1). (ii) For this, let $t \in (t_0, t_1)$. We define $G : (t_0, t) \to X$ by $G(s) := U(t,s)v(s)$ for all $s \in (t_0, t)$. Then it follows for every $s \in (t_0, t)$ and $h \in \mathbb{R}^*$ such that $s + h \in (t_0, t)$

$$
\begin{aligned}
\frac{1}{h}\,[G(s+h) - G(s)] &= \frac{1}{h}\,[U(t,s+h)v(s+h) - U(t,s)v(s)] \\
&= U(t,s+h)\frac{1}{h}\,[v(s+h) - v(s)] + \frac{1}{h}\,[U(t,s+h) - U(t,s)]v(s)
\end{aligned}
$$

and hence by Lemma 8.5.2

$$
\lim_{h \to 0} \frac{1}{h}\,[G(s+h) - G(s)] = U(t,s)[v'(s) + A(s)\,v(s)] = U(t,s)f(s)
$$

for all continuity points s of A in (t_0, t_1). Altogether, it follows the differentiability of G and $G'(s) = U(t,s)f(s)$ for all continuity points s of A in (t_0, t_1). Note that there is a continuous extension $\hat{G} : [t_0, t] \to X$ of G such that $\hat{G}(t_0) = U(t, t_0)\,\xi$ and $\hat{G}(t) = v(t)$. Finally, note that $F : [t_0, t] \to X$ defined by $F(s) := U(t, s) f(s)$ for every $s \in [t_0, t]$ is continuous by the proof of (i). Hence it follows the weak Lebesgue integrability of F with respect to $L(X, \mathbb{K})$ and by the fundamental theorem of calculus

$$v(t) - U(t, t_0) = \hat{G}(t) - \hat{G}(t_0) = \int_{t_0}^{t} F(s)\,ds$$
$$= \int_{t_0}^{t} U(t, s) f(s)\,ds = \int_{t_0}^{t} U(t, s) f(s)\,ds$$

and hence $v(t) = u(t)$ for every $t \in [t_0, t_1)$. □

Lemma 9.0.8. In addition to the assumptions of Theorem 9.0.6, let $U \in C_*(\triangle(I), L(X, X))$ be the Y/X-evolution operator for A. Further, let

(iv) Z_0 be a dense subspace of Z contained in $D(\Phi(A(t)))$ for all $t \in I$, $\| \ \|_0$ a norm for Z_0 such that $(Z_0, \| \ \|_0)$ is a $\mathbb{K}$-Banach space and such that the inclusion $\iota_{Z_0 \hookrightarrow Z}$ of Z_0 into Z is continuous. Finally, let $\Phi(A(t))|_{Z_0} \in L((Z_0, \| \ \|_0), Z)$ for all $t \in I$ and $\Phi A := (I \to L((Z_0, \| \ \|_0), Z), t \mapsto \Phi(A(t))|_{Z_0}) \in PC_*(I, L((Z_0, \| \ \|_0), Z))$.

Then

(i)

$$\| (U(t, r) - \mathrm{id}_X)\,\xi\|_Y \leqslant (1 + \|\hat{U}\|_{\infty,Z,Z}) \cdot \|S\xi - \xi_0\|_Z$$
$$+ \left[\|U\|_{\infty,X,X}\,\|A\|_{\infty,Y,X}\,\|\xi\|_Y + \|\hat{U}\|_{\infty,Z,Z}\,\|\hat{A}\xi_0\|_{\infty,Z}\right] \cdot (t - r)\ .$$

for all $\xi \in Y$, $\xi_0 \in Z_0$ and $(t, r) \in \triangle(I)$ where $\hat{A} \in PC_*(I\ , L((Z_0, \| \ \|_0), Z))$ is defined by $\hat{A}(t) := (\Phi A)(t) + B(t)$ for all $t \in I$ and $\hat{A}\xi_0 := (I \to Z, t \mapsto \hat{A}(t)\xi_0)$.

(ii) For every compact subset K of $(Y, \| \ \|_Y)$ and each $\varepsilon > 0$, there is $R \in [0, \infty)$ such that

$$\| (U(t, r) - \mathrm{id}_X)\,\xi\|_Y \leqslant \varepsilon + R \cdot (t - r)$$

for all $\xi \in K$ and all $(t, r) \in \triangle(I)$.

Proof. '(i)': For this, let $\xi \in Y$ and $(t, r) \in \triangle(I)$. Then it follows for every $s, s' \in [a, t]$

$$\|U(t, s')A(s')\xi - U(t, s)A(s)\xi\|_X$$
$$\leqslant \|U\|_{\infty,X,X} \cdot \|A(s')\xi - A(s)\xi\|_X + \|U(t, s')A(s)\xi - U(t, s)A(s)\xi\|_X$$

and hence that $([a, t] \to X, s \mapsto U(t, s)A(s)\xi)$ is bounded by $\|U\|_{\infty,X,X}\,\|A\|_{\infty,Y,X}\,\|\xi\|_Y$, continuous in the points of strong continuity of A in $[a, t]$ and hence also almost everywhere continuous. In particular, it follows by Theorem 3.2.4 (ii) and Theorem 3.2.5 from (9.0.14) that

$$\|U(t, r)\xi - \xi\|_X = \left\| \int_r^t U(t, s)A(s)\xi\,ds \right\|_X \leqslant \int_r^t \|U(t, s)A(s)\xi\|_X\,ds$$
$$\leqslant \|U\|_{\infty,X,X}\,\|A\|_{\infty,Y,X}\,\|\xi\|_Y \cdot (t - r)\ .$$

Further, by Lemma 8.5.4 it follows, in the following we use the notation from the proof of Theorem 9.0.6, that for every $\xi_0 \in Z_0$ and $\nu \in \mathbb{N}^*$

$$\int_r^t \hat{U}_\nu(t,s)\hat{A}_\nu(s)\xi_0\,ds = \xi_0 - \hat{U}_\nu(t,r)\xi_0 \tag{9.0.16}$$

where integration denotes weak Lebesgue integration with respect to $L(Z,\mathbb{K})$ and $\hat{A}_\nu : I \to L((Z_0,\|\ \|_0),Z)$ is defined by $\hat{A}_\nu(a_i) := (\Phi(A(a_i)) + B(a_i))|_{Z_0}$ for all $i \in \{0,\dots,n\}$ and for all $i \in \{0,\dots,n-1\}$

$$\hat{A}_\nu(t) := (\Phi(A(a_{i\nu k})) + B(a_{i\nu k}))\,|_{Z_0}\ ,\ a_{i\nu k} := a_i + \frac{k}{\nu}\,(a_{i+1} - a_i)$$

for all $t \in I_{i\nu k}$, $k \in \{1,\dots,\nu-1\}$,

$$\hat{A}_\nu(t) := (\Phi(A(a_{i\nu 0})) + B(a_{i\nu 0}))\,|_{Z_0}\ ,\ a_{i\nu 0} := a_i + \frac{1}{\nu}\,(a_{i+1} - a_i)$$

for all $t \in I_{i\nu 0}$. In the following, let $\hat{A} := \Phi A + B_0$ where $B_0 := (I \to L((Z_0,\|\ \|_0),Z),\ t \mapsto B(t)|_{Z_0})$. Note that $B_0 \in PC_*(I, L((Z_0,\|\ \|_0),Z))$ and hence that $\hat{A} \in PC_*(I, L((Z_0,\|\ \|_0),Z))$. In addition, $\hat{U}_\nu(t,\cdot) \circ \hat{A}_\nu\xi_0, \hat{U}(t,\cdot) \circ \hat{A}\xi_0$ are almost everywhere continuous where $\hat{A}_\nu\xi_0 := (I \to Z, t \mapsto \hat{A}_\nu(t)\xi_0)$ and $\hat{A}\xi_0 := (I \to Z, t \mapsto \hat{A}(t)\xi_0)$. Hence it follows by Theorem 3.2.5 that

$$\left\| \int_r^t \hat{U}_\nu(t,s)\hat{A}_\nu(s)\xi_0\,ds - \int_r^t \hat{U}(t,s)\hat{A}(s)\xi_0\,ds \right\|_Z$$
$$\leqslant \int_r^t \|\hat{U}_\nu(t,s)\hat{A}_\nu(s)\xi_0 - \hat{U}(t,s)\hat{A}(s)\xi_0\|_Z\,ds\ .$$

Because of

$$\|\hat{U}_\nu(t,s)\hat{A}_\nu(s)\xi_0 - \hat{U}(t,s)\hat{A}(s)\xi_0\|_Z \leqslant c\,e^{(|\mu|+c\|B\|_\infty)(b-a)}\,\|\hat{A}_\nu(s)\xi_0 - \hat{A}(s)\xi_0\|_Z$$
$$+ \|\hat{U}_\nu(t,s)\hat{A}(s)\xi_0 - \hat{U}(t,s)\hat{A}(s)\xi_0\|_Z \leqslant 4c\,e^{(|\mu|+c\|B\|_\infty)(b-a)}\,\|\hat{A}\|_{\infty,Z_0,Z}\,\|\xi_0\|_0\ ,$$

the uniform convergence of $\hat{A}_\nu\xi_0$ to $\hat{A}\xi_0$ and Lebesgue's dominated convergence Theorem, this implies that

$$\lim_{\nu\to\infty} \int_r^t \hat{U}_\nu(t,s)\hat{A}_\nu(s)\xi_0\,ds = \int_r^t \hat{U}(t,s)\hat{A}(s)\xi_0\,ds$$

and hence, finally, by (9.0.8) and (9.0.16) that

$$\int_r^t \hat{U}(t,s)\hat{A}(s)\xi_0\,ds = \xi_0 - \hat{U}(t,r)\xi_0\ .$$

Further, it follows for every $s, s' \in [a,t]$

$$\|\hat{U}(t,s')\hat{A}(s')\xi_0 - \hat{U}(t,s)\hat{A}(s)\xi_0\|_Z$$
$$\leqslant \|\hat{U}\|_{\infty,Z,Z} \cdot \|\hat{A}(s')\xi_0 - \hat{A}(s)\xi_0\|_Z + \|\hat{U}(t,s')\hat{A}(s)\xi_0 - \hat{U}(t,s)\hat{A}(s)\xi_0\|_Z$$

and hence that $([a,t] \to X, s \mapsto \hat{U}(t,s)\hat{A}(s)\xi_0)$ is bounded by

$$\|\hat{U}\|_{\infty,Z,Z}\, \|\hat{A}\|_{\infty,Z_0,Z}\, \|\xi_0\|_0 \ ,$$

is continuous in the points of strong continuity of $\hat{A}$ in $[a,t]$ and hence also almost everywhere continuous. In particular, it follows by Theorem 3.2.4 (ii) and Theorem 3.2.5

$$\begin{aligned} \|\hat{U}(t,r)\xi_0 - \xi_0\|_Z &= \left\| \int_r^t \hat{U}(t,s)\hat{A}(s)\xi_0\, ds \right\|_Z \leqslant \int_r^t \|\hat{U}(t,s)\hat{A}(s)\xi_0\|_Z\, ds \\ &\leqslant \|\hat{U}\|_{\infty,Z,Z}\, \|\hat{A}\xi_0\|_{\infty,Z} \cdot (t-r) \end{aligned}$$

where $\hat{A}\xi_0 := (I \to Z, t \mapsto \hat{A}(t)\xi_0)$. Hence it follows that

$$\begin{aligned} \|U(t,r)\xi - \xi\|_Y &\leqslant \|U(t,r)\xi - \xi\|_X + \|S\,U(t,r)\xi - S\xi\|_Z \\ &= \|U(t,r)\xi - \xi\|_X + \|\hat{U}(t,r)S\xi - S\xi\|_Z \\ &\leqslant \|U(t,r)\xi - \xi\|_X + \|(\hat{U}(t,r) - \mathrm{id}_Z)(S\xi - \xi_0)\|_Z + \|\hat{U}(t,r)\xi_0 - \xi_0\|_Z \\ &\leqslant (1 + \|\hat{U}\|_{\infty,Z,Z}) \cdot \|S\xi - \xi_0\|_Z \\ &\quad + \left[\|U\|_{\infty,X,X}\, \|A\|_{\infty,Y,X}\, \|\xi\|_Y + \|\hat{U}\|_{\infty,Z,Z}\, \|\hat{A}\xi_0\|_{\infty,Z}\right] \cdot (t-r)\ . \end{aligned}$$

'(ii)': For this, let K be a non-empty compact subset of $(Y, \| \ \|_Y)$, $\varepsilon > 0$ be given and $(t,r) \in \triangle(I)$. Then K is in particular bounded, i.e, there is $M > 0$ such that $\|\xi\|_Y \leqslant M$ for all $\xi \in K$. Hence by (i)

$$\begin{aligned} \|\,(U(t,r) - \mathrm{id}_X)\,\xi\|_Y &\leqslant (1 + \|\hat{U}\|_{\infty,Z,Z}) \cdot \|S\xi - \xi_0\|_Z \\ &\quad + \left[M\, \|U\|_{\infty,X,X}\, \|A\|_{\infty,Y,X} + \|\hat{U}\|_{\infty,Z,Z}\, \|\hat{A}\xi_0\|_{\infty,Z}\right] \cdot (t-r)\ . \end{aligned}$$

for every $\xi \in K$ and $\xi_0 \in Z_0$. Since $S : (Y, \| \ \|_Y) \to (Z, \| \ \|_Z)$ is continuous, $S(K)$ is a compact subset of Z. Hence for $\delta_1 > 0$, there are $n \in \mathbb{N}^*$ and $\xi_1, \ldots, \xi_n \in K$ such that the union of $U_{\delta_1,Z}(S\xi_1), \ldots, U_{\delta_1,Z}(S\xi_n)$ is covering $S(K)$. Since Z_0 is dense in Z, for every $\delta_2 > 0$ and every $i \in \{1,\ldots,n\}$, there is $\xi_{i0} \in Z_0$ such that $\|S\xi_i - \xi_{i0}\|_Z < \delta_2$. Then

$$\|S\xi - \xi_{i0}\|_Z < \delta_1 + \delta_2$$

and

$$\begin{aligned} \|\,(U(t,r) - \mathrm{id}_X)\,\xi\|_Y &\leqslant (1 + \|\hat{U}\|_{\infty,Z,Z}) \cdot (\delta_1 + \delta_2) \\ &\quad + \left[M\, \|U\|_{\infty,X,X}\, \|A\|_{\infty,Y,X} + \|\hat{U}\|_{\infty,Z,Z}\, \|\hat{A}\xi_{0i}\|_{\infty,Z}\right] \cdot (t-r) \\ &\leqslant (1 + \|\hat{U}\|_{\infty,Z,Z}) \cdot (\delta_1 + \delta_2) + [M\, \|U\|_{\infty,X,X}\, \|A\|_{\infty,Y,X} \\ &\quad + \|\hat{U}\|_{\infty,Z,Z} \max\{\|\hat{A}\xi_{0i}\|_{\infty,Z} : i \in \{1,\ldots,n\}\}] \cdot (t-r) \end{aligned}$$

for all $\xi \in S^{-1}(U_{\delta_i,Z}(\xi_i))$. □

10

Examples of Linear Evolution Equations

This chapter gives examples of applications of Theorem 9.0.6 to linear evolution equations from General Relativity and Astrophysics. For examples of applications of Kato's older results [107, 108] to such problems, see, for instance, [9, 10, 14, 66, 67, 80, 83, 84, 93, 94, 100, 115, 136, 143, 145] and to non-autonomous wave equations, see, for instance, [41, 173, 174].

The first example considers a scalar field of mass $m_0 \geqslant 0$ in the gravitational field of a spherical black hole. The field is required to vanish on a spherical surface which is in a spherically symmetric collapse into the black hole.[1] For this, Kruskal coordinates are used because of the singular behaviour of the Schwarzschild coordinates on the horizon. Due to its spherical symmetry, the problem can be separated by decomposition into spherical harmonic functions. Note that the situation considered by the example is at the basis of the analysis of Hawking radiation from a Schwarzschild black holes [96]. Theorem 10.1.6 is a new result. The second example considers non-autonomous Hermitian hyperbolic systems.

In both sections below, for the convenience of the reader, material is developed which is needed in the application of Theorem 9.0.6, but can also be found in some form in other sources. This includes fractional powers of infinitesimal generators [39,47,52,57,90,99,106,120,168,179,224], results from the theory of Sobolev spaces [2, 227], harmonic analysis [140, 141, 200, 201] and from the spectral theory of self-adjoint linear operators [53, 106, 179, 216]. Lemma 10.2.1 (vii) gives a result which is weaker than Calderon's first commutator [40, 140, 141, 201], but can be derived without the use of the methods singular integral operators. It has been included to keep the course self-contained and at the same time to limit the size of the section on non-autonomous linear Hermitian hyperbolic systems. The developed material is also used in Chapter 12 that gives examples of the application of Theorem 11.0.7 to quasi-linear evolution equations.

[1] For instance, the surface of a spherical star.

10.1 Scalar Fields in the Gravitational Field of a Spherical Black Hole

The reduced wave equation for a scalar field $\psi(v,u)Y_{lm}(\theta,\varphi)$ of mass $m_0 \geqslant 0$, where $l \in \mathbb{N}$, $m \in \{-l,-l+1,\dots,l\}$ and Y_{lm} is the corresponding spherically harmonic function, in the gravitational field of a spherical symmetric black hole of mass $M > 0$ is given by

$$\frac{\partial^2 \hat{\psi}}{\partial v^2} - \frac{\partial^2 \hat{\psi}}{\partial u^2} + \frac{32M^3}{r}\left[\frac{l(l+1)}{r^2} + \frac{2M}{r^3} + m_0^2\right] e^{-r/(2M)}\hat{\psi} = 0 \,. \tag{10.1.1}$$

Here (v,u,θ,φ), where

$$(v,u) \in \Omega := \{\mathbb{R}^2 : u^2 - v^2 > -1\} \,,\; \theta \in (0,\pi) \,,\; \varphi \in (-\pi,\pi) \,,$$

are the so called 'Kruskal coordinates', $r : \Omega \to (0,\infty)$ is defined by

$$r(u,v) := h^{-1}(u^2 - v^2)$$

for all $(v,u) \in \Omega$ where $h : (0,\infty) \to (-1,\infty)$ is defined by

$$h(x) := \left(\frac{x}{2M} - 1\right) e^{x/(2M)}$$

for all $x \in (0,\infty)$ and $\hat{\psi} := r\psi$. In particular, geometrical units are used where the speed of light and the gravitational constant have the value 1. From

$$h(x) = -1 + \frac{1}{4M^2}\int_0^x y\, e^{y/(2M)}\, dy \geqslant -1 + \frac{1}{4M^2}\int_0^x y\, dy = \frac{x^2}{8M^2} - 1 \,,$$

valid for all $x > 0$, it follows

$$r(u,v) = h^{-1}(u^2 - v^2) \leqslant 2^{3/2} M (1 + u^2 - v^2)^{1/2} \tag{10.1.2}$$

and from

$$\begin{aligned} h(x) &= -1 + \frac{1}{4M^2}\int_0^x y\, e^{y/(2M)}\, dy \leqslant -1 + \frac{1}{2M\delta}\int_0^x e^{(1+\delta)y/(2M)}\, dy \\ &= \frac{1}{\delta(1+\delta)}\left[e^{(1+\delta)x/(2M)} - 1\right] - 1 \,, \end{aligned}$$

valid for all $x > 0$, it follows

$$\begin{aligned} r(u,v) = h^{-1}(u^2 - v^2) &\geqslant \frac{2M}{1+\delta} \ln[1 + \delta(1+\delta)(1 + u^2 - v^2)] \\ &\geqslant \frac{2M\delta(1 + u^2 - v^2)}{1 + \delta(1+\delta)(1 + u^2 - v^2)} \end{aligned}$$

for every $\delta > 0$ and all $u, v \in \Omega$ where

$$\ln(1+x) = \int_0^x \frac{dy}{1+y} \geqslant \frac{x}{1+x}\,,$$

valid for $x > 0$, has been used. The last inequality gives for $\delta = (1+u^2-v^2)^{-1/2}$

$$r(u,v) \geqslant \frac{2M(1+u^2-v^2)^{1/2}}{2+(1+u^2-v^2)^{1/2}} \tag{10.1.3}$$

and for $\delta = 1$

$$e^{-r(u,v)/(2M)} \leqslant \frac{1}{[1+2(1+u^2-v^2)]^{1/2}}$$

for all $(u,v) \in \Omega$ and as a consequence

$$0 < \frac{1}{r(u,v)} \leqslant \frac{1}{M}\left[\frac{1}{2} + \frac{1}{(1-v^2)^{1/2}}\right]\,,\quad e^{-r(u,v)/(2M)} \leqslant \frac{1}{(1+u^2)^{1/2}}$$

for all $(u,v) \in \mathbb{R} \times (-1,1)$. In addition, note that

$$\frac{\partial r}{\partial v}(u,v) = -\frac{8M^2 v}{r(u,v)}\,e^{-r(u,v)/(2M)}\,,\quad \frac{\partial r}{\partial u}(u,v) = \frac{8M^2 u}{r(u,v)}\,e^{-r(u,v)/(2M)}\,,$$

$$\frac{\partial^2 r}{\partial v \partial u}(u,v) = -\frac{8M^2 u}{r(u,v)}\left(\frac{1}{r(u,v)} + \frac{1}{2M}\right)\frac{\partial r}{\partial v}(u,v)\,e^{-r(u,v)/(2M)}\,,$$

$$\frac{\partial^2 r}{\partial^2 u}(u,v) = \frac{8M^2}{r(u,v)}\,e^{-r(u,v)/(2M)} - \frac{8M^2 u}{r(u,v)}\left(\frac{1}{r(u,v)} + \frac{1}{2M}\right)\frac{\partial r}{\partial u}(u,v)\,e^{-r(u,v)/(2M)}$$

for all $(u,v) \in \Omega$.

We demand that $\hat{\psi}$ vanishes on a radially into the black hole falling trajectory

$$S := \{(f(v), v) : v \in (-1,1)\}$$

(boundary of a star) where $f \in C^2((-1,1), \mathbb{R})$ is such that

$$|f'(v)| < 1$$

for all $v \in (-1,1)$. From (10.1.1), it follows by the coordinate transformation

$$\bar{v}(u,v) := v\,,\ \ \bar{u}(u,v) := u - f(v)$$

for all $u, v \in \Omega$ the equation

$$\frac{\partial^2 \bar{\psi}}{\partial \bar{v}^2} - 2f'\,\frac{\partial^2 \bar{\psi}}{\partial \bar{u}\,\partial \bar{v}} - (1-f'^2)\,\frac{\partial^2 \bar{\psi}}{\partial \bar{u}^2} - f''\,\frac{\partial \bar{\psi}}{\partial \bar{u}} + \frac{32M^3}{\bar{r}}\left[\frac{l(l+1)}{\bar{r}^2} + \frac{2M}{\bar{r}^3} + m_0^2\right] e^{-\bar{r}/(2M)}\bar{\psi} = 0 \tag{10.1.4}$$

where

$$\bar{r}(\bar{u},\bar{v}) := r(\bar{u}+f(\bar{v}),\bar{v}) \ , \ \bar{\psi}(\bar{u},\bar{v}) := r(\bar{u}+f(\bar{v}),\bar{v})\,\psi(\bar{u}+f(\bar{v}),\bar{v})$$

for all $(\bar{u},\bar{v}) \in (0,\infty) \times (-1,1)$. Note that

$$\begin{aligned}
\frac{\partial \bar{r}}{\partial \bar{v}}(\bar{u},\bar{v}) &= f'(\bar{v})\,\frac{\partial r}{\partial u}(\bar{u}+f(\bar{v}),\bar{v}) + \frac{\partial r}{\partial v}(\bar{u}+f(\bar{v}),\bar{v}),\\
\frac{\partial \bar{r}}{\partial \bar{u}}(\bar{u},\bar{v}) &= \frac{\partial r}{\partial u}(\bar{u}+f(\bar{v}),\bar{v}),\\
\frac{\partial^2 \bar{r}}{\partial \bar{v}\partial \bar{u}}(\bar{u},\bar{v}) &= f'(\bar{v})\,\frac{\partial^2 r}{\partial^2 u}(\bar{u}+f(\bar{v}),\bar{v}) + \frac{\partial^2 r}{\partial v \partial u}(\bar{u}+f(\bar{v}),\bar{v})\ .
\end{aligned}$$

Lemma 10.1.1. Let $(X,\langle\,|\,\rangle_X)$ be a Hilbert space over $\mathbb{K} \in \{\mathbb{R},\mathbb{C}\}$, $A : D(A) \to X$ a densely-defined, linear, self-adjoint and positive operator and $B \in L(X,X)$ self-adjoint and positive. Then

$$D((A+B)^{1/2}) = D(A^{1/2})\ . \tag{10.1.5}$$

Proof. First, we notice that Theorem 3.1.9 implies that $D(A)$ is a core for $A^{1/2}$ and $(A+B)^{1/2}$ and hence that $D(A)$ is dense in $(D(A^{1/2}),\|\ \|_{A^{1/2}})$ and in $(D((A+B)^{1/2}),$ $\|\ \|_{D((A+B)^{1/2})})$. In particular, for $\xi \in D(A)$,

$$\begin{aligned}
\|A^{1/2}\xi\|^2_{A^{1/2}} &= \|\xi\|^2 + \langle\xi|A\xi\rangle \leqslant \|(A+B)^{1/2}\xi\|^2_{(A+B)^{1/2}}\\
&= \|\xi\|^2 + \langle\xi|(A+B)\xi\rangle \leqslant (1+\|B\|)\cdot(\|\xi\|^2+\langle\xi|A\xi\rangle)\\
&= (1+\|B\|)\cdot\|A^{1/2}\xi\|^2_{A^{1/2}}\ .
\end{aligned}$$

Hence the restrictions of $\|\ \|_{A^{1/2}}$ and $\|\ \|_{D((A+B)^{1/2})}$ to $D(A)$ are equivalent. Finally, from this follows (10.1.5) by the denseness of $D(A)$ in $(D(A^{1/2}),\|\ \|_{A^{1/2}})$, $(D((A+B)^{1/2}),\|\ \|_{D((A+B)^{1/2})})$, the completeness of these spaces and the continuity of their inclusions into X. □

Theorem 10.1.2. (Fractional powers of generators) Let $\mathbb{K} \in \{\mathbb{R},\mathbb{C}\}$, $(X,\|\ \|)$ a $\mathbb{K}$-Banach space and A the generator of a strongly continuous semigroup $T : [0,\infty) \to L(X,X)$ for which there are $c \in [1,\infty)$, $\mu < 0$ such that

$$\|T(t)\| \leqslant c\,e^{\mu t}$$

for all $t \in [0,\infty)$.

(i) Then by

$$A^{-\alpha} := \frac{1}{\Gamma(\alpha)}\int_0^\infty t^{\alpha-1}T(t)\,dt\ , \tag{10.1.6}$$

$\alpha > 0$, there is defined a bounded linear operator on X such that

$$\|A^{-\alpha}\| \leqslant \frac{c}{|\mu|^\alpha}\ .$$

In addition,

$$A^{-\alpha}A^{-\beta} = A^{-(\alpha+\beta)}$$

for all $\alpha, \beta > 0$,

$$\text{s}-\lim_{\alpha\to 0+} A^{-\alpha} = \text{id}_X \tag{10.1.7}$$

and

$$A^{-n} = [R_A(0)]^n \tag{10.1.8}$$

for every $n \in \mathbb{N}^*$. Hence by $P_A := ([0,\infty) \to L(X,X), \alpha \mapsto A^{-\alpha})$ defined by $P_A(0) := \text{id}_X$ and $P_A(\alpha) := A^{-\alpha}$ for $\alpha > 0$, there is given a strongly continuous semigroup whose infinitesimal generator will be denoted by $\ln(A)$.

(ii)

$$A^{-\alpha} = \frac{\sin(\pi\alpha)}{\pi}\int_0^\infty \lambda^{-\alpha}(A+\lambda)^{-1}\,d\lambda$$

for all $\alpha \in (0,1)$.

Proof. '(i)': By the strong continuity of $((0,\infty) \to L(X,X), t \mapsto t^{\alpha-1}T(t))$ and

$$\|t^{\alpha-1}T(t)\| \leqslant c\,t^{\alpha-1}\,e^{-|\mu|t},$$

for all $t > 0$ and Theorem 3.2.11, it follows that by (10.1.6) there is defined a bounded linear operator $A^{-\alpha}$ on X such that

$$\|A^{-\alpha}\| \leqslant \frac{c}{\Gamma(\alpha)}\int_0^\infty t^{\alpha-1}\,e^{-|\mu|t}\,dt = \frac{c}{|\mu|^\alpha}$$

for every $\alpha > 0$. For the following let $\alpha, \beta > 0$. By Lemma 8.1.1 (ii), it follows the strong continuity of $((0,\infty)^2 \to L(X,X), (t,s) \mapsto t^{\alpha-1}T(t)s^{\beta-1}T(s) = t^{\alpha-1}s^{\beta-1}T(t+s))$. Further,

$$\|t^{\alpha-1}s^{\beta-1}T(t+s)\| \leqslant t^{\alpha-1}s^{\beta-1} \leqslant t^{\alpha-1}s^{\beta-1}e^{-|\mu|(t+s)}$$

for all $t, s > 0$ and hence it follows by Theorem 3.2.11 that by

$$\int_{(0,\infty)^2} t^{\alpha-1}s^{\beta-1}T(t+s)\,dt\,ds$$

there is given a bounded linear operator $A^{-\alpha}$ on X. In particular, it follows by Fubini's theorem and Theorem 3.2.10 that

$$\begin{aligned}
A^{-\alpha}A^{-\beta}\xi &= \frac{1}{\Gamma(\beta)}\int_0^\infty s^{\beta-1}A^{-\alpha}T(s)\xi\,ds \\
&= \frac{1}{\Gamma(\alpha)\,\Gamma(\beta)}\int_0^\infty s^{\beta-1}\left(\int_0^\infty t^{\alpha-1}T(t+s)\xi\,dt\right)ds \\
&= \frac{1}{\Gamma(\alpha)\,\Gamma(\beta)}\int_{(0,\infty)^2} t^{\alpha-1}s^{\beta-1}T(t+s)\xi\,dt\,ds
\end{aligned}$$

$$= \frac{1}{\Gamma(\alpha)\,\Gamma(\beta)} \int_{s'>t'} t'^{\alpha-1}(s'-t')^{\beta-1} T(s')\xi\,dt'\,ds'$$

$$= \frac{1}{\Gamma(\alpha)\,\Gamma(\beta)} \int_0^\infty \left(\int_0^{s'} t'^{\alpha-1}(s'-t')^{\beta-1}\,dt' \right) T(s')\xi\,ds'$$

$$= \frac{1}{\Gamma(\alpha)\,\Gamma(\beta)} \int_0^\infty \left(s'^{\alpha+\beta-1} \int_0^1 t''^{\alpha-1}(1-t'')^{\beta-1}\,dt'' \right) T(s')\xi\,ds'$$

$$= \frac{1}{\Gamma(\alpha+\beta)} \int_0^\infty s'^{\alpha+\beta-1} T(s')\xi\,ds' = A^{-(\alpha+\beta)}\xi$$

for every $\xi \in X$. Further, the validness of (10.1.8) for every $n \in \mathbb{N}^*$ follows from Theorem 4.1.1 (iv). Since

$$\|A^{-\alpha}\| \leqslant \frac{c}{|\mu|^\alpha} \leqslant \frac{c}{M^\alpha} = c\,e^{-\alpha \ln(M)} \leqslant c\,e^{|\ln(M)|\,\alpha}$$

where $M := \min\{|\mu|, 1\} \in (0,1]$, it follows the uniform boundedness of $(A^{-\alpha})_{\alpha\in(0,1]}$. Therefore, since $D(A)$ is dense in X, (10.1.7) follows if it can be shown that

$$\lim_{\alpha\to 0+} A^{-\alpha}A^{-1}\xi = \lim_{\alpha\to 0+} A^{-(\alpha+1)}\xi = A^{-1}\xi \tag{10.1.9}$$

for all $\xi \in X$. Since for every $\xi \in X$

$$\|A^{-(\alpha+1)}\xi - A^{-1}\xi\|$$
$$\leqslant \frac{1}{\Gamma(\alpha+1)} \int_0^\infty |t^\alpha - 1|\,\|T(t)\xi\|\,dt + \left| \frac{1}{\Gamma(\alpha+1)} - \frac{1}{\Gamma(\alpha)} \right| \int_0^\infty \|T(t)\xi\|\,dt$$
$$\leqslant \frac{c\|\xi\|}{\Gamma(\alpha+1)} \int_0^\infty |t^\alpha - 1|\,e^{-|\mu|t}\,dt + \frac{c\|\xi\|}{|\mu|}\,\frac{|\Gamma(\alpha+1)-\Gamma(\alpha)|}{\Gamma(\alpha)\Gamma(\alpha+1)}\,,$$

(10.1.9) follows by an application Lebesgue's dominated convergence theorem.

'(ii)': For this, let $\alpha \in (0,1)$. By the strong continuity of $((0,\infty) \to L(X,X), \lambda \mapsto \lambda^{-\alpha}(A+\lambda)^{-1})$ and

$$\|\lambda^{-\alpha}(A+\lambda)^{-1}\| \leqslant \frac{c\,\lambda^{-\alpha}}{\lambda + |\mu|}\,,$$

for all $\lambda > 0$ and Theorem 3.2.11, it follows that by

$$B_\alpha := \frac{\sin(\pi\alpha)}{\pi} \int_0^\infty \lambda^{-\alpha}(A+\lambda)^{-1}\,d\lambda$$

there is defined a bounded linear operator B_α on X. By Lemma 8.1.1 (ii), it follows the strong continuity of $((0,\infty)^2 \to L(X,X), (t,\lambda) \mapsto \lambda^{-\alpha}e^{-\lambda t}\,T(t))$. Further,

$$\|\lambda^{-\alpha}e^{-\lambda t}\,T(t)\| \leqslant c\,\lambda^{-\alpha}e^{-(\lambda+|\mu|)t}$$

for all $t, \lambda > 0$, and hence it follows by Tonelli's theorem along with Theorem 3.2.11 that by

$$\int_{(0,\infty)^2} \lambda^{-\alpha} e^{-\lambda t} T(t)\, dt\, d\lambda$$

there is given a bounded linear operator on X. Finally, it follows by Fubini's theorem, Theorem 3.2.10 and Theorem 4.1.1 (iv) that

$$\begin{aligned}
B_\alpha \xi &:= \frac{\sin(\pi\alpha)}{\pi} \int_0^\infty \lambda^{-\alpha}(A+\lambda)^{-1}\xi\, d\lambda \\
&= \frac{\sin(\pi\alpha)}{\pi} \int_0^\infty \lambda^{-\alpha} \left(\int_0^\infty e^{-\lambda t} T(t)\xi\, dt \right) d\lambda \\
&= \frac{\sin(\pi\alpha)}{\pi} \int_{(0,\infty)^2} \lambda^{-\alpha} e^{-\lambda t} T(t)\xi\, dt\, d\lambda = \int_0^\infty \left(\int_0^\infty e^{-t\lambda}\lambda^{-\alpha}\, d\lambda \right) T(t)\xi\, dt\,, \\
& \frac{\sin(\pi\alpha)}{\pi} \int_0^\infty \left(\int_0^\infty e^{-\lambda'} \lambda'^{1-\alpha-1}\, d\lambda' \right) t^{\alpha-1} T(t)\xi\, dt \\
&= \frac{\Gamma(1-\alpha)}{\Gamma(\alpha)\Gamma(1-\alpha)} \int_0^\infty t^{\alpha-1} T(t)\xi\, dt = A^{-\alpha}\xi
\end{aligned}$$

for every $\xi \in X$. □

Lemma 10.1.3. Let $(X, \langle\,|\,\rangle_X)$ be a complex Hilbert space, $A : D(A) \to X$ a densely-defined, linear, self-adjoint and positive operator in X. Then

$$[(A+\delta)^{1/2}]^{-1} = \frac{1}{\pi} \int_0^\infty \lambda^{-1/2}(A+\lambda+\delta)^{-1}\, d\lambda$$

for all $\delta > 0$ and

$$A^{1/2}\xi = \frac{1}{\pi} \int_0^\infty \lambda^{-1/2}(A+\lambda)^{-1}A\xi\, d\lambda$$

for all $\xi \in D(A)$.

Proof. For this, let $\delta > 0$. Then $A + \delta$ is a densely-defined, linear, self-adjoint operator in X which is semibounded from below with lower bound δ. Hence it follows by Theorem 4.2.6 that $A + \delta$ is the infinitesimal generator of a strongly continuous semigroup $T : [0,\infty) \to L(X,X)$ such that

$$\|T(t)\| \leqslant \exp(-t\delta)$$

for all $t \in [0,\infty)$. Further, $(A+\delta)^{-1} \in L(X,X)$ is in particular positive self-adjoint and $(A+\delta)^{-1/2} \in L(X,X)$, defined as in Theorem 10.1.2, is positive symmetric and hence also self-adjoint. In addition,

$$[(A+\delta)^{-1/2}]^2 = (A+\delta)^{-1}$$

and hence

$$(A+\delta)^{-1/2} = [(A+\delta)^{-1}]^{1/2} = f(A+\delta) = \left(f \circ \mathrm{id}_{\mathbb{R}}^2\right)((A+\delta)^{1/2}) = [(A+\delta)^{1/2}]^{-1}$$

where f is the real-valued function defined on the spectrum $\sigma(A+\delta)$ of A by $f(\lambda) := \lambda^{-1/2}$ for all $\lambda \in \sigma(A+\delta)$. Hence it follows for every $\xi \in D(A)$ that

$$(A+\delta)^{1/2}\xi = (A+\delta)^{-1/2}(A+\delta)\xi = \frac{1}{\pi}\int_0^\infty \lambda^{-1/2}(A+\lambda+\delta)^{-1}(A+\delta)\xi \, d\lambda \, .$$

In the next step, it will we proved that

$$\lim_{\delta\to 0}(A+\delta)^{1/2}\xi = A^{1/2}\xi \, , \tag{10.1.10}$$

for every $\xi \in D(A^2)$ where $\nu \in \mathbb{N}^*$. First, we notice that

$$A^{1/2}(A+\delta)^{-1/2} \supset (A+\delta)^{-1/2}A^{1/2}$$

since A and $A^{1/2}$ commute and also that A and $A+\delta$ commute since the operators of their associated one-parameter unitary groups commute. Hence it follows for $\xi \in D(A^2)$ that

$$\begin{aligned} A^{1/2}(A+\delta)^{1/2}\xi &= A^{1/2}(A+\delta)^{-1/2}(A+\delta)\xi = (A+\delta)^{-1/2}A^{1/2}(A+\delta)\xi \\ &= (A+\delta)^{-1/2}(A+\delta)A^{1/2}\xi = (A+\delta)^{1/2}A^{1/2}\xi \end{aligned}$$

and hence

$$\begin{aligned} &[(A+\delta)^{1/2}+A^{1/2}][(A+\delta)^{1/2}-A^{1/2}]\xi \\ &\quad = (A+\delta)\xi - (A+\delta)^{1/2}A^{1/2}\xi + A^{1/2}(A+\delta)^{1/2}\xi - A\xi = \delta\xi \, . \end{aligned} \tag{10.1.11}$$

Further, $(A+\delta)^{1/2}+A^{1/2}$ is a densely-defined, linear and positive symmetric operator in X such that

$$\langle \xi | [(A+\delta)^{1/2}+A^{1/2}]\xi \rangle \geqslant \delta^{1/2}\|\xi\|^2 \tag{10.1.12}$$

for all $\xi \in D(A^{1/2})$. Since $A^{1/2}$ and $(A+\delta)^{1/2}$ commute, also their associated one-parameter unitary groups $U, U_\delta : \mathbb{R} \to L(X,X)$ commute. Hence $(\mathbb{R} \to L(X,X), t \mapsto U_\delta(t)U(t))$ is a strongly continuous unitary one-parameter group whose generator A_δ is an extension of $(A+\delta)^{1/2}+A^{1/2}$ and has $D(A^{1/2})$ as a core. Hence it follows by (10.1.12) that the spectrum of A_δ is contained in $[\delta^{1/2}, \infty)$ and therefore that

$$\|A_\delta^{-1}\| \leqslant \delta^{-1/2} \, .$$

As a consequence, it follows from (10.1.11) that

$$\|[(A+\delta)^{1/2}-A^{1/2}]\xi\| = \delta \, \|A_\delta^{-1}\xi\| \leqslant \delta^{1/2} \, \|\xi\| \tag{10.1.13}$$

for all $\xi \in D(A^2)$ and hence that (10.1.10) holds for every $\xi \in D(A)$. For later use we note that, since $D(A^2)$ is a core for $A^{1/2}$ and $(A+\delta)^{1/2}$, (10.1.13) is true for all $\xi \in D(A^{1/2})$. Further, let be $\xi \in D(A)$ and $\varepsilon \geqslant 0$. Then $h_{\xi,\varepsilon} : (0,\infty) \to D(A)$ defined by

$$h_{\xi,\varepsilon}(\lambda) := \pi^{-1}\,\lambda^{-1/2}(A+\lambda+\varepsilon)^{-1}A\xi$$

is continuous and satisfies

$$\|h_{\xi,\varepsilon}(\lambda)\| = \lambda^{-1/2}\|(A+\lambda+\varepsilon)^{-1}A\xi\| \leqslant \pi^{-1}\,\lambda^{-1/2}\,\frac{\|A\xi\|}{\lambda+\varepsilon} \leqslant \pi^{-1}\,\|A\xi\|\cdot\lambda^{-3/2}$$
$$\leqslant \pi^{-1}\,\|\xi\|_{\mathrm{A}}\,\lambda^{-3/2}$$
$$\|h_{\xi,\varepsilon}(\lambda)\| \leqslant \pi^{-1}\,\lambda^{-1/2}\,\|(A+\lambda+\varepsilon)^{-1}A\xi\|$$
$$\leqslant \pi^{-1}\,\lambda^{-1/2}\,\|\xi-(\lambda+\varepsilon)(A+\lambda+\varepsilon)^{-1}\xi\| \leqslant 2\pi^{-1}\,\lambda^{-1/2}\,\|\xi\|$$
$$\leqslant 2\pi^{-1}\,\lambda^{-1/2}\,\|\xi\|_{\mathrm{A}}$$

for every $\lambda > 0$. Hence $h_{\xi,\varepsilon}$ is weakly summable and by $A_{\varepsilon,1/2} : D(A) \to X$ defined by

$$A_{\varepsilon,1/2}\xi := \int_0^\infty h_{\xi,\varepsilon}(\lambda)\,d\lambda$$

for every $\xi \in D(A)$ there is given an element of $L((D(A), \|\ \|_{\mathrm{A}}), X)$ satisfying

$$\|A_{\varepsilon,1/2}\|_{\mathrm{A,X}} \leqslant C := \pi^{-1}\int_0^\infty \min\{\lambda^{-3/2}, 2\lambda^{-1/2}\}\,d\lambda\,.$$

Further, it follows for $\xi \in D(A)$, $\lambda > 0$ that

$$(A+\lambda+\delta)^{-1}(A+\delta)\xi = (A+\lambda)^{-1}(A+\lambda)\xi - \lambda(A+\lambda+\delta)^{-1}\xi$$
$$= (A+\lambda)^{-1}A\xi + \lambda\,[(A+\lambda)^{-1}\xi - (A+\lambda+\delta)^{-1}\xi]$$
$$= (A+\lambda)^{-1}A\xi - \lambda\delta\,(A+\lambda)^{-1}(A+\lambda+\delta)^{-1}\xi$$

and hence that

$$\|(A+\lambda+\delta)^{-1}(A+\delta)\xi - (A+\lambda)^{-1}A\xi\| \leqslant \delta\,(\lambda+\delta)^{-1}\|\xi\|$$

$$\|(A+\delta)^{1/2}\xi - A_{0,1/2}\xi\|$$
$$\leqslant \pi^{-1}\int_0^\infty \lambda^{-1/2}\,\|(A+\lambda+\delta)^{-1}(A+\delta)\xi - (A+\lambda)^{-1}A\xi\|\,d\lambda$$
$$\leqslant \pi^{-1}\delta\,\|\xi\|\int_0^\infty \lambda^{-1/2}\,(\lambda+\delta)^{-1}\,d\lambda = \pi^{-1}\delta^{1/2}\,\|\xi\|\int_0^\infty \bar{\lambda}^{-1/2}\,(\bar{\lambda}+1)^{-1}\,d\bar{\lambda}$$

and therefore that

$$\lim_{\delta\to 0}\|(A+\delta)^{1/2}\xi - A_{0,1/2}\xi\| = 0\,. \tag{10.1.14}$$

From (10.1.10), (10.1.14), it follows that

$$A^{1/2}\xi = A_{0,1/2}\xi$$

for every $\xi \in D(A^2)$. Finally, it follows by an application of Theorem 3.1.6 that also $A^{1/2}|_{D(A)} \in L((D(A), \|\ \|_{\mathrm{A}}), X)$ and therefore, since $D(A^2)$ is according to Theorem 6.1.1 a core for A, that

$$A^{1/2}|_{D(A)} = A_{0,1/2}\,.$$

□

Lemma 10.1.4. Let $I := (0, \infty)$. For every $k \in \mathbb{N}^*$, we define the derivative operators D_I^k, the Sobolev spaces $W_{\mathbb{C}}^k(I)$ with corresponding scalar products $\langle\,,\rangle_k$ and their induced norms $|||\;|||_k$ as in Definition 5.3.1. In addition, we define $W_{0,\mathbb{C}}^k(I)$ for every $k \in \mathbb{N}^*$ as the closure of $C_0^k(I, \mathbb{C})$ in $(W_{\mathbb{C}}^k(I), |||\;|||_k)$. Then

(i)
$$A := -D_I^{2*}|_{W_{0,\mathbb{C}}^1(I) \cap W_{\mathbb{C}}^2(I)}$$
is a densely-defined, linear, self-adjoint and positive operator in $L_{\mathbb{C}}^2(I)$.

(ii) For every $\lambda > 0$, $f \in L_{\mathbb{C}}^2(I)$
$$[R_{\mathrm{A}}(-\lambda)f]\,(x) = \frac{1}{\sqrt{\lambda}}\left[e^{-\sqrt{\lambda}x}\int_0^x \sinh(\sqrt{\lambda}y)f(y)\,dy \right.$$
$$\left. + \sinh(\sqrt{\lambda}x)\int_x^\infty e^{-\sqrt{\lambda}y}f(y)\,dy\right] = \int_0^\infty K_{-\lambda}(x,y)f(y)\,dy$$
for all $x \in I$ where
$$K_{-\lambda}(x,y) := \frac{1}{\sqrt{\lambda}}\begin{cases} e^{-\sqrt{\lambda}x}\sinh(\sqrt{\lambda}y) & \text{if } y < x \\ \sinh(\sqrt{\lambda}x)\,e^{-\sqrt{\lambda}y} & \text{if } y \geqslant x\,. \end{cases}$$
for all $(x, y) \in I^2$.

(iii)
$$D_I^{1*}|_{W_{0,\mathbb{C}}^1(I)}$$
is a densely-defined, linear and closed operator in $L_{\mathbb{C}}^2(I)$ such that
$$\langle f|D_I^* g\rangle_2 = -\langle D_I^* f|g\rangle_2$$
for all $(f, g) \in (W_{0,\mathbb{C}}^1(I))^2$.

(iv) Further, $W_{0,\mathbb{C}}^1(I) \cap W_{\mathbb{C}}^2(I)$ is a core for $D_I^{1*}|_{W_{0,\mathbb{C}}^1(I)}$ and $A^{1/2}$,
$$D(A^{1/2}) = W_{0,\mathbb{C}}^1(I) \quad \text{and} \quad \|A^{1/2}f\|_2 = \|D_I^{1*}f\|_2$$
for all $f \in W_{0,\mathbb{C}}^1(I)$.

(v) For every $\lambda > 0$,
$$D_I^{1*}R_{\mathrm{A}}(-\lambda) \supset R_{\mathrm{A}}(-\lambda)D_I^{1*} \tag{10.1.15}$$
and for every $\delta > 0$
$$D_I^{1*}[(A+\delta)^{1/2}]^{-1} \supset [(A+\delta)^{1/2}]^{-1}D_I^{1*}\,. \tag{10.1.16}$$
as well as
$$D_I^{1*}(A+\delta)^{1/2}f = (A+\delta)^{1/2}D_I^{1*}f\,. \tag{10.1.17}$$
for all $f \in D(A)$.

(vi) (**A commutator estimate**) Let $g \in C^1(I, \mathbb{C}) \cap L^\infty_{\mathbb{C}}(I)$ such that $g' \in L^\infty_{\mathbb{C}}(I)$, and denote by T_g the corresponding maximal multiplication operator in $L^2_{\mathbb{C}}(I)$ which is a bounded linear operator on $L^2_{\mathbb{C}}(I)$. Finally, let $\delta > 0$. Then

$$(A+\delta)^{1/2} T_g (A+\delta)^{-1/2} \in L(L^2_{\mathbb{C}}(I), L^2_{\mathbb{C}}(I))$$

and in particular

$$\|(A+\delta)^{1/2} T_g (A+\delta)^{-1/2}\| \leqslant 3\|g\|_\infty + \delta^{-1/2}\|g'\|_\infty . \tag{10.1.18}$$

Proof. '(i)': First, obviously, $A_0 : C^2_0(I, \mathbb{C}) \to L^2_{\mathbb{C}}(I)$ defined by $A_0 f := -f''$ for every $f \in C^2_0(I, \mathbb{C})$ is a densely-defined, linear, symmetric and positive operator in $L^2_{\mathbb{C}}(I)$. In addition, the deficiency subspaces of A_0 are given by

$$[\mathrm{Ran}(A_0 - i)]^\perp = \mathbb{C}.f_i \ , \ [\mathrm{Ran}(A_0 + i)]^\perp = \mathbb{C}.f_{-i}$$

where $f_i : (0, \infty) \to \mathbb{R}$, $f_{-i} : (0, \infty) \to \mathbb{R}$ are defined by $f_i(x) := \exp(-(1+i)\, x/\sqrt{2}))$ and $f_{-i}(x) := \exp(-(1-i)x/\sqrt{2}))$ for all $x \in (0, \infty)$. (For example, see [179], Volume II, X.1, Example 2.) Further, A is a linear extension of A_0. We note that

$$\langle f | D^{2*}_I g \rangle_2 = -\langle D^*_I f | D^*_I g \rangle_2 \tag{10.1.19}$$

for all $(f, g) \in W^1_{0,\mathbb{C}}(I) \times W^2_{\mathbb{C}}(I)$, which can be seen as follows. For this, we define the sesquilinear form $s_1 : W^1_{0,\mathbb{C}}(I) \times W^2_{\mathbb{C}}(I) \to \mathbb{C}$ by

$$s_1(f, g) := \langle f | D^{2*}_I g \rangle_2 + \langle D^*_I f | D^*_I g \rangle_2$$

for all $(f, g) \in W^1_{0,\mathbb{C}}(I) \times W^2_{\mathbb{C}}(I)$. By the continuity of D^*_I, D^{2*}_I follows the continuity of s_1. Further,

$$\langle f | D^{2*}_I g \rangle_2 = \langle f'' | g \rangle_2 = \langle f' | D^*_I g \rangle_2 = -\langle D^*_I f | D^*_I g \rangle_2$$

for all $f \in C^\infty_0(I, \mathbb{C})$ and $g \in W^2_{\mathbb{C}}(I)$. Since $C^\infty_0(I, \mathbb{C}) \times W^2_{\mathbb{C}}(I)$ is dense in $W^1_{0,\mathbb{C}}(I) \times W^2_{\mathbb{C}}(I)$, this implies that s_1 vanishes and hence that (10.1.19) holds for all $(f, g) \in W^1_{0,\mathbb{C}}(I) \times W^2_{\mathbb{C}}(I)$. Obviously, (10.1.19) implies that A is a symmetric and positive. Further, we note that

$$\langle f | D^*_I g \rangle_2 = -\langle D^*_I f | g \rangle_2 \tag{10.1.20}$$

for all $(f, g) \in W^1_{0,\mathbb{C}}(I) \times W^1_{\mathbb{C}}(I)$, which can be seen as follows. For this, we define the sesquilinear form $s_2 : W^1_{0,\mathbb{C}}(I) \times W^1_{\mathbb{C}}(I) \to \mathbb{C}$ by

$$s_2(f, g) := \langle f | D^*_I g \rangle_2 + \langle D^*_I f | g \rangle_2$$

for all $(f, g) \in W^1_{0,\mathbb{C}}(I) \times W^1_{\mathbb{C}}(I)$. By the continuity of D^*_I, it follows the continuity of s_2 and by partial integration that $s_2(f, g) = 0$ for all $f \in C^\infty_0(I, \mathbb{C})$ and $f \in C^\infty(I, \mathbb{C}) \cap W^1_{\mathbb{C}}(I)$. Since $C^\infty_0(I, \mathbb{C}) \times ((C^\infty(I, \mathbb{C}) \cap W^1_{\mathbb{C}}(I))$ is dense in $W^1_{0,\mathbb{C}}(I) \times W^1_{\mathbb{C}}(I)$, this implies the vanishing of s_2 and hence the validity of (10.1.20) for all

$(f, g) \in W^1_{0,\mathbb{C}}(I) \times W^1_{\mathbb{C}}(I)$. In the next step, we conclude that A is closed. For this, let $(f, g) \in G(\bar{A})$ and $f_0, f_1, \dots$ a sequence in $D(A)$ converging to f and such that $Af_1, Af_2, \dots$ is converging to g. Since

$$-\langle f_\nu | D_I^{2*} f_\nu \rangle_2 = \| D_I^* f_\nu \|_2^2$$

for all $\nu \in \mathbb{N}$, this implies that $f_0, f_1, \dots$ is a Cauchy sequence in $(W^1_{0,\mathbb{C}}(I), ||| \, |||_1)$, $(W^2_{\mathbb{C}}(I), ||| \, |||_2)$, and hence by the completeness of $(W^1_{0,\mathbb{C}}(I), ||| \, |||_1)$, $(W^2_{\mathbb{C}}(I), ||| \, |||_2)$, the continuity of the inclusions of $(W^1_{\mathbb{C}}(I), ||| \, |||_1)$ $(W^2_{\mathbb{C}}(I), ||| \, |||_2)$ into $L^2_{\mathbb{C}}(I)$ and the continuity of D_I^{2*} that $f \in D(A)$, $Af = g$ and hence $(f, g) \in G(A)$. In the final step, we prove that the deficiency subspaces of A are trivial and hence that A is self-adjoint. For this, we define the function $f \in C_0^\infty(I, \mathbb{R})$ by

$$f(x) := \sin(x/\sqrt{2})\, e^{-x/\sqrt{2}}$$

for all $x \in I$. Then

$$f'(x) = \frac{1}{\sqrt{2}} \left(\cos(x/\sqrt{2}) - \sin(x/\sqrt{2}) \right) e^{-x/\sqrt{2}}, \; f''(x) = -\cos(x/\sqrt{2})\, e^{-x/\sqrt{2}}$$

for all $x \in \mathbb{R}$ and hence $f \in W^2_{\mathbb{C}}(I)$. In addition, $f \in W^1_{0,\mathbb{C}}(I)$. For the proof let $h \in C^\infty(\mathbb{R}, \mathbb{R})$ be an auxiliary function such that

$$h(x) \begin{cases} = 0 & \text{if } x < -2 \\ \in [0,1] & \text{if } -2 \leqslant x \leqslant -1 \\ = 1 & \text{if } x > -1 \, . \end{cases}$$

Obviously, such a function is easy to construct. We define $h_\nu \in C^\infty(I, \mathbb{R})$ by

$$h_\nu(x) := h\left(-(x - x^{-1})^2/\nu^2\right)$$

for all $x \in I$ and $\nu \in \mathbb{N}^*$. Then

$$h_\nu(x) = \begin{cases} 0 & \text{if } x < \left(1 + \sqrt{2}\,\nu\right)^{-1} \text{ or if } x > 1 + \sqrt{2}\,\nu \\ 1 & \text{if } \nu^{-1} < x < \nu \end{cases}$$

and hence $h_\nu \in C_0^\infty(I, \mathbb{R})$ as well as $\operatorname{Ran} h_\nu \subset [0, 1]$ for all $\nu \in \mathbb{N}^*$. In particular,

$$\begin{aligned} |x\, h_\nu'(x)| &\leqslant \frac{2(x^4+1)}{\nu^2 x^2} \left| h'\left(-(x - x^{-1})^2/\nu^2\right) \right| \\ &\leqslant \|h'\|_\infty \frac{2(x^4+1)}{\nu^2 x^2} \left(\chi_{[(1+\sqrt{2}\nu)^{-1}, \nu^{-1}]}(x) + \chi_{[\nu, 1+\sqrt{2}\nu]}(x) \right) \\ &\leqslant 2\,(\sqrt{2} + \nu^{-1})^2 \, \|h'\|_\infty \, (x^4+1)\, \chi_{[(1+\sqrt{2}\nu)^{-1}, \nu^{-1}]}(x) \\ &\quad + 2\,\nu^{-4} \, \|h'\|_\infty \, (x^4+1)\, \chi_{[\nu, 1+\sqrt{2}\nu]}(x) \\ &\leqslant 2\,(1 + \sqrt{2})^2 \, \|h'\|_\infty \, (x^4+1) \end{aligned} \tag{10.1.21}$$

for all $x \in I$ and $\nu \in \mathbb{N}^*$. An application of Lebesgue's dominated convergence leads to

$$\lim_{\nu\to\infty} \|h_\nu f - f\|_2 = \lim_{\nu\to\infty} \|(h_\nu f)' - f'\|_2 = 0$$

and hence to the fact that $f \in W^1_{0,\mathbb{C}}(I)$. Finally,

$$-f''(x) \pm i f(x) = f_{\mp i}(x)$$

for all $x \in I$ implies that the deficiency subspaces of A are trivial and hence that A is also self-adjoint.

'(ii)': For this, let $\lambda > 0$. Then it follows that $K_\lambda \in C(I^2, \mathbb{R})$ and hence that K_λ is measurable. In addition, $K(x,\cdot) \in L^1_{\mathbb{C}}(I)$ is such that

$$\|K(x,\cdot)\|_1 = \frac{1}{\lambda}\left(1 - e^{-\sqrt{\lambda}x}\right) \leqslant \frac{1}{\lambda}$$

for all $x \in I$. Since K is in particular symmetric, this implies that by

$$[B_\lambda f](x) := \int_0^\infty K_\lambda(x,y) f(y)\, dy$$

for every $x > 0$ and $f \in L^2_{\mathbb{C}}(I)$, there is defined a bounded linear operator B_λ on $L^2_{\mathbb{C}}(I)$ satisfying

$$\|B_\lambda\| \leqslant \frac{1}{\lambda}\,.$$

In addition, it follows for every $f \in C_0(I, \mathbb{C})$ that $B_\lambda f \in C^2(I, \mathbb{C})$ and

$$[B_\lambda f]'(x) = \int_0^\infty G_\lambda(x,y) f(y)\, dy\,,$$

for every $x > 0$ where

$$G_{-\lambda}(x,y) := \begin{cases} -e^{-\sqrt{\lambda}x}\sinh(\sqrt{\lambda}y) & \text{if } y < x \\ \cosh(\sqrt{\lambda}x)\, e^{-\sqrt{\lambda}y} & \text{if } y \geqslant x\,. \end{cases}$$

for all $(x,y) \in I^2$ as well as that

$$[B_\lambda f]''(x) = \lambda[B_\lambda f](x) - f(x) \tag{10.1.22}$$

for all $x > 0$. Further, G_λ is measurable, $G(x,\cdot), G(\cdot,y) \in L^1_{\mathbb{C}}(I)$ and

$$\|G(x,\cdot)\|_1 = \frac{1}{\sqrt{\lambda}} e^{-2\sqrt{\lambda}x} \leqslant \frac{1}{\sqrt{\lambda}}\,,\ \|G(\cdot,y)\|_1 = \frac{1}{\sqrt{\lambda}}\left(1 - e^{-2\sqrt{\lambda}y}\right) \leqslant \frac{1}{\sqrt{\lambda}}$$

for all $x, y \in I$. This implies that by

$$[C_\lambda f](x) := \int_0^\infty G_\lambda(x,y) f(y)\, dy$$

for every $x > 0$ and $f \in L^2_{\mathbb{C}}(I)$ there is defined a bounded linear operator C_λ on $L^2_{\mathbb{C}}(I)$ satisfying

$$\|C_\lambda\| \leqslant \frac{1}{\sqrt{\lambda}} .$$

Hence it follows for every $f \in C_0(I, \mathbb{C})$ that $B_\lambda f \in W^2_{\mathbb{C}}(I)$ and by (10.1.21) that $B_\lambda f \in W^1_{0,\mathbb{C}}(I)$ and therefore by (10.1.22) that

$$(A + \lambda)B_\lambda f = f .$$

Finally, since $A + \lambda$ is bijective and $C_0(I, \mathbb{C})$ is dense in $L^2_{\mathbb{C}}(I)$, this leads to

$$R_A(-\lambda) = B_\lambda .$$

'(iii)': First, it follows by the completeness of $W^1_{0,\mathbb{C}}(I)$, the continuity of the canonical imbedding of $W^1_{0,\mathbb{C}}(I)$ into $L^2_{\mathbb{C}}(I)$, the continuity of D^*_I and the denseness of $W^1_{0,\mathbb{C}}(I)$ in $L^2_{\mathbb{C}}(I)$ that the restriction D^*_{I0} of D^*_I to $W^1_{0,\mathbb{C}}(I)$ defines a densely-defined, linear and closed operator in $L^2_{\mathbb{C}}(I)$. Further, it follows by (10.1.20) that

$$2\,\mathrm{Re}\,(\langle f|D^*_I f\rangle_2) = \langle f|D^*_I f\rangle_2 + \langle D^*_I f|f\rangle_2 = -\langle D^*_I f|f\rangle_2 + \langle f|D^*_I f\rangle_2 = 0$$

for all $f \in W^1_{0,\mathbb{C}}(I)$.
'(iv)': By Theorem 3.1.9, it follows that $D(A)$ is a core for $A^{1/2}$ and hence that $D(A)$ is dense in $(D(A^{1/2}), \| \|_{A^{1/2}})$. Further, by (10.1.19), it follows

$$\|A^{1/2} f\|_2 = (\langle f|Af\rangle_2)^{1/2} = \|D^*_I f\|_2$$

for all $f \in D(A)$. In addition, the denseness of $D(A)$ in $W^1_{0,\mathbb{C}}(I)$ and the continuity of D^*_I imply that $D(A)$ is a core for D^*_I, too. Hence it follows that $D(A^{1/2}) = W^1_{0,\mathbb{C}}(I)$ and that $\|A^{1/2} f\|_2 = \|D^{1*}_I f\|_2$ for all $f \in W^1_{0,\mathbb{C}}(I)$.
'(v)': For this, let $\lambda > 0$. Then

$$R_A(-\lambda) D^{1*}_I \in L(W^1_{0,\mathbb{C}}(I), L^2_{\mathbb{C}}(I)) .$$

Further, it follows by direct calculation that

$$R_A(-\lambda) D^{1*}_I f = -R_A(-\lambda) f' = -\int_0^\infty G_\lambda(x, y) f(y)\, dy$$

for $f \in C^\infty_0(I, \mathbb{C})$ and therefore by the denseness of $C^\infty_0(I, \mathbb{C})$ in $W^1_{0,\mathbb{C}}(I)$, the continuity of the inclusion of $W^1_{0,\mathbb{C}}(I)$ into $L^2_{0,\mathbb{C}}(I)$ and $C_\lambda \in L(L^2_{0,\mathbb{C}}(I), L^2_{0,\mathbb{C}}(I))$ that

$$R_A(-\lambda) D^{1*}_I f = -R_A(-\lambda) f' = -\int_0^\infty G_\lambda(x, y) f(y)\, dy$$

for all $f \in W^1_{0,\mathbb{C}}(I)$. Finally, since $W^1_{\mathbb{C}}(I) \subset C(I, \mathbb{C})$, it follows by the proof of (ii) that

$$D^{1*}_I R_A(-\lambda) f = -\int_0^\infty G_\lambda(x, y) f(y)\, dy$$

for all $f \in W^1_{0,\mathbb{C}}(I)$ and therefore (10.1.15). In the following, let $f \in W^1_{0,\mathbb{C}}(I)$. Then $h_f := (I \to L^2_{0,\mathbb{C}}(I), \lambda \mapsto R_{\mathrm{A}}(-\lambda)f)$ is continuous and satisfies

$$\|h_f\|_2 \leqslant \frac{\|f\|_2}{\lambda}$$

for all $\lambda > 0$. In addition, it follows for for $\lambda, \mu > 0$ that

$$\begin{aligned}\|D_I^{1*}(h_f(\lambda) - (h_f(\mu))\|_2 &= \|D_I^{1*}R_{\mathrm{A}}(-\lambda)f - D_I^{1*}R_{\mathrm{A}}(-\mu)f\|_2 \\ &= \|R_{\mathrm{A}}(-\lambda)D_I^{1*}f - R_{\mathrm{A}}(-\mu)D_I^{1*}f\|_2 \\ &= |\mu - \lambda|\,\|R_{\mathrm{A}}(-\lambda)R_{\mathrm{A}}(-\mu)D_I^{1*}f\|_2 \leqslant \|D_I^{1*}f\|_2 \left|\frac{1}{\lambda} - \frac{1}{\mu}\right|\end{aligned}$$

and

$$\|D_I^{1*}h_f(\lambda)\|_2 = \|D_I^{1*}R_{\mathrm{A}}(-\lambda)f\|_2 = \|R_{\mathrm{A}}(-\lambda)D_I^{1*}f\|_2 \leqslant \frac{\|D_I^{1*}f\|_2}{\lambda} .$$

As a consequence, it follows for $\delta > 0$ that $k_f := (I \to W^1_{0,\mathbb{C}}(I), \lambda \mapsto \lambda^{-1/2}R_{\mathrm{A}}(-(\lambda + \varepsilon))f)$ is continuous such that

$$|||k_f(\lambda)|||_1 \leqslant \lambda^{-1/2}\,(\lambda + \delta)^{-1}\,|||f|||_1$$

and hence the summability of k_f. Hence it follows by the continuity of D_I^{1*} that

$$\begin{aligned}[(A+\delta)^{1/2}]^{-1}D_I^{1*}f &= \frac{1}{\pi}\int_0^\infty \lambda^{-1/2}(A + \lambda + \delta)^{-1}\,D_I^{1*}f\,d\lambda \\ &= \frac{1}{\pi}\int_0^\infty \lambda^{-1/2}D_I^{1*}(A + \lambda + \delta)^{-1}\,f\,d\lambda \\ &= D_I^{1*}\,\frac{1}{\pi}\int_0^\infty \lambda^{-1/2}(A + \lambda + \delta)^{-1}\,f\,d\lambda = D_I^{1*}[(A+\delta)^{1/2}]^{-1}f\end{aligned}$$

and therefore (10.1.16). Finally, it follows by (10.1.16) for every $f \in D(A)$ that

$$D_I^{1*}(A+\delta)^{1/2}f = (A+\delta)^{1/2}\,[(A+\delta)^{1/2}]^{-1}D_I^{1*}(A+\delta)^{1/2}f = (A+\delta)^{1/2}D_I^{1*}f$$

and hence (10.1.17).
'(vi)': For this, let $g \in C^1(I, \mathbb{C}) \cap L^\infty_{\mathbb{C}}(I)$ such that $g' \in L^\infty_{\mathbb{C}}(I)$ and $f \in W^1_{0,\mathbb{C}}(I)$. Then

$$gf \in W^1_{\mathbb{C}}(I)$$

and

$$D_I^{1*}gf = gD_I^{1*}f - g'f .$$

Further, if $f_1, f_2, \ldots$ is a sequence in $C^1_0(I, \mathbb{C})$ such that

$$\lim_{\nu \to \infty} |||f_\nu - f|||_1 = 0 ,$$

then $gf_1, gf_2, \dots$ is a sequence in $C_0^1(I, \mathbb{C})$ such that

$$\lim_{\nu\to\infty} |||gf_\nu - gf|||_1 = 0$$

since $(W^1_{\mathbb{C}}(I) \to W^1_{\mathbb{C}}(I), h \mapsto gh)$ is a continuous linear map. Hence $gf \in W^1_{0,\mathbb{C}}(I)$. Further, it follows for $\delta > 0$ by (10.1.13) that

$$\begin{aligned}\|(A+\delta)^{1/2}gf\|_2 &\leqslant \|A^{1/2}gf\|_2 + \delta^{1/2}\|gf\|_2 \leqslant \|D_I^{1*}gf\|_2 + \delta^{1/2}\|g\|_\infty\|f\|_2\\ &\leqslant \|g\|_\infty\|A^{1/2}f\|_2 + \left(\|g'\|_\infty + \delta^{1/2}\|g\|_\infty\right)\|f\|_2\\ &\leqslant \|g\|_\infty\|(A+\delta)^{1/2}f\|_2 + \left(\|g'\|_\infty + 2\delta^{1/2}\|g\|_\infty\right)\|f\|_2\end{aligned}$$

and hence for every $f \in L^2_{0,\mathbb{C}}(I)$ that

$$\begin{aligned}&\|(A+\delta)^{1/2}g(A+\delta)^{-1/2}f\|_2\\ &\leqslant \|g\|_\infty\|f\|_2 + \left(\|g'\|_\infty + 2\delta^{1/2}\|g\|_\infty\right)\|(A+\delta)^{-1/2}f\|_2\\ &\leqslant \left[3\|g\|_\infty + \delta^{-1/2}\|g'\|_\infty\right]\|f\|_2 .\end{aligned}$$

□

Lemma 10.1.5. Let $n \in \mathbb{N}^*$ and J, U be non-empty open subsets of $\mathbb{R}$ and $\mathbb{R}^n$, respectively. In addition, let $f \in C^1(J \times U, \mathbb{C})$, $h \in C(J, \mathbb{R})$ such that $f(t,\cdot), f_{,1}(t,\cdot) \in BC(U, \mathbb{C})$ and

$$\|f_{,1}(t,\cdot)\|_\infty \leqslant h(t)$$

for all $t \in J$. Then $(J \to L^\infty_{\mathbb{C}}(U), t \mapsto f(t,\cdot))$ is continuous.

Proof. For this, let $t_1, t_2 \in J$, $u \in U$. Then it follows by the mean value theorem that

$$\begin{aligned}|f(t_1,u) - f(t_2,u)| &\leqslant \sup\{|f_{,1}(t_1 + \lambda(t_2 - t_1), u)| : \lambda \in [0,1]\} \cdot |t_1 - t_2|\\ &\leqslant \max\{|h(t_1 + \lambda(t_2 - t_1))| : \lambda \in [0,1]\} \cdot |t_1 - t_2|\end{aligned}$$

and hence that

$$\|f(t_1,\cdot) - f(t_2,\cdot)\|_\infty \leqslant \max\{|h(t_1 + \lambda(t_2 - t_1))| : \lambda \in [0,1]\} \cdot |t_1 - t_2| .$$

□

Theorem 10.1.6. Let $\varepsilon > 0$, $I := (0, \infty)$, $f : (-1,1) \to \mathbb{R}$ twice differentiable with a piecewise continuous f'', $f'(0) = 0$ and

$$|f'(\bar{v})| < \frac{1}{5}\sqrt{5}$$

for every $\bar{v} \in (-1,1)$, $U : (-1,1) \to L^\infty_{\mathbb{C}}(I)$ piecewise continuous such that, $U(\bar{v}) \in C^1(I, \mathbb{C})$ and $(U(\bar{v}))' \in L^\infty_{\mathbb{C}}(I)$ for every $\bar{v} \in (-1,1)$ and such that $U_{,\bar{u}} := ((-1,1) \to L^\infty_{\mathbb{C}}(I), \bar{v} \mapsto (U(\bar{v}))')$ is piecewise continuous,

$$A_{\bar{v}} := -(1 - f'^2(\bar{v}))\, D_I^{2*}|_{W^1_{0,\mathbb{C}}(I)\cap W^2_{\mathbb{C}}(I)} + \varepsilon\ ,\ B_{\bar{v}} := -2if'(\bar{v}))\, D_I^{1*}|_{W^1_{0,\mathbb{C}}(I)}\ ,$$
$$C_{\bar{v}} := -f''(\bar{v})\, D_I^{1*}|_{W^1_{0,\mathbb{C}}(I)} + U(\bar{v}) - \varepsilon$$

for every $\bar{v} \in (-1, 1)$ and

$$X := W^1_{0,\mathbb{C}}(I) \times L^2_{\mathbb{C}}(I)\ .$$

(i) Let $\bar{v} \in (-1, 1)$. Then X equipped with $(\,|\,)_{\bar{v}} : X^2 \to \mathbb{C}$ defined by

$$((g_1, g_2)|(g_3, g_4))_{\bar{v}} := \langle A_{\bar{v}}^{1/2} g_1 | A_{\bar{v}}^{1/2} g_3\rangle + \langle g_2 | g_4\rangle_2$$

for all $(g_1, g_2), (g_3, g_4) \in X$ is a complex Hilbert space. Further,

$$G_{\bar{v}} : Y \to X\ ,$$

where

$$Y := \left(W^1_{0,\mathbb{C}}(I) \cap W^2_{\mathbb{C}}(I)\right) \times W^1_{0,\mathbb{C}}(I)\ ,$$

defined by

$$G_{\bar{v}}(g_1, g_2) := (-g_2, (A_{\bar{v}} + C_{\bar{v}})\, g_1 + iB_{\bar{v}} g_2)$$

for all $(g_1, g_2) \in Y$, is the infinitesimal generator of a strongly continuous semigroup $T(\bar{v}) : [0, \infty) \to L((X, (\,|\,)_{\bar{v}}), (X, (\,|\,)_{\bar{v}}))$ such that

$$\|T(\bar{v})(t)\|_{\bar{v}} \leqslant e^{\mu_{\bar{v}} t}$$

for all $t \in [0, \infty)$ where

$$\mu_{\bar{v}} := \left(\frac{5}{2}\right)^{1/2} |f''(\bar{v})| + \left(\frac{2}{\varepsilon}\right)^{1/2} \|U(\bar{v}) - \varepsilon\|_\infty .$$

(ii) Let I be some closed subinterval of $(-1, 1)$, $l(I) \geqslant 0$ the length of I,

$$C_{f,I} := \sup\{|f''(\bar{v})| : \bar{v} \in I\}\ ,$$

$C_{U,I} \geqslant 0$ such that

$$\|U(\bar{v}) - \varepsilon\|_\infty \leqslant C_{U,I}\ .$$

Then $(G_{\bar{v}})_{\bar{v}\in I}$ is a stable family of infinitesimal generators of strongly continuous semigroups on $(X, (\,|\,))$, where $(\,|\,) := (\,|\,)_0$, with constants

$$\mu_I := \left(\frac{5}{2}\right)^{1/2} C_{f,I} + \left(\frac{2}{\varepsilon}\right)^{1/2} C_{U,I}\ ,\quad e^{2\sqrt{5}\, l(I)\, C_{f,I}}\ .$$

(iii) Let $A := A_0$ and $(\,|\,) := (\,|\,)_0$. Then $S : Y \to (X, (\,|\,))$ defined by

$$S\,(g_1, g_2) := (A^{1/2} g_1, A^{1/2} g_2)$$

for every $(g_1, g_2) \in Y$ is a densely-defined linear and closed operator in $(X, (\,|\,))$.

(iv)

$$G_{\bar{v}} \in L((Y, \| \|_s), X)$$

for every $\bar{v} \in (-1, 1)$. In addition,

$$G := ((-1,1) \to L((Y, \| \|_s), X), \bar{v} \mapsto G_{\bar{v}})$$

is piecewise norm continuous.

(v) Let

$$B(\bar{v}) := \left(X \to X, (g_1, g_2) \mapsto (0, (A^{1/2}U(\bar{v})A^{-1/2} - U(\bar{v}))g_1)\right)$$

for every $\bar{v} \in (-1, 1)$. Then $B \in PC((-1,1), L(X, X))$. In addition, for every $\bar{v} \in (-1, 1)$ and $(g_1, g_2) \in \tilde{D} := D(A^{3/2}) \times D(A)$,

$$S G_{\bar{v}}(g_1, g_2) = (G_{\bar{v}} + B(\bar{v}))\, S\,(g_1, g_2)\ .$$

Finally, for every $\bar{v} \in (-1, 1)$ there is $\lambda < 0$ such that

$$(G_{\bar{v}} - \lambda)\tilde{D} = Y\ .$$

(vi) Let J be some closed subinterval of $(-1, 1)$ and assume that Y is equipped with $\| \|_s$. Then there is a unique Y/X-evolution operator $U \in C_*(\triangle(J), L(X, X))$ for $G|_J$.

Proof. '(i)': For this, let $\bar{v} \in (-1, 1)$. It follows by Lemma 10.1.4 that $A_{\bar{v}}$ is densely-defined, linear, self-adjoint operator in $L^2_{\mathbb{C}}(I)$ which is semibounded from below with lower bound ε, that $B_{\bar{v}}$ is a densely-defined, linear and symmetric operator in $L^2_{\mathbb{C}}(I)$, that $C_{\bar{v}}$ is a densely-defined, linear and closed operator in $L^2_{\mathbb{C}}(I)$ and by Lemma 10.1.1 that $D(A_{\bar{v}})$ is a core for $A_{\bar{v}}^{1/2}$, $B_{\bar{v}}$ and $C_{\bar{v}}$ as well as that

$$D(A_{\bar{v}}^{1/2}) = W^1_{0,\mathbb{C}}(I)\ .$$

In addition, it follows for $g \in D(A_{\bar{v}})$ that

$$\begin{aligned}\|B_{\bar{v}}g\|_2^2 &= 4\,|f'(\bar{v})|^2\,\|D_I^{1*}g\|_2^2 = 4\,|f'(\bar{v})|^2\,\langle g| - D_I^{2*}g\rangle_2 = \frac{4\,|f'(\bar{v})|^2}{1 - f'^2(\bar{v})}\,\langle g|A_{\bar{v}}g\rangle_2 \\ &= \frac{4\,|f'(\bar{v})|^2}{1 - f'^2(\bar{v})}\,\|A_{\bar{v}}^{1/2}g\|_2^2\end{aligned}$$

and

$$\begin{aligned}\|C_{\bar{v}}g\|_2^2 &\leqslant \left(\,|f''(\bar{v})|\,\|D_I^{1*}g\|_2 + \|U(\bar{v}) - \varepsilon\|_\infty\,\|g\|_2\,\right)^2 \leqslant 2|f''(\bar{v})|^2\,\|D_I^{1*}g\|_2^2 \\ &\quad + 2\|U(\bar{v}) - \varepsilon\|_\infty^2\,\|g\|_2^2 \leqslant \frac{2|f''(\bar{v})|^2}{1 - f'^2(\bar{v})}\,\|A_{\bar{v}}^{1/2}g\|_2^2 + 2\|U(\bar{v}) - \varepsilon\|_\infty^2\,\|g\|_2^2 \\ &\leqslant \frac{5}{2}\,|f''(\bar{v})|^2\,\|A_{\bar{v}}^{1/2}g\|_2^2 + 2\|U(\bar{v}) - \varepsilon\|_\infty^2\,\|g\|_2^2\end{aligned}$$

and hence also that

$$\|B_{\bar{v}}g\|_2^2 \leqslant \frac{4\,|f'(\bar{v})|^2}{1-f'^2(\bar{v})}\,\|A_{\bar{v}}^{1/2}g\|_2^2\,,$$
$$\|C_{\bar{v}}g\|_2^2 \leqslant \frac{5}{2}\,|f''(\bar{v})|^2\,\|A_{\bar{v}}^{1/2}g\|_2^2 + 2\|U(\bar{v})-\varepsilon\|_\infty^2\,\|g\|_2^2$$

for all $g \in D(A_{\bar{v}}^{1/2})$. Since

$$\frac{4\,|f'(\bar{v})|^2}{1-f'^2(\bar{v})} < 1\,,$$

the statement (i) follows by application of Theorem 5.4.3, Lemma 5.4.6 and Theorem 5.4.7.
'(ii)': For $\bar{v}_1, \bar{v}_2 \in I$ and $(g_1, g_2) \in \left(W^1_{0,\mathbb{C}}(I) \cap W^2_{\mathbb{C}}(I)\right) \times L^2_{\mathbb{C}}(I)$, it follows

$$\begin{aligned}
&\|(g_1,g_2)\|_{\bar{v}_1}^2 \\
&= \|A_{\bar{v}_1}^{1/2}g_1\|_2^2 + \|g_2\|_2^2 = (1-f'^2(\bar{v}_1))\langle g_1| - D_I^{2*}g_1\rangle_2 + \varepsilon\|g_1\|_2^2 + \|g_2\|_2^2 \\
&= \frac{1-f'^2(\bar{v}_1)}{1-f'^2(\bar{v}_2)}\left[\|A_{\bar{v}_2}^{1/2}g_1\|_2^2 - \varepsilon\|g_1\|_2^2\right] + \varepsilon\|g_1\|_2^2 + \|g_2\|_2^2 \\
&= \frac{1-f'^2(\bar{v}_1)}{1-f'^2(\bar{v}_2)}\,\|(g_1,g_2)\|_{\bar{v}_2}^2 + \left[1 - \frac{1-f'^2(\bar{v}_1)}{1-f'^2(\bar{v}_2)}\right](\varepsilon\,\|g_1\|_2^2 + \|g_2\|_2^2) \\
&\leqslant \left(1 + 2\,\frac{|f'^2(\bar{v}_1) - f'^2(\bar{v}_2)|}{1-f'^2(\bar{v}_2)}\right)\|(g_1,g_2)\|_{\bar{v}_2}^2 \\
&\leqslant \left(1 + \sqrt{5}\,|f'(\bar{v}_1) - f'(\bar{v}_2)|\right)\|(g_1,g_2)\|_{\bar{v}_2}^2 \\
&\leqslant \left(1 + \sqrt{5}\,C_{f,I}\,|\bar{v}_1 - \bar{v}_2|\right)\|(g_1,g_2)\|_{\bar{v}_2}^2 \leqslant e^{\sqrt{5}\,C_{f,I}\,|\bar{v}_1-\bar{v}_2|}\,\|(g_1,g_2)\|_{\bar{v}_2}^2\,.
\end{aligned}$$

Since $W^1_{0,\mathbb{C}}(I) \cap W^2_{\mathbb{C}}(I)$ is a core for $A_{\bar{v}_1}^{1/2}$ and $A_{\bar{v}_2}^{1/2}$, it follows that

$$\|(g_1,g_2)\|_{\bar{v}_1}^2 \leqslant e^{\sqrt{5}\,C_{f,I}\,|\bar{v}_1-\bar{v}_2|}\,\|(g_1,g_2)\|_{\bar{v}_2}^2$$

for all $(g_1, g_2) \in X$. Hence the statement follows by using part (i) and Theorem 8.6.4.
'(iii)': First, it follows that S is a densely-defined linear operator in X and

$$\|(g_1,g_2)\|_S^2 = \|A^{1/2}g_1\|_2^2 + \|Ag_1\|_2^2 + \|g_2\|_2^2 + \|A^{1/2}g_2\|_2^2$$

for all $(g_1, g_2) \in Y$. Further, if $(g_{11}, g_{21}), (g_{12}, g_{22}), \dots$ is a Cauchy sequence in $(Y, \|\ \|_S)$, the closedness of $A^{1/2}$ implies the existence of $g_2, h \in W^1_{0,\mathbb{C}}(I)$ such that

$$\lim_{\nu\to\infty}\|g_{2\nu} - g_2\|_2 = 0\,,\ \lim_{\nu\to\infty}\|A^{1/2}g_{2\nu} - A^{1/2}g_2\|_2 = 0\,,$$
$$\lim_{\nu\to\infty}\|A^{1/2}g_{1\nu} - h\|_2 = 0\,,\ \lim_{\nu\to\infty}\|Ag_{1\nu} - A^{1/2}h\|_2 = 0\,.$$

Further, since A is bijective, it follows that

$$\lim_{\nu\to\infty} \|g_{1\nu} - g_1\|_2 = 0\ ,\ \lim_{\nu\to\infty} \|A^{1/2}g_{1\nu} - A^{1/2}g_1\|_2 = 0\ ,\ \lim_{\nu\to\infty} \|Ag_{1\nu} - Ag_1\|_2 = 0\ ,$$

where

$$g_1 := A^{-1}A^{1/2}h = (A^{1/2})^{-1}(A^{1/2})^{-1}A^{1/2}h = (A^{1/2})^{-1}h \in W^1_{0,\mathbb{C}}(I) \cap W^2_{\mathbb{C}}(I)\ ,$$

and hence that $(g_1, g_2) \in Y$ as well as that

$$\lim_{\nu\to\infty} \|(g_{1\nu}, g_{2\nu}) - (g_1, g_2)\|_{\text{s}} = 0\ .$$

Hence $(Y, \|\ \|_{\text{s}})$ is complete and therefore S is closed.
'(iv)': For this, let $\bar{v} \in (-1, 1)$. Then

$$\begin{aligned}
G_{\bar{v}}(g_1, g_2) &= (-g_2, (A_{\bar{v}} + C_{\bar{v}})\, g_1 + iB_{\bar{v}}g_2) \\
&= (-g_2, \big[-(1 - f'^2(\bar{v}))\, D_I^{2*} + \varepsilon - f''(\bar{v})\, D_I^{1*} + U(\bar{v}) - \varepsilon\big]\, g_1 \\
&\quad + i(-2i) f'(\bar{v})\, D_I^{1*} g_2) \\
&= (-g_2, \big[(1 - f'^2(\bar{v}))\, A - f''(\bar{v})\, D_I^{1*} + U(\bar{v}) - \varepsilon\,(1 - f'^2(\bar{v}))\big]\, g_1 \\
&\quad + 2f'(\bar{v})\, D_I^{1*} g_2)
\end{aligned}$$

for all $(g_1, g_2) \in Y$. Hence it follows by using Lemma 10.1.4 (iii) that

$$\begin{aligned}
&\|G_{\bar{v}}(g_1, g_2)\|_2^2 = \|A^{1/2}g_2\|_2^2 \\
&\quad + \big\|\big[(1 - f'^2(\bar{v}))\, A - f''(\bar{v})\, D_I^{1*} + U(\bar{v}) - \varepsilon\,(1 - f'^2(\bar{v}))\big]\, g_1 + 2f'(\bar{v})\, D_I^{1*} g_2\big\|_2^2 \\
&\leqslant \|A^{1/2}g_2\|_2^2 + 5\Big[|1 - f'^2(\bar{v})|^2\, \|Ag_1\|_2^2 + |f''(\bar{v})|^2\, \|A^{1/2}g_1\|_2^2 + \|U(\bar{v})\|_\infty^2\, \|g_1\|_2^2 \\
&\quad + \varepsilon^2\, |1 - f'^2(\bar{v})|^2\, \|g_1\|_2^2 + 4\, |f'(\bar{v})|^2\, \|A^{1/2}g_2\|_2^2\Big] \\
&= 5\, |f''(\bar{v})|^2\, \|A^{1/2}g_1\|_2^2 + 5\, |1 - f'^2(\bar{v})|^2\, \|Ag_1\|_2^2 + (1 + 20\, |f'(\bar{v})|^2)\, \|A^{1/2}g_2\|_2^2 \\
&\quad + \big[\|U(\bar{v})\|_\infty^2 + \varepsilon^2\, |1 - f'^2(\bar{v})|^2\big]\, \|g_1\|_2^2 \\
&\leqslant \Big[5\, |f''(\bar{v})|^2 + \big(\|U(\bar{v})\|_\infty^2 + \varepsilon^2\, |1 - f'^2(\bar{v})|^2\big)\, \|(A^{1/2})^{-1}\|\Big]\, \|A^{1/2}g_1\|_2^2 \\
&\quad + 5\, |1 - f'^2(\bar{v})|^2\, \|Ag_1\|_2^2 + (1 + 20\, |f'(\bar{v})|^2)\, \|A^{1/2}g_2\|_2^2 \leqslant C^2(\bar{v})\, \|(g_1, g_2)\|_{\text{s}}^2
\end{aligned}$$

for all $(g_1, g_2) \in Y$ where

$$\begin{aligned}
C(\bar{v}) := \Big(&\max\{5\, |f''(\bar{v})|^2 + \big(\|U(\bar{v})\|_\infty^2 + \varepsilon^2\, |1 - f'^2(\bar{v})|^2\big)\, \|(A^{1/2})^{-1}\|, \\
&5\, |1 - f'^2(\bar{v})|^2, 1 + 20\, |f'(\bar{v})|^2\}\Big)^{1/2}\ .
\end{aligned}$$

Further, let $\bar{v}_1, \bar{v}_2 \in (-1,1)$ and $(g_1, g_2) \in Y$. Then

$$\begin{aligned}
&\|G_{\bar{v}_1}(g_1,g_2) - G_{\bar{v}_2}(g_1,g_2)\|^2 \\
&= \| [(f'^2(\bar{v}_2)) - f'^2(\bar{v}_1))A + (f''(\bar{v}_2) - f''(\bar{v}_1))\, D_I^{1*} + U(\bar{v}_1) - U(\bar{v}_2) \\
&\quad -\varepsilon\,(f'^2(\bar{v}_2) - f'^2(\bar{v}_1))]\, g_1 + 2(f'(\bar{v}_1) - f'(\bar{v}_2))\, D_I^{1*} g_2)\|_2^2 \\
&\leqslant 5\,|f'^2(\bar{v}_2)) - f'^2(\bar{v}_1)|^2\, \|A g_1\|_2^2 + 5\,|f''(\bar{v}_2) - f''(\bar{v}_1)|^2\, \|A^{1/2} g_1\|_2^2 \\
&\quad + 5\,\left[\|U(\bar{v}_1) - U(\bar{v}_2)\|_\infty^2 + \varepsilon^2\,|f'^2(\bar{v}_2) - f'^2(\bar{v}_1)|^2\right] \|g_1\|_2^2 \\
&\quad + 20\,|f'(\bar{v}_1) - f'(\bar{v}_2)|^2\, \|A^{1/2} g_2\|_2^2 \\
&\leqslant 5\,\|(g_1,g_2)\|_S^2\, \left[|f'^2(\bar{v}_2)) - f'^2(\bar{v}_1)|^2 + |f''(\bar{v}_2) - f''(\bar{v}_1)|^2\right. \\
&\quad + \left[\|U(\bar{v}_1) - U(\bar{v}_2)\|_\infty^2 + \varepsilon^2\,|f'^2(\bar{v}_2) - f'^2(\bar{v}_1)|^2\right] \|(A^{1/2})^{-1}\| \\
&\quad \left. + 4\,|f'(\bar{v}_1) - f'(\bar{v}_2)|^2\right].
\end{aligned}$$

'(v)': By (10.1.18), it follows that

$$\begin{aligned}
&\|B(\bar{v})(g_1,g_2)\| = \|(0, (A^{1/2} U(\bar{v}) A^{-1/2} - U(\bar{v}))g_1\| \\
&\quad = \|(A^{1/2} U(\bar{v}) A^{-1/2} - U(\bar{v}))g_1\|_2 \leqslant \left[4\|U(\bar{v})\|_\infty + \varepsilon^{-1/2}\|[U(\bar{v})]'\|_\infty\right] \|g_1\|_2 \\
&\quad \leqslant \|A^{-1/2}\| \left[4\|U(\bar{v})\|_\infty + \varepsilon^{-1/2}\|[U(\bar{v})]'\|_\infty\right] \|A^{1/2} g_1\|_2 \\
&\quad \leqslant \|A^{-1/2}\| \left[4\|U(\bar{v})\|_\infty + \varepsilon^{-1/2}\|[U(\bar{v})]'\|_\infty\right] \|(g_1,g_2)\|
\end{aligned}$$

and

$$\begin{aligned}
&\|B(\bar{v}_1)(g_1,g_2) - B(\bar{v}_2)(g_1,g_2)\| \\
&\quad \leqslant \|A^{-1/2}\| \left[4\,\|U(\bar{v}_1) - U(\bar{v}_2)\|_\infty + \varepsilon^{-1/2}\,\|[U(\bar{v}_1)]' - [U(\bar{v}_2)]'\|_\infty\right] \|(g_1,g_2)\|
\end{aligned}$$

for all $(g_1, g_2) \in X$ and $\bar{v}, \bar{v}_1, \bar{v}_2 \in (-1,1)$. Further,

$$\begin{aligned}
&S G_{\bar{v}}(g_1,g_2) \\
&= S\,(-g_2, [(1 - f'^2(\bar{v}))\,A - f''(\bar{v})\, D_I^{1*} + U(\bar{v}) - \varepsilon\,(1 - f'^2(\bar{v}))]\, g_1 \\
&\quad + 2f'(\bar{v})\, D_I^{1*} g_2) \\
&= \left(-A^{1/2} g_2, \left[(1 - f'^2(\bar{v}))\, A^{1/2} A - f''(\bar{v})\, A^{1/2} D_I^{1*} + A^{1/2} U(\bar{v})\right.\right. \\
&\quad \left.\left. -\varepsilon\,(1 - f'^2(\bar{v})) A^{1/2}\right] g_1 + 2f'(\bar{v})\, A^{1/2} D_I^{1*} g_2\right) \\
&= \left(-A^{1/2} g_2, \left[(1 - f'^2(\bar{v}))\, A - f''(\bar{v})\, D_I^{1*} + A^{1/2} U(\bar{v}) A^{-1/2}\right.\right. \\
&\quad \left.\left. -\varepsilon\,(1 - f'^2(\bar{v}))\right] A^{1/2} g_1 + 2f'(\bar{v})\, D_I^{1*} A^{1/2} g_2\right)
\end{aligned}$$

$$= \Big(-A^{1/2}g_2, \big[(1-f'^2(\bar{v}))\,A - f''(\bar{v})\,D_I^{1*} + U(\bar{v})$$
$$-\varepsilon\,(1-f'^2(\bar{v}))\big]\,A^{1/2}g_1 + 2f'(\bar{v})\,D_I^{1*}A^{1/2}g_2\Big)$$
$$+\,(0, [A^{1/2}U(\bar{v})A^{-1/2} - U(\bar{v})]A^{1/2}g_1) = (G_{\bar{v}} + B(\bar{v}))S\,(g_1, g_2)$$

for all $(g_1, g_2) \in \tilde{D}$ and $\bar{v} \in (-1,1)$. Further, let $\bar{v} \in (-1,1)$. According to Theorem 6.1.1, the restriction of $G_{\bar{v}}$ in domain to $D(G_{\bar{v}}^2)$ and in image to Y is the infinitesimal generator of a strongly continuous semigroup. Hence there is $\lambda < 0$ such that

$$(G_{\bar{v}} - \lambda)D(G_{\bar{v}}^2) = Y\,.$$

In the following, it will be shown that

$$D(G_{\bar{v}}^2) = \tilde{D}\,.$$

First, remembering that

$$Y = D(A) \times D(A^{1/2}) = \big(W^1_{0,\mathbb{C}}(I) \cap W^2_{\mathbb{C}}(I)\big) \times W^1_{0,\mathbb{C}}(I)\,,$$

it follows that

$$G_{\bar{v}}(g_1, g_2) = (-g_2, \big[(1-f'^2(\bar{v}))\,A - f''(\bar{v})\,D_I^{1*} + U(\bar{v}) - \varepsilon\,(1-f'^2(\bar{v}))\big]\,g_1$$
$$+\,2f'(\bar{v})\,D_I^{1*}g_2) \in Y$$

for all $(g_1, g_2) \in \tilde{D} = D(A^{3/2}) \times D(A)$ and hence that

$$\tilde{D} \subset D(G_{\bar{v}}^2)\,.$$

On the other hand, if $(h_1, h_2) \in Y$ and $(g_1, g_2) \in Y$ is such that

$$(G_{\bar{v}} - \lambda)(g_1, g_2) = (-g_2 - \lambda g_1, \big[(1-f'^2(\bar{v}))\,A - f''(\bar{v})\,D_I^{1*} + U(\bar{v})$$
$$-\varepsilon\,(1-f'^2(\bar{v}))\big]\,g_1 + 2f'(\bar{v})\,D_I^{1*}g_2 - \lambda g_2) = (h_1, h_2)\,,$$

then

$$g_2 = -(h_1 + \lambda g_1) \in D(A)$$

and

$$(1-f'^2(\bar{v}))\left(A - \varepsilon + \frac{\lambda^2}{1-f'^2(\bar{v})}\right)g_1$$
$$= h_2 + 2f'(\bar{v})D_I^{1*}h_1 - \lambda h_1 + f''(\bar{v})\,D_I^{1*}g_1 - U(\bar{v})g_1 + 2\lambda f'(\bar{v})g_1 \in D(A^{1/2})$$

and hence $g_1 \in D(A^{3/2})$. Therefore $(g_1, g_2) \in \tilde{D}$ and

$$\tilde{D} \supset D(G_{\bar{v}}^2)\,.$$

'(vi)': The statement (vi) is an immediate consequence of Theorem 9.0.6 and the statements (i)-(v). □

10.2 Non-Autonomous Linear Hermitian Hyperbolic Systems

In the following, we consider the solutions of the system

$$\frac{\partial u}{\partial t}(t,x) + \sum_{j=1}^{n} A^j(t,x) \cdot \frac{\partial u}{\partial x^j}(t,x) + \sum_{j=1}^{n} B^j(t,x) \cdot u(t,x) = 0$$

for $t \in \mathbb{R}$, $x \in \mathbb{R}^n$, $\mathbb{C}^p$-valued u, $A^1, \dots, A^n$ assuming values in the set of Hermitian $p \times p$-matrices and $\mathbb{C}^p$-valued f where $p \in \mathbb{N}^*$, $\cdot$ denotes matrix multiplication and partial differentiation is to be interpreted component-wise.[2] The main sources for this section are the papers [107, 109]. Here these results are 'adapted' to the late Kato's most recent approach to abstract quasi-linear evolution equations [114] given in this second part of the notes. For additional material on initial boundary value problems for non-autonomous linear symmetric hyperbolic systems see, for instance, [158, 161, 177, 193, 194].

Lemma 10.2.1. (A commutator estimate) Let $n \in \mathbb{N}^*$, $a \in C^1(\mathbb{R}^n, \mathbb{C})$ such that a, $a_{,1}, \dots, a_{,n}$ are bounded and $b \in C^2(\mathbb{R}^n, \mathbb{C})$ such that $b, b_{,j}, b_{,jk}$ are bounded for all $j, k, l \in \{1, \dots, n\}$. Further, for every complex-valued function g which is a.e. defined on $\mathbb{R}^n$ and measurable, we denote by T_g the corresponding maximal multiplication operator in $L^2_{\mathbb{C}}(\mathbb{R}^n)$. In addition, let $F_2 : L^2_{\mathbb{C}}(\mathbb{R}^n) \to L^2_{\mathbb{C}}(\mathbb{R}^n)$ be the unitary Fourier transformation determined by

$$(F_2 f)(x) := (2\pi)^{-n/2} \int_{\mathbb{R}^n} e^{-ixy} f(y)\, dy$$

for all $x \in \mathbb{R}^n$, for every $f \in C_0^\infty(\mathbb{R}^n, \mathbb{C})$. Finally, we denote for every $j \in \{1, \dots, n\}$ by p_j the projection of $\mathbb{R}^n$ onto the j-th coordinate.

(i) For every $j \in \{1, \dots, n\}$ and $f \in W^2_{\mathbb{C}}(\mathbb{R}^n)$

$$(-\Delta + 1)^{1/2} \partial^j f = \partial^j (-\Delta + 1)^{1/2} f \; .$$

(ii) For every $f \in W^1_{\mathbb{C}}(\mathbb{R}^n)$, $g \in L^2_{\mathbb{C}}(\mathbb{R}^n)$

$$(-\Delta + \lambda)^{-1} \partial^j f = \partial^j (-\Delta + \lambda)^{-1} f \; , \; \|\partial^j (-\Delta + \lambda)^{-1} g\|_2 \leqslant \frac{1}{2\sqrt{\lambda}} \|g\|_2 \; .$$

(iii) For every $f \in W^2_{\mathbb{C}}(\mathbb{R}^n)$,

$$T_b(-\Delta) f - (-\Delta) T_b f = 2 \sum_{j=1}^{n} \partial^j b_{,j} f - \left(\sum_{j=1}^{n} b_{,jj} \right) f \; .$$

[2] The elements of $\mathbb{C}^p$ are considered as column vectors.

(iv) For every $f \in L^2_{\mathbb{C}}(\mathbb{R}^n)$,

$$\|(-\Delta + 1)^{1/2}\, T_a\, [\,(-\Delta + 1)^{1/2}\,]^{-1} f\|_2 \leqslant C_n(a)\, \|f\|_2 \tag{10.2.1}$$

where

$$C_n(a) := (2n + 1)\, \|a\|_\infty + \sum_{j=1}^{n} \|a_{,j}\|_\infty \,.$$

(v) In addition, let $j \in \{1, \dots, n\}$. Then

$$\left((0, \infty) \to L^2_{\mathbb{C}}(\mathbb{R}^n, \lambda \mapsto \lambda^{1/2} \partial^j (-\Delta + \lambda + 1)^{-2} f\right)$$

is weakly summable and

$$\int_0^\infty \lambda^{1/2} \partial^j (-\Delta + \lambda + 1)^{-2} f \, d\lambda$$
$$= \left(\int_0^\infty \frac{\lambda^{1/2}}{(\lambda + 1)^2} \, d\lambda \right) F_2^{-1} T_{i p_j \cdot (|\ |^2 + 1)^{-1/2}} F_2 f \tag{10.2.2}$$

for every $f \in L^2_{\mathbb{C}}(\mathbb{R}^n)$.

(vi) There is $K \geqslant 0$ such that for every $h \in BC(\mathbb{R}^n, \mathbb{C}) \cap \mathrm{Lip}(\mathbb{R}, \mathbb{C})$ and every $C \geqslant 0$ such that $|\, h(x) - h(y)\,| \leqslant C\, |\, x - y\,|$ for all $x, y \in \mathbb{R}^n$

$$\|T_h (-\Delta + \lambda)^{-1} - (-\Delta + \lambda)^{-1} T_h\| \leqslant \frac{KC}{\lambda^{3/2}} \,.$$

(vii)

$$\|T_b (-\Delta + 1)^{1/2} - (-\Delta + 1)^{1/2} T_b\|$$
$$\leqslant \frac{2}{\pi} \left(\sum_{j=1}^{n} [\, 2\, \|b_{,j}\|_\infty + \|b_{,jj}\|_\infty] + K \sum_{j,k=1}^{n} \|b_{,jk}\|_\infty \right) .$$

Proof. For this, note that $F_2 W^k_{\mathbb{C}}(\mathbb{R}^n)$ coincides with the domain of the maximal multiplication operator in $L^2_{\mathbb{C}}(\mathbb{R}^n)$ by $(1 + |\ |^2)^{k/2}$ for every $k \in \mathbb{N}$.

'(i)': Let $j \in \{1, \dots, n\}$ and $f \in W^2_{\mathbb{C}}(\mathbb{R}^n)$. Then

$$\begin{aligned} (-\Delta + 1)^{1/2} \partial^j f &= i F_2^{-1} T_{(1+|\ |^2)^{1/2}} F_2 F_2^{-1} T_{p_j} F_2 f = i F_2^{-1} T_{p_j (1+|\ |^2)^{1/2}} F_2 \\ &= i F_2^{-1} T_{p_j} F_2 F_2^{-1} T_{(1+|\ |^2)^{1/2}} F_2 f = \partial^j (-\Delta + 1)^{1/2} f \,. \end{aligned}$$

'(ii)': For this, let $\lambda > 0$ and $j \in \{1, \dots, n\}$. Then

$$\begin{aligned} \|\partial^j (-\Delta + \lambda)^{-1} g\|_2 &= \|\, i F_2^{-1} T_{p_j} F_2 F_2^{-1} T_{(|\ |^2 + \lambda)^{-1}} F_2 g\|_2 = \|T_{p_j \cdot (|\ |^2 + \lambda)^{-1}} F_2 g\|_2 \\ &\leqslant \|p_j \cdot (|\ |^2 + \lambda)^{-1}\|_\infty \, \|g\|_2 \leqslant \frac{1}{2\sqrt{\lambda}} \, \|g\|_2 \end{aligned}$$

for every $g \in L^2_{\mathbb{C}}(\mathbb{R}^n)$. Finally,

$$(-\triangle + \lambda)^{-1}\partial^j f = iF_2^{-1}T_{p_j(1+|\ |^2)^{1/2}}F_2 f = \partial^j(-\triangle + \lambda)^{-1} f$$

for every $f \in W^1_{\mathbb{C}}(\mathbb{R}^n)$.
'(iii)': For this, let $f \in W^2_{\mathbb{C}}(\mathbb{R}^n)$. Then

$$\begin{aligned}\triangle T_b f &= \sum_{j=1}^n \partial^j \partial^j b f = 2\sum_{j=1}^n \partial^j b_{,j} f + \sum_{j=1}^n \partial^j (b\,\partial^j f - b_{,j} f)\\ &= T_b \triangle f + 2\sum_{j=1}^n \partial^j b_{,j} f - \left(\sum_{j=1}^n b_{,jj}\right) f\end{aligned}$$

'(iv)': It follows for $j \in \{1, \dots, n\}$ and $f \in W^1_{\mathbb{C}}(\mathbb{R}^n)$ that

$$\begin{aligned}\|(-\triangle)^{1/2} f\|_2 &= \|F_2(-\triangle)^{1/2} f\|_2 = \|\,|\ | F_2 f\,\|_2 \leqslant \|\sum_{j=1}^n |p_j|\, F_2 f\,\|_2 \leqslant \sum_{j=1}^n \||p_j|\, F_2 f\|_2\\ &= \sum_{j=1}^n \|p_j F_2 f\|_2 = \sum_{j=1}^n \|F_2\,\partial^j f\|_2 = \sum_{j=1}^n \|\partial^j f\|_2 \qquad (10.2.3)\end{aligned}$$

and

$$\begin{aligned}\|\partial^j f\|_2 &= \|F_2\,\partial^j f\|_2 = \|p_j F_2 f\|_2 = \||p_j| F_2 f\|_2 \leqslant \|\,|\ | F_2 f\,\|_2 \qquad (10.2.4)\\ &= \|(-\triangle)^{1/2} f\|_2\,.\end{aligned}$$

Further, by using Lemma 10.1.1, (10.1.13) and $af \in W^1_{\mathbb{C}}(\mathbb{R}^n)$, it follows that

$$\begin{aligned}&\|(-\triangle + 1)^{1/2}\, T_a f\|_2 \leqslant \|(-\triangle)^{1/2}\, af\|_2 + \|af\|_2 \leqslant \sum_{j=1}^n \|\partial^j af\|_2 + \|a\|_\infty \|f\|_2\\ &\leqslant \sum_{j=1}^n \|a_{,j} f + a\,\partial^j f\|_2 + \|a\|_\infty \|f\|_2 \leqslant \sum_{j=1}^n \|a_{,j} f\|_2 + \sum_{j=1}^n \|a\,\partial^j f\|_2 + \|a\|_\infty \|f\|_2\\ &\leqslant [\,\|a\|_\infty + \sum_{j=1}^n \|a_{,j}\|_\infty\,]\,\|f\|_2 + \|a\|_\infty \sum_{j=1}^n \|\partial^j f\|_2 \leqslant [\,\|a\|_\infty + \sum_{j=1}^n \|a_{,j}\|_\infty\,]\,\|f\|_2\\ &\quad + n\|a\|_\infty\,\|(-\triangle)^{1/2} f\|_2 \leqslant [\,(n+1)\,\|a\|_\infty + \sum_{j=1}^n \|a_{,j}\|_\infty\,]\,\|f\|_2\\ &\quad + n\|a\|_\infty\,\|(-\triangle + 1)^{1/2} f\|_2 \leqslant [\,(2n+1)\,\|a\|_\infty + \sum_{j=1}^n \|a_{,j}\|_\infty\,]\cdot\|(-\triangle + 1)^{1/2} f\|_2\end{aligned}$$

and hence, finally, (10.2.1).

'(v)': First, it follows for $\lambda_1, \lambda_2 > 0$, $f \in L^2_{\mathbb{C}}(\mathbb{R}^n)$ that

$$\begin{aligned}
&\|\partial^j(-\Delta + \lambda_1 + 1)^{-1} f - \partial^j(-\Delta + \lambda_2 + 1)^{-1} f\|_2 \\
&= |\lambda_1 - \lambda_2| \cdot \|\partial^j(-\Delta + \lambda_1 + 1)^{-1}(-\Delta + \lambda_2 + 1)^{-1} f\|_2 \\
&\leqslant |\lambda_1 - \lambda_2| \cdot \|T_{p_j \cdot (|\ |^2 + \lambda_1 + 1)^{-1}}\| \cdot \|(-\Delta + \lambda_2 + 1)^{-1}\| \cdot \|f\|_2 \\
&\leqslant \frac{|\lambda_1 - \lambda_2|}{2(\lambda_2 + 1)\sqrt{\lambda_1 + 1}} \|f\|_2
\end{aligned}$$

and hence the strong continuity of

$$\left((0,\infty) \to L(L^2_{\mathbb{C}}(\mathbb{R}^n), L^2_{\mathbb{C}}(\mathbb{R}^n)), \lambda \mapsto \partial^j(-\Delta + \lambda + 1)^{-1}\right)$$

and therefore by Lemma 8.1.1 (ii) also the strong continuity of

$$h := \left((0,\infty) \to L(L^2_{\mathbb{C}}(\mathbb{R}^n), L^2_{\mathbb{C}}(\mathbb{R}^n)), \lambda \mapsto \lambda^{1/2}\, \partial^j(-\Delta + \lambda + 1)^{-2}\right).$$

Further, it follows for $f, g \in L^2_{\mathbb{C}}(\mathbb{R}^n)$ and $\lambda > 0$ that

$$\begin{aligned}
\langle f | h(\lambda) g \rangle &= i\lambda^{1/2} \langle F_2 f |\, p_j \cdot (|\ |^2 + \lambda + 1)^{-2} F_2 g \rangle \\
&= \int_{\mathbb{R}^n} \frac{i\lambda^{1/2} k_j}{(|k|^2 + \lambda + 1)^2} (F_2 f)^*(k) (F_2 g)(k)\, dk\, .
\end{aligned}$$

In addition, by change of variables

$$\begin{aligned}
&\int_0^\infty \frac{\lambda^{1/2}\, |k_j|}{(|k|^2 + \lambda + 1)^2} \cdot |(F_2 f)(k)| \cdot |(F_2 g)(k)|\, d\lambda \\
&= \left(\int_0^\infty \frac{\lambda^{1/2}}{(\lambda + 1)^2}\, d\lambda \right) \frac{|k_j|}{(|k|^2 + 1)^{1/2}} \cdot |(F_2 f)(k)| \cdot |(F_2 g)(k)|
\end{aligned}$$

for almost all $k \in \mathbb{R}^n$. Since

$$\left(D(F_2 f) \cap D(F_2 g) \to \mathbb{R}, k \mapsto \frac{|k_j|}{(|k|^2 + 1)^{1/2}} \cdot |(F_2 f)(k)| \cdot |(F_2 g)(k)| \right)$$

is summable, it follows by Tonelli's theorem the summability of

$$\left((0,\infty) \times D(F_2 f) \cap D(F_2 g) \to \mathbb{C}, (\lambda, k) \mapsto \frac{i\lambda^{1/2} k_j}{(|k|^2 + \lambda + 1)^2} \cdot (F_2 f)^*(k) \cdot (F_2 g)(k) \right)$$

and therefore by Fubini's theorem and change of variables the summability of

$$\left((0,\infty) \to \mathbb{C}, \lambda \mapsto \langle f | h(\lambda) g \rangle\right)$$

and

$$\int_0^\infty \langle f|h(\lambda)g\rangle \, d\lambda = \left(\int_0^\infty \frac{\lambda^{1/2}}{(\lambda+1)^2}\, d\lambda\right) \int_{\mathbb{R}^n} \frac{ik_j}{(\,|k|^2+1)^{1/2}} \cdot (F_2 f)^*(k) \cdot (F_2 g)(k)\, dk$$

$$= \left(\int_0^\infty \frac{\lambda^{1/2}}{(\lambda+1)^2}\, d\lambda\right) \langle f|\, F_2^{-1} T_{ip_j \cdot (|\ |^2+1)^{-1/2}} F_2 g\rangle \, .$$

Finally, since $(X, \langle\,|\,\rangle)$ is reflexive, from this follows by Theorem 3.2.2 the weak integrability of

$$\left((0,\infty) \to L^2_{\mathbb{C}}(\mathbb{R}^n), \lambda \mapsto h(\lambda) f\right)$$

and (10.2.2).

'(vi)': In a first step, we find a representation of the operator $(-\Delta + \lambda)^{-1}$ where $\lambda > 0$. In this, we consider the cases $n = 1$, $n = 3$ and $n \in \mathbb{R}^* \backslash \{1,3\}$. For $n = 1$, we define $B_{1\lambda} : \mathbb{R}^* \to \mathbb{R}$ by

$$B_{1\lambda}(x) := \frac{1}{2\sqrt{\lambda}}\, e^{-\sqrt{\lambda}\,|x|}$$

for every $x \in \mathbb{R}^*$. Then

$$\begin{aligned}
(F_1 B_{1\lambda})(k) &= 2\lambda^{-1/2} \int_{-\infty}^{\infty} \exp(-(ikx + \sqrt{\lambda}\,|x|\,))\, dx \\
&= 2^{-1}\lambda^{-1/2} \int_0^\infty \exp(-(ik + \sqrt{\lambda}\,)\, x)\, dx \\
&\quad + 2^{-1}\lambda^{-1/2} \int_0^\infty \exp(-(-ik + \sqrt{\lambda}\,)\, x)\, dx \\
&= 2^{-1}\lambda^{-1/2} \frac{1}{ik + \sqrt{\lambda}} + 2^{-1}\lambda^{-1/2} \frac{1}{-ik + \sqrt{\lambda}} = \frac{1}{k^2 + \lambda}
\end{aligned}$$

for every $k \in \mathbb{R}$. For $n = 3$, we define $B_{3\lambda} : \mathbb{R}^3 \backslash \{0\} \to \mathbb{R}$ by

$$B_{3\lambda}(x) := \frac{1}{4\pi} \frac{e^{-\sqrt{\lambda}\,|x|}}{|x|}$$

for every $x \in \mathbb{R}^3 \backslash \{0\}$. The calculation in this case employs change of variables to spherical coordinates, $r \in (0,\infty)$, $\theta \in (0,\pi)$, $\phi \in (-\pi,\pi)$. Then

$$\begin{aligned}
&(F_1 B_{3\lambda})(0,0,|k|) \\
&= (4\pi)^{-1} \int_{(0,\infty)\times(-\pi,\pi)\times(0,\pi)} \exp(-i\,|k|\, r\cos(\theta)) \exp(-\sqrt{\lambda} r)\, r \sin(\theta)\, dr\, d\varphi\, d\theta \\
&= \frac{1}{2i|k|} \int_0^\infty [\exp(-(-i|k| + \sqrt{\lambda}\,)\, r) - \exp(-(i\,|k| + \sqrt{\lambda}\,)\, r)]\, dr \\
&= \frac{1}{2i|k|} \left[\frac{1}{-i|k| + \sqrt{\lambda}} - \frac{1}{i|k| + \sqrt{\lambda}}\right] = \frac{1}{|k|^2 + \lambda}
\end{aligned}$$

for every $k \in \mathbb{R}^3\backslash\{0\}$ and hence it follows because of the spherical symmetry and continuity of the Fourier transform that

$$(F_1 B_{3\lambda})(k) = \frac{1}{|k|^2 + \lambda}$$

for every $k \in \mathbb{R}^3$. The calculation in the remaining cases $n \in \mathbb{N}^*\backslash\{1,3\}$ uses the facts that

$$F_2\, \varepsilon^{-n} \exp(-|\ |^2/(2\varepsilon^2)) = \exp(-\varepsilon^2|\ |^2/2)\ ,$$

where $\varepsilon > 0$, as well as that

$$\int_{\mathbb{R}} e^{-\alpha(x+i\beta)^2} dx = \int_{\mathbb{R}} e^{-\alpha x^2} dx = \left(\frac{\pi}{\alpha}\right)^{1/2}$$

for all $\alpha > 0$ and $\beta \in \mathbb{R}$. Then

$$\begin{aligned} \frac{1}{|k|^2 + \lambda} \exp(-\varepsilon^2|k|^2/2) &= \exp(-\varepsilon^2|k|^2/2) \int_0^\infty e^{-(|k|^2+\lambda)\,t}\, dt \\ &= \int_0^\infty e^{-\lambda t} e^{-[t+(\varepsilon^2/2)]\,|k|^2}\, dt \end{aligned}$$

for every $k \in \mathbb{R}^n$. Since $(|\ |^2 + \lambda)^{-1} \exp(-\varepsilon^2|\ |^2/2) \in L^1_{\mathbb{C}}(\mathbb{R}^n) \cap L^2_{\mathbb{C}}(\mathbb{R}^n)$, it follows by Fubini's theorem

$$\begin{aligned} &[F_2^{-1}(|\ |^2 + \lambda)^{-1} \exp(-\varepsilon^2(|\ |^2 + \lambda)/2)](x) \\ &= (2\pi)^{-n/2} \exp(-\lambda\varepsilon^2/2) \int_0^\infty e^{-\lambda t} \left[\int_{\mathbb{R}^n} e^{ixk} e^{-[t+(\varepsilon^2/2)]\,|k|^2}\, dk\right] dt \\ &= (2\pi)^{-n/2} \exp(-\lambda\varepsilon^2/2) \int_0^\infty e^{-\lambda t} \\ &\left[\int_{\mathbb{R}^n} \exp\left(-[t+(\varepsilon^2/2)]\left(k - \frac{i}{2[t+(\varepsilon^2/2)]}x\right)^2\right) \exp\left(-\frac{x^2}{4[t+(\varepsilon^2/2)]}\right) dk\right] dt \\ &= 2^{-n/2} \exp(-\lambda\varepsilon^2/2) \int_0^\infty (t + (\varepsilon^2/2))^{-n/2} \exp\left(-\lambda t - \frac{x^2}{4[t+(\varepsilon^2/2)]}\right) dt \\ &= 2^{-n/2} \int_{\varepsilon^2/2}^\infty t^{-n/2} \exp\left(-\lambda t - \frac{x^2}{4t}\right) dt \end{aligned} \tag{10.2.5}$$

for every $x \in \mathbb{R}^n$ where we used that

$$h := ((0,\infty) \times \mathbb{R}^n \to \mathbb{C}, (t,k) \mapsto \exp(-\lambda t + ixk - [t + (\varepsilon^2/2)]\,|k|^2))$$

is summable since h is continuous and hence measurable and that $|h|$ is majorized by the summable function

$$((0,\infty) \times \mathbb{R}^n \to \mathbb{R}, (t,k) \mapsto \exp(-\lambda t - (\varepsilon^2/2)\,|k|^2))\ .$$

We define $B_{n\lambda} : \mathbb{R}^n\backslash\{0\} \to \mathbb{R}$ by

$$B_{n\lambda}(x) := \frac{1}{(4\pi)^{n/2}} \int_0^\infty t^{-n/2} \exp\left(-\lambda t - \frac{x^2}{4t}\right) dt$$

for all $x \in \mathbb{R}^n \backslash \{0\}$. Note that

$$\begin{aligned}
B_{n\lambda}(x) &= \frac{1}{(4\pi)^{n/2}} \int_{-\infty}^{\infty} e^{(\frac{n}{2}-1)\tau} \exp\left(-\left[\frac{x^2}{4} e^{\tau} + \lambda e^{-\tau}\right]\right) d\tau \\
&= \frac{1}{(4\pi)^{n/2}} \int_{-\infty}^{\infty} e^{(\frac{n}{2}-1)\tau} \exp\left(-|x| \sqrt{\lambda} \cosh\left[\tau + \ln\left(\frac{|x|}{2\sqrt{\lambda}}\right)\right]\right) d\tau \\
&= \frac{1}{(4\pi)^{n/2}} \left(\frac{|x|}{2\sqrt{\lambda}}\right)^{1-\frac{n}{2}} \int_{-\infty}^{\infty} e^{(\frac{n}{2}-1)u} \exp(-|x| \sqrt{\lambda} \cosh(u))\, du \\
&= \frac{2}{(4\pi)^{n/2}} \left(\frac{|x|}{2\sqrt{\lambda}}\right)^{1-\frac{n}{2}} K_{\frac{n}{2}-1}(\sqrt{\lambda}|x|)
\end{aligned}$$

for all $x \in \mathbb{R}^n \backslash \{0\}$ where the modified Bessel function $K_{(n/2)-1}$ is defined according to [1]. [3] Further, it follows for every $x \in \mathbb{R}^n \backslash \{0\}$ that

$$\begin{aligned}
B_{n\lambda}(x) &= \frac{1}{(4\pi)^{n/2}} \int_0^{\infty} t^{-n/2} \exp\left(-\lambda t - \frac{x^2}{4t}\right) dt \\
&= \frac{1}{(4\pi)^{n/2}} \left(\frac{|x|}{2\sqrt{\lambda}}\right)^{1-(n/2)} \int_0^{\infty} \tau^{-n/2} \exp\left(-\frac{|x|\sqrt{\lambda}}{2}\left(\tau + \frac{1}{\tau}\right)\right) d\tau \\
&= \frac{1}{(4\pi)^{n/2}} \left(\frac{|x|}{2\sqrt{\lambda}}\right)^{1-(n/2)} \left[\int_0^{1} \tau^{-n/2} \exp\left(-\frac{|x|\sqrt{\lambda}}{2}\left(\tau + \frac{1}{\tau}\right)\right) d\tau\right. \\
&\quad \left. + \int_1^{\infty} \tau^{-n/2} \exp\left(-\frac{|x|\sqrt{\lambda}}{2}\left(\tau + \frac{1}{\tau}\right)\right) d\tau\right] \\
&= \frac{1}{(4\pi)^{n/2}} \left(\frac{|x|}{2\sqrt{\lambda}}\right)^{1-(n/2)} \int_0^{\infty} \tau^{-n/2} \exp\left(-\frac{|x|\sqrt{\lambda}}{2}\left(\tau + \frac{1}{\tau}\right)\right) d\tau \\
&= \frac{1}{(4\pi)^{n/2}} \left(\frac{|x|}{2\sqrt{\lambda}}\right)^{1-(n/2)} \left[\int_1^{\infty} u^{(n/2)-2} \exp\left(-\frac{|x|\sqrt{\lambda}}{2}\left(u + \frac{1}{u}\right)\right) du\right. \\
&\quad \left. + \int_1^{\infty} \tau^{-n/2} \exp\left(-\frac{|x|\sqrt{\lambda}}{2}\left(\tau + \frac{1}{\tau}\right)\right) d\tau\right] \\
&\leqslant \frac{1}{(4\pi)^{n/2}} \left(\frac{|x|}{2\sqrt{\lambda}}\right)^{1-(n/2)} \left[\int_1^{\infty} u^{(n/2)-2} \exp\left(-\frac{|x|\sqrt{\lambda}}{2} u\right) du\right. \\
&\quad \left. + \int_1^{\infty} \tau^{-n/2} \exp\left(-\frac{|x|\sqrt{\lambda}}{2}\tau\right) d\tau\right]
\end{aligned}$$

[3] See [1] formula 9.6.24.

$$\leqslant \frac{1}{(4\pi)^{n/2}} \left(\frac{|x|}{2\sqrt{\lambda}}\right)^{1-(n/2)} \exp(-|x|\sqrt{\lambda}/2)$$
$$\left[\int_1^\infty u^{(n/2)-2} \exp\left(-\frac{|x|\sqrt{\lambda}}{2}(u-1)\right) du + \frac{2}{|x|\sqrt{\lambda}}\right]$$

and hence in the case $n = 2$ that

$$B_{2\lambda}(x) \leqslant \frac{1}{4\pi} \exp(-|x|\sqrt{\lambda}/2) \left[\int_1^\infty \exp\left(-\frac{|x|\sqrt{\lambda}}{2}(u-1)\right) du + \frac{2}{|x|\sqrt{\lambda}}\right]$$
$$= \frac{1}{\pi} \frac{1}{|x|\sqrt{\lambda}} \exp(-|x|\sqrt{\lambda}/2) ,$$

whereas it follows in the case $n \neq 2$ that

$$B_{n\lambda}(x) \leqslant \frac{1}{(4\pi)^{n/2}} \left(\frac{|x|}{2\sqrt{\lambda}}\right)^{1-(n/2)} \exp(-|x|\sqrt{\lambda}/2)$$
$$\cdot \left[\int_0^\infty (y+1)^{(n/2)-2} \exp\left(-\frac{|x|\sqrt{\lambda}}{2} y\right) dy + \frac{2}{|x|\sqrt{\lambda}}\right]$$
$$\leqslant \frac{1}{(4\pi)^{n/2}} \left(\frac{|x|}{2\sqrt{\lambda}}\right)^{1-(n/2)} \exp(-|x|\sqrt{\lambda}/2)$$
$$\cdot \left[\int_0^\infty y^{(n/2)-2} \exp\left(-\frac{|x|\sqrt{\lambda}}{2} y\right) dy + \frac{2}{|x|\sqrt{\lambda}}\right]$$
$$= \frac{1}{(4\pi)^{n/2}} \left(\frac{|x|}{2\sqrt{\lambda}}\right)^{1-(n/2)} \exp(-|x|\sqrt{\lambda}/2) \left[\frac{\Gamma((n/2)-1)}{(|x|\sqrt{\lambda}/2)^{(n/2)-1}} + \frac{2}{|x|\sqrt{\lambda}}\right]$$
$$= \frac{1}{(4\pi)^{n/2}} \exp(-|x|\sqrt{\lambda}/2) \left[\Gamma((n/2)-1) \left(\frac{|x|}{2}\right)^{2-n} + \frac{1}{\lambda}\left(\frac{|x|}{2\sqrt{\lambda}}\right)^{-(n/2)}\right].$$

As a consequence, $h_n \in L^1_{\mathbb{C}}(\mathbb{R}^n)$. Further, it follows for every $x \in \mathbb{R}^n \backslash \{0\}$ that

$$\int_{\mathbb{R}^n} \left|(4\pi)^{-n/2} \int_{\varepsilon^2/2}^\infty t^{-n/2} \exp\left(-\lambda t - \frac{x^2}{4t}\right) dt - B_n(x)\right| dx$$
$$= (4\pi)^{-n/2} \int_{\mathbb{R}^n} \left[\int_0^{\varepsilon^2/2} t^{-n/2} \exp\left(-\lambda t - \frac{x^2}{4t}\right) dt\right] dx$$
$$= \int_0^{\varepsilon^2/2} \exp(-\lambda t)\, dt = \frac{1}{\lambda}\left[1 - \exp(-\lambda\varepsilon^2/2)\right] .$$

Note that in this Fubini's theorem can been applied since the continuous and hence measurable function $((0, \varepsilon^2/2) \times \mathbb{R}^n, (t, x) \mapsto t^{-n/2} \exp(-\lambda t - (x^2/(4t)))$ is integrable by Tonelli's theorem. Finally, since

$$\lim_{\varepsilon \to 0} \| \exp(-\varepsilon^2(|\ |^2 + \lambda)/2) - \chi_{\mathbb{R}^n} \|_\infty = 0 ,$$

it follows from (10.2.5) that

$$(F_1 B_{n\lambda})(k) = \frac{1}{|k|^2 + \lambda} \, .$$

for every $k \in \mathbb{R}^n$. In the following, we drop additional restrictions of n. We note that it follows by a standard criterion for integral operators [4] that for every $G \in L^1_{\mathbb{C}}(\mathbb{R}^n)$ by

$$\text{Int}\,(\text{G})f := G * f$$

for every $f \in L^1_{\mathbb{C}}(\mathbb{R}^n)$ there is given a bounded linear operator Int (G) of norm equal or smaller than $\|G\|_1$. The representation of the operator $(-\triangle + \lambda)^{-1}$ follows by the convolution theorem for elements of $L^1_{\mathbb{C}}(\mathbb{R}^n)$. For this, let $f \in L^1_{\mathbb{C}}(\mathbb{R}^n) \cap L^2_{\mathbb{C}}(\mathbb{R}^n)$. Then

$$\begin{aligned}(-\triangle + \lambda)^{-1} f &= F_2^{-1} \frac{1}{|\ |^2 + \lambda} F_2 f = (2\pi)^{-n/2} F_2^{-1} \frac{1}{|\ |^2 + \lambda} F_1 f \\ &= (2\pi)^{-n/2} F_2^{-1} F_1 (B_{n\lambda} * f) = B_{n\lambda} * f \, .\end{aligned}$$

Since $L^1_{\mathbb{C}}(\mathbb{R}^n) \cap L^2_{\mathbb{C}}(\mathbb{R}^n)$ is dense in $L^2_{\mathbb{C}}(\mathbb{R}^n)$, from this it follows that

$$(-\triangle + \lambda)^{-1} f = B_{n\lambda} * f$$

for all $f \in L^2_{\mathbb{C}}(\mathbb{R}^n)$. We note that

$$B_{n\lambda}(\lambda^{-1/2}\, x) = \lambda^{\frac{n}{2}-1} B_{n1}(x)$$

for all $\lambda > 0$ and $x \in \mathbb{R}^n$. For the cases $n = 1, 3$, this is obvious. For the cases $n \in \mathbb{N}^* \backslash \{1, 3\}$, this follows from

$$\begin{aligned}B_{n\lambda}(\lambda^{-1/2}\, x) &= \frac{1}{(4\pi)^{n/2}} \int_0^\infty t^{-n/2} \exp\left(-\lambda t - \frac{x^2}{4\lambda t}\right) dt \\ &= \frac{\lambda^{\frac{n}{2}-1}}{(4\pi)^{n/2}} \int_0^\infty \tau^{-n/2} \exp\left(-\tau - \frac{x^2}{4\tau}\right) d\tau = \lambda^{\frac{n}{2}-1} B_{n1}(x)\end{aligned}$$

for all $\lambda > 0$ and $x \in \mathbb{R}^n$. Further, it follows for $h \in BC(\mathbb{R}^n, \mathbb{C})$ for which there is $C \geqslant 0$ such that $|h(x) - h(y)| \leqslant C\, |x - y|$ for all $x, y \in \mathbb{R}^n$ and $f \in L^2_{\mathbb{C}}(\mathbb{R}^n)$ that

$$\begin{aligned}&\left(\left[T_h(-\triangle + \lambda)^{-1} - (-\triangle + \lambda)^{-1} T_h\right] f\right)(x) \\ &= \int_{\mathbb{R}^n} B_{n\lambda}(x - y)(h(x) - h(y)) f(y)\, dy\end{aligned}$$

for every $x, y \in \mathbb{R}^n$ and hence

$$\begin{aligned}\|T_h(-\triangle + \lambda)^{-1} - (-\triangle + \lambda)^{-1} T_h\| &\leqslant C \int_{\mathbb{R}^n} |x|\, B_{n\lambda}(x)\, dx \\ = C\, \lambda^{\frac{n}{2}-1} \int_{\mathbb{R}^n} |x|\, B_{n\lambda}(\lambda^{1/2}\, x)\, dx &= C\, \lambda^{-\frac{3}{2}} \int_{\mathbb{R}^n} |x|\, B_{n1}(x)\, dx \, .\end{aligned}$$

---

[4] Sometimes referred to as 'Schur's Lemma' (for integral operators). For instance, see the Corollary of Theorem 6.24 in [216] or Theorem 6.18 in [68].

Finally, note for future use that

$$\begin{aligned}
-\frac{1}{\pi}\int_0^\infty \lambda^{1/2} B_{n\lambda}(x-y)\,d\lambda &= -\frac{1}{\pi}\int_0^\infty \lambda^{\frac{n-1}{2}} B_{n1}(\lambda^{1/2}(x-y))\,d\lambda \\
&= -\frac{1}{\pi}\int_0^\infty \lambda^{\frac{n-1}{2}} b_n(\lambda^{1/2}\,|x-y|\,)\,d\lambda \\
&= -\frac{1}{\pi}\,\frac{1}{|x-y|^{n+1}}\int_0^\infty \bar{\lambda}^{\frac{n-1}{2}} b_n(\bar{\lambda}^{1/2})\,d\bar{\lambda} \\
&= -\frac{2}{\pi}\left(\int_0^\infty \tau^n b_n(\tau)\,d\tau\right)\frac{1}{|x-y|^{n+1}}
\end{aligned}$$

for all $x, y \in \mathbb{R}^n$ such that $x \neq y$ where $b_n : (0,\infty) \to \mathbb{R}$ is well-defined by $b_n(\,|x|\,) := B_{n1}(x)$ for all $x \in \mathbb{R}^n$.

'(vii)': For this, let $f \in W^2_{\mathbb{C}}(\mathbb{R}^n)$. By Lemma 10.1.3

$$\begin{aligned}
(-\triangle+1)^{1/2} f &= \frac{1}{\pi}\int_0^\infty \lambda^{-1/2}(-\triangle+\lambda+1)^{-1}(-\triangle+1)f\,d\lambda \\
&= \frac{1}{\pi}\int_0^\infty \lambda^{-1/2}\,[\,f-\lambda(-\triangle+\lambda+1)^{-1}f\,]\,d\lambda\,.
\end{aligned}$$

Further, since $T_b \in L(L^2_{\mathbb{C}}(\mathbb{R}^n), L^2_{\mathbb{C}}(\mathbb{R}^n))$

$$\begin{aligned}
&[T_b(-\triangle+1)^{1/2} - (-\triangle+1)^{1/2}T_b]f \\
&= -\frac{1}{\pi}\int_0^\infty \lambda^{1/2}\,[\,T_b(-\triangle+\lambda+1)^{-1}f - (-\triangle+\lambda+1)^{-1}T_b f\,]\,d\lambda \\
&= \frac{1}{\pi}\int_0^\infty \lambda^{1/2}\,(-\triangle+\lambda+1)^{-1}(\triangle T_b - T_b\triangle)(-\triangle+\lambda+1)^{-1}f\,d\lambda \\
&= \frac{1}{\pi}\int_0^\infty \sum_{j=1}^n \lambda^{1/2}\,(-\triangle+\lambda+1)^{-1}(\partial^j T_{b,j} + T_{b,j}\partial^j)(-\triangle+\lambda+1)^{-1}f\,d\lambda \\
&= \frac{1}{\pi}\int_0^\infty \sum_{j=1}^n \lambda^{1/2}\,(-\triangle+\lambda+1)^{-1}(T_{b,jj} + 2\,T_{b,j}\partial^j)(-\triangle+\lambda+1)^{-1}f\,d\lambda\,.
\end{aligned}$$

In addition, it follows for $\lambda > 0$ that

$$\begin{aligned}
&2\lambda^{1/2}\,(-\triangle+\lambda+1)^{-1}T_{b,j}\partial^j(-\triangle+\lambda+1)^{-1}f \\
&\quad = 2\lambda^{1/2}\,T_{b,j}(-\triangle+\lambda+1)^{-1}\partial^j(-\triangle+\lambda+1)^{-1}f \\
&\qquad + 2\lambda^{1/2}\,[(-\triangle+\lambda+1)^{-1}T_{b,j} - T_{b,j}(-\triangle+\lambda+1)^{-1}]\partial^j(-\triangle+\lambda+1)^{-1}f
\end{aligned}$$

and by (vi)

$$\|2\lambda^{1/2}\,[(-\triangle+\lambda+1)^{-1}T_{b,j} - T_{b,j}(-\triangle+\lambda+1)^{-1}]\partial^j(-\triangle+\lambda+1)^{-1}f\|_2$$

$$\leqslant 2\lambda^{1/2} K \left(\sum_{k=1}^{n} \|b_{,jk}\|_\infty \right) (\lambda+1)^{-3/2} \, 2^{-1} (\lambda+1)^{-1/2} \, \|f\|_2$$

$$\leqslant K \left(\sum_{k=1}^{n} \|b_{,jk}\|_\infty \right) (\lambda+1)^{-3/2} \, \|f\|_2$$

$$\|\lambda^{1/2} (-\triangle + \lambda + 1)^{-1} T_{b_{,jj}} (-\triangle + \lambda + 1)^{-1} f\|_2$$

$$\leqslant \|b_{,jj}\|_\infty \, \|f\|_2 \, \lambda^{1/2} (\lambda+1)^{-2}$$

and hence by (v)

$$\|[T_b(-\triangle+1)^{1/2} - (-\triangle+1)^{1/2} T_b] f\|_2$$

$$\leqslant \frac{2}{\pi} \left(\sum_{j=1}^{n} [\, 2\, \|b_{,j}\|_\infty + \|b_{,jj}\|_\infty] + K \sum_{j,k=1}^{n} \|b_{,jk}\|_\infty \right) \|f\|_2 \, .$$

From this follows (vi), since $W^2_{\mathbb{C}}(\mathbb{R}^n)$ is dense in $(W^1_{\mathbb{C}}(\mathbb{R}^n), \|\;\|_1)$ and $T_b, (-\triangle+1)^{1/2}$ define continuous operators from $(W^1_{\mathbb{C}}(\mathbb{R}^n), \|\;\|_1)$ to $L^2_{\mathbb{C}}(\mathbb{R}^n)$.

□

Theorem 10.2.2. Let $n, p \in \mathbb{N}^*$, $X := (L^2_{\mathbb{C}}(\mathbb{R}^n))^p$, $Y := (W^1_{\mathbb{C}}(\mathbb{R}^n))^p$,

$$S := \bigtimes_{j=1}^{p} (-\triangle + 1)^{1/2}$$

and J be a non-empty open interval of $\mathbb{R}$. Further, let I be some closed subinterval of J, $l(I) \geqslant 0$ be the length of I, $A^1, \dots, A^n : J \times \mathbb{R}^n \to M(p \times p, \mathbb{C})$, $B : J \times \mathbb{R}^n \to M(p \times p, \mathbb{C})$ such that $A^j(t,x)$ is Hermitian for all $(t,x) \in J \times \mathbb{R}^n$ and $j \in \{1,\dots,n\}$ as well as such that

a) For all $j \in \{1,\dots,n\}$, $k,l \in \{1,\dots,p\}$ and $t \in I$: $A^j_{kl}(t,\cdot) \in C^2(\mathbb{R}^n, \mathbb{C})$ such that all partial derivatives from zeroth up to second order, inclusively, are bounded. $B_{kl}(t,\cdot) \in C^1(\mathbb{R}^n, \mathbb{C})$ such that all partial derivatives from zeroth up to first order, inclusively, are bounded.
b) For all $j \in \{1,\dots,n\}$, $k,l \in \{1,\dots,p\}$, $m \in \{1,\dots,n\}$: The restrictions of $A^j_{kl,m}$, B_{kl} to $I \times \mathbb{R}^n$ are bounded.
c) For all $j, m_1, m_2 \in \{1,\dots,n\}$, $k,l \in \{1,\dots,p\}$: A^j_{kl}, A^j_{kl,m_1}, $A^j_{kl,m_1 m_2}$, B_{kl}, $B_{kl,m}$ are continuously partially differentiable with respect to time with derivative defining a bounded function on $I \times \mathbb{R}^n$.

Note that here we associate the index 0 to 'time', i.e., the coordinate projection of $J \times \mathbb{R}^n$ onto the first coordinate and the indices $1,\dots,n$ to the coordinate projections of $\mathbb{R}^n$. Finally, for every complex-valued function g which is a.e. defined on $\mathbb{R}^n$ and measurable, we denote by T_g the corresponding maximal multiplication operator in $L^2_{\mathbb{C}}(\mathbb{R}^n)$. Then

(i) S is a densely-defined linear, self-adjoint and bijective operator in X. In particular,

$$n^{-1/2}\left(\sum_{k=1}^{p}|||u_k|||_1^2\right)^{1/2} \leqslant \|u\|_Y \leqslant (2n+1)^{1/2}\left(\sum_{k=1}^{p}|||u_k|||_1^2\right)^{1/2} \tag{10.2.6}$$

for all $u = (u_1, \ldots, u_p) \in Y$.

(ii) For every $t \in I$, by

$$\left(Y \to X, u \mapsto \sum_{j=1}^{n} A^j(t,\cdot)\partial^j u + B(t,\cdot)u\right),$$

there is defined a densely-defined linear and quasi-accretive operator in X with bound

$$-\frac{1}{2}\sum_{j=1}^{n}\left(\sum_{k,l=1}^{p}\|A^j_{kl,j}(t,\cdot)\|_\infty^2\right)^{1/2} - \left(\sum_{k,l=1}^{p}\|B_{kl}(t,\cdot)\|_\infty^2\right)^{1/2} \tag{10.2.7}$$

whose closure $A(t)$ is the infinitesimal generator of a strongly continuous group on X.

(iii) $(A(t))_{t\in I}$ is a stable family of infinitesimal generators of strongly continuous semigroups on X with constants

$$\mu_I := \frac{1}{2}\left\{\sum_{j=1}^{n}\left(\sum_{k,l=1}^{p}\|A^j_{kl,j}|_{I\times\mathbb{R}^n}\|_\infty^2\right)^{1/2} + \left(\sum_{k,l=1}^{p}\|B_{kl}|_{I\times\mathbb{R}^n}\|_\infty^2\right)^{1/2}\right\},\ 1.$$

(iv) For every $t \in I$,

$$A(t)|_Y \in L((Y, \|\ \|_s), X). \tag{10.2.8}$$

Further,

$$A := (I \to L((Y, \|\ \|_s), X), t \mapsto A(t)|_Y)$$

is continuous.

(v) Define for every $t \in I$

$$\begin{aligned}
&\mathcal{B}_0(t)u := \\
&\Bigg(\sum_{j=1}^{n}\sum_{l=1}^{p}[\,(-\Delta+1)^{1/2}T_{A^j_{1l}(t,\cdot)} - T_{A^j_{1l}(t,\cdot)}(-\Delta+1)^{1/2}\,]\,\partial^j(-\Delta+1)^{-1/2}u_l \\
&\quad+\sum_{l=1}^{p}[\,(-\Delta+1)^{1/2}T_{B_{1l}(t,\cdot)}(-\Delta+1)^{-1/2} - T_{B_{1l}(t,\cdot)}\,]\,u_l, \ldots, \\
&\sum_{j=1}^{n}\sum_{l=1}^{p}[\,(-\Delta+1)^{1/2}T_{A^j_{pl}(t,\cdot)} - T_{A^j_{pl}(t,\cdot)}(-\Delta+1)^{1/2}\,]\,\partial^j(-\Delta+1)^{-1/2}u_l
\end{aligned}$$

$$+\sum_{l=1}^{p}[\,(-\triangle+1)^{1/2}T_{B_{pl}(t,\cdot)}(-\triangle+1)^{-1/2}-T_{B_{pl}(t,\cdot)}\,]\,u_l\Bigg)\in X$$

for every $u=(u_1,\dots,u_l)\in Y$. Then

$$SA(t)u=(A(t)+\mathcal{B}_0(t))Su$$

for all $u\in\tilde{D}:=(W^2_{\mathbb{C}}(\mathbb{R}^n))^p$. Further, $\mathcal{B}_0(t)$ has an extension to an element of $L(X,X)$ which we denote by $\mathcal{B}(t)$ and $\mathcal{B}:=(I\to L(X,X),t\mapsto\mathcal{B}(t))\in C(I,L(X,X))$. Finally, for every $t\in I$, there is $\lambda<0$ such that

$$(A(t)+\mathcal{B}(t)-\lambda)S\tilde{D}$$

is dense in X.

(vi) Assume that Y is equipped with $\|\ \|_S$. Then there is a unique Y/X-evolution operator $U\in C_*(\triangle(I),L(X,X))$ for A.

Proof. '(i)': Obviously, S is a densely-defined linear, self-adjoint operator in X as direct sum of the densely-defined linear and self-adjoint operator $(-\triangle+1)^{1/2}$ operator in $L^2_{\mathbb{C}}(\mathbb{R}^n)$. Also is S bijective as a consequence of the bijectivity of $(-\triangle+1)^{1/2}$. In particular, it follows by Lemma 10.1.1, (10.1.13), (10.2.3) that

$$\begin{aligned}\|Su\|_X^2&=\sum_{k=1}^{p}\|(-\triangle+1)^{1/2}u_k\|_2^2\leqslant 2\,\|u\|_X^2+2\sum_{k=1}^{p}\|(-\triangle)^{1/2}u_k\|_2^2\\&\leqslant 2\,\|u\|_X^2+2\sum_{k=1}^{p}\left(\sum_{j=1}^{n}\|\partial^j u_k\|_2\right)^2\leqslant 2\,\|u\|_X^2+2n\sum_{k=1}^{p}\sum_{j=1}^{n}\|\partial^j u_k\|_2^2\\&\leqslant 2n\sum_{k=1}^{p}\|u_k\|_1^2\end{aligned}$$

for all $u=(u_1,\dots,u_p)\in Y$. Further, it follows for every $f\in W^2_{\mathbb{C}}(\mathbb{R}^n)$ that

$$\|(-\triangle+1)^{1/2}f\|_2^2=\langle f|(-\triangle+1)f\rangle_2=\|(-\triangle)^{1/2}f\|_2^2+\|f\|_2^2$$

and since $W^2_{\mathbb{C}}(\mathbb{R}^n)$ is dense in $(W^1_{\mathbb{C}}(\mathbb{R}^n),\|\ \|_1)$ that

$$\|(-\triangle+1)^{1/2}f\|_2^2=\|(-\triangle)^{1/2}f\|_2^2+\|f\|_2^2 \tag{10.2.9}$$

for all $f\in W^1_{\mathbb{C}}(\mathbb{R}^n)$. Using this along with (10.2.3), we conclude that

$$\|Su\|_X^2\geqslant\sum_{k=1}^{p}\|(-\triangle)^{1/2}u_k\|_2^2\geqslant\sum_{k=1}^{p}\|\partial^j u_k\|_2^2$$

for every $j\in\{1,\dots,n\}$ and hence that

$$\|u\|_Y^2=\frac{1}{n}\,n\,[\,\|u\|_X^2+\|Su\|_X^2\,]\geqslant\frac{1}{n}\sum_{j=1}^{n}\sum_{k=1}^{p}[\,\|u_k\|_2^2+\|\partial^j u_k\|_2^2\,]\geqslant\frac{1}{n}\sum_{k=1}^{p}\|u_k\|_1^2\,.$$

'(ii)': For this, let $t \in I$. First, since $B_{kl}(t,\cdot)$ are bounded for all $k,l \in \{1,\dots,p\}$, it follows for every $u \in X$ that

$$B(t,\cdot)u := {}^t\Big(\sum_{l=1}^{p} B_{1l}(t,\cdot)u_l,\dots,\sum_{l=1}^{p} B_{pl}(t,\cdot)u_l\Big) \in X$$

and

$$\begin{aligned}\|B(t,\cdot)u\|_X^2 &= \sum_{k=1}^{p}\|\sum_{l=1}^{p} B_{kl}(t,\cdot)u_l\|_2^2 \leqslant \sum_{k=1}^{p}\left(\sum_{l=1}^{p}\|B_{kl}(t,\cdot)\|_\infty\|u_l\|_2\right)^2 \\ &\leqslant \left(\sum_{k,l=1}^{p}\|B_{kl}(t,\cdot)\|_\infty^2\right)\|u\|_X^2 .\end{aligned}$$

Hence, obviously, it follows that by $(X \to X, u \mapsto B(t,\cdot)u)$ there is given a bounded linear operator $B(t,\cdot)$ on X such that

$$\|B(t,\cdot)\| \leqslant \left(\sum_{k,l=1}^{p}\|B_{kl}(t,\cdot)\|_\infty^2\right)^{1/2} .$$

Further, according to Corollary 5.5.3, by $A_0(t) := (Y \to X, u \mapsto \sum_{j=1}^{n} A^j(t,\cdot)\partial^j u)$ there is defined a densely-defined, linear operator in X whose closure is the infinitesimal generator of a strongly continuous group on X. In addition, defining $C(t) \in L(X,X)$ by

$$C(t)u := \frac{1}{2}\sum_{j=1}^{n} A^j_{,j}(t,\cdot)u$$

for every $u \in X$, it follows that

$$\begin{aligned}\mathrm{Re}\,\langle u|\overline{A_0(t)}\,u\rangle_X &= \mathrm{Re}\,\langle iu\,|\,\overline{i(A_0(t)+C(t))}u\rangle_X - \mathrm{Re}\,\langle u\,|\,C(t)u\rangle_X \\ &= -\frac{1}{2}\sum_{j=1}^{n}\mathrm{Re}\,\langle u\,|\,A^j_{,j}(t,\cdot)u\rangle_X \geqslant -\frac{1}{2}\sum_{j=1}^{n}\|A^j_{,j}(t,\cdot)u\|_X\,\|u\|_X \\ &\geqslant -\frac{1}{2}\sum_{j=1}^{n}\left(\sum_{k,l=1}^{p}\|A^j_{kl,j}(t,\cdot)\|_\infty^2\right)^{1/2}\|u\|_X^2\end{aligned}$$

for all $u \in Y$ and hence that $\overline{A_0(t)+B(t,\cdot)}$ is quasi-accretive with bound given by (10.2.7).
'(iii)': The statement is a simple consequence of (ii), the boundedness of the restrictions of $A^j_{kl}, A^j_{kl,m}, B_{kl}$, $k,l \in \{1,\dots,p\}$ and $j,m \in \{1,\dots,n\}$, to $I \times \mathbb{R}^n$ and Lemma 8.6.4.
'(iv)': For this, let $t \in I$. Then

$$\|\sum_{j=1}^{n} A^j(t,\cdot)\,\partial^j u\,\|_X^2 \leqslant n\sum_{j=1}^{n}\|A^j(t,\cdot)\,\partial^j u\,\|_X^2$$

$$\leqslant n \sum_{j=1}^{n} \left(\sum_{k,l=1}^{p} \|A^j_{kl}(t,\cdot)\|^2_\infty \right) \|\partial^j u\|^2_X$$

$$\leqslant n \left(\sum_{j=1}^{n} \sum_{k,l=1}^{p} \|A^j_{kl}(t,\cdot)\|^2_\infty \right) \sum_{k=1}^{p} \|\!|u_k|\!\|^2_1$$

for every $u \in Y$ and hence by using the continuity of the inclusion of $(Y, \| \|_S)$ into X the validity of (10.2.8). Further,

$$\| \sum_{j=1}^{n} A^j(t_2,\cdot)\,\partial^j u - \sum_{j=1}^{n} A^j(t_1,\cdot)\,\partial^j u \|^2_X$$

$$\leqslant n \left(\sum_{j=1}^{n} \sum_{k,l=1}^{p} \|A^j_{kl}(t_2,\cdot) - A^j_{kl}(t_1,\cdot)\|^2_\infty \right) \sum_{k=1}^{p} \|\!|u_k|\!\|^2_1 \,,$$

$$\|B(t_2,\cdot)u - B(t_1,\cdot)u\|^2_X \leqslant \left(\sum_{k,l=1}^{p} \|B_{kl}(t_2,\cdot) - B_{kl}(t_1,\cdot)\|^2_\infty \right) \sum_{k=1}^{p} \|\!|u_k|\!\|^2_1 .$$

for all $t_1, t_2 \in I$ and $u \in Y$. Since by the mean value theorem

$$\|A^j_{kl}(t_2,\cdot) - A^j_{kl}(t_1,\cdot)\|_\infty \leqslant \|A^j_{kl,0}|_{I\times\mathbb{R}^n}\|_\infty \, |t_2 - t_1| \,,$$
$$\|B_{kl}(t_2,\cdot) - B_{kl}(t_1,\cdot)\|_\infty \leqslant \|B_{kl,0}|_{I\times\mathbb{R}^n}\|_\infty \, |t_2 - t_1| \tag{10.2.10}$$

for all $k, l \in \{1,\dots,p\}$, $j \in \{1,\dots,n\}$, it follows also the continuity of A.

'(v)': For this, let $t \in I$, $u \in \tilde{D}$. Then $S\,u \in Y$,

$$A(t)u = \left(\sum_{j=1}^{n}\sum_{l=1}^{p} A^j_{1l}(t,\cdot)\,\partial^j u_l + \sum_{l=1}^{p} B_{1l}(t,\cdot)u_l, \dots, \right.$$
$$\left. \sum_{j=1}^{n}\sum_{l=1}^{p} A^j_{pl}(t,\cdot)\,\partial^j u_l + \sum_{l=1}^{p} B_{pl}(t,\cdot)u_l \right) \in Y$$

and

$$(S A(t) - A(t) S)u$$
$$= \left(\sum_{j=1}^{n}\sum_{l=1}^{p} (-\triangle + 1)^{1/2} A^j_{1l}(t,\cdot)\,\partial^j u_l + \sum_{l=1}^{p} (-\triangle + 1)^{1/2} B_{1l}(t,\cdot)u_l, \dots, \right.$$
$$\left. \sum_{j=1}^{n}\sum_{l=1}^{p} (-\triangle + 1)^{1/2} A^j_{pl}(t,\cdot)\,\partial^j u_l + \sum_{l=1}^{p} (-\triangle + 1)^{1/2} B_{pl}(t,\cdot)u_l \right)$$
$$- \left(\sum_{j=1}^{n}\sum_{l=1}^{p} A^j_{1l}(t,\cdot)\,\partial^j (-\triangle + 1)^{1/2} u_l + \sum_{l=1}^{p} B_{1l}(t,\cdot)(-\triangle + 1)^{1/2} u_l, \dots, \right.$$
$$\left. \sum_{j=1}^{n}\sum_{l=1}^{p} A^j_{pl}(t,\cdot)\,\partial^j (-\triangle + 1)^{1/2} u_l + \sum_{l=1}^{p} B_{pl}(t,\cdot)(-\triangle + 1)^{1/2} u_l \right)$$

$$= \left(\sum_{j=1}^{n} \sum_{l=1}^{p} [\,(-\triangle+1)^{1/2} A^{j}_{1l}(t,\cdot) - A^{j}_{1l}(t,\cdot)(-\triangle+1)^{1/2}\,]\,\partial^{j}(-\triangle+1)^{-1/2}(Su)_l \right.$$
$$+ \sum_{l=1}^{p} [\,(-\triangle+1)^{1/2} B_{1l}(t,\cdot)(-\triangle+1)^{-1/2} - B_{1l}(t,\cdot)\,](Su)_l, \dots,$$
$$\sum_{j=1}^{n} \sum_{l=1}^{p} [\,(-\triangle+1)^{1/2} A^{j}_{pl}(t,\cdot) - A^{j}_{pl}(t,\cdot)(-\triangle+1)^{1/2}\,]\,\partial^{j}(-\triangle+1)^{-1/2}(Su)_l$$
$$\left. + \sum_{l=1}^{p} [\,(-\triangle+1)^{1/2} B_{pl}(t,\cdot)(-\triangle+1)^{-1/2} - B_{pl}(t,\cdot)\,](Su)_l \right) = \mathcal{B}_0(t) S u\,.$$

Further, by Lemma 10.2.1 (ii), (iv), (vi), it follows that $\mathcal{B}_0(t)$ is a densely-defined and bounded linear operator on X, which hence can be uniquely extended to a bounded linear operator $\mathcal{B}(t)$ on X of the same norm as $\mathcal{B}_0(t)$. Noting that

$$\begin{aligned}
&[\,(-\triangle+1)^{1/2} T_{b_1} - T_{b_1}(-\triangle+1)^{1/2}\,]\,\partial^{j}(-\triangle+1)^{-1/2} \\
&\quad - [\,(-\triangle+1)^{1/2} T_{b_2} - T_{b_2}(-\triangle+1)^{1/2}\,]\,\partial^{j}(-\triangle+1)^{-1/2} \\
&= [\,(-\triangle+1)^{1/2} T_{b_1-b_2} - T_{b_1-b_2}(-\triangle+1)^{1/2}\,]\,\partial^{j}(-\triangle+1)^{-1/2} \\
&\quad [\,(-\triangle+1)^{1/2} T_{b_1}(-\triangle+1)^{-1/2} - T_{b_1}\,] - [\,(-\triangle+1)^{1/2} T_{b_2}(-\triangle+1)^{-1/2} - T_{b_2}\,] \\
&= [\,(-\triangle+1)^{1/2} T_{b_1-b_2}(-\triangle+1)^{-1/2} - T_{b_1-b_2}\,]
\end{aligned}$$

for all $b_1, b_2 \in C^1(\mathbb{R}^n, \mathbb{C})$ such that $b_1, b_{1,k}, b_2, b_{2,k}$ are bounded for all $k \in \{1, \dots, n\}$, it follows by Lemma 10.2.1 (ii), (iv), (vi) along with the estimates for the higher derivatives analogous to (10.2.10) the continuity of $\mathcal{B}$. Finally, for every $t \in I$, the closure of the operator $A(t)+\mathcal{B}_0(t)$ is the generator of strongly continuous semigroup on X. Hence $S\tilde{D} = Y$ is a core for that operator and as a consequence it follows the existence of $\lambda < 0$ such that $(A(t) + \mathcal{B}_0(t) - \lambda)S\tilde{D}$ is dense in X.
'(vi)': The statement (vi) is an immediate consequence of Lemma 6.2.2 (ii), Theorem 9.0.6 and the statements (i)-(v). □

Remark 10.2.3. By the stronger commutator estimate given in [113], it follows that all statements in the previous theorem are true with conditions a) and c) replaced by the following weaker conditions:

a) For all $j \in \{1, \dots, n\}$, $k, l \in \{1, \dots, p\}$ and $t \in I$: $A^{j}_{kl}(t,\cdot) \in C^1(\mathbb{R}^n, \mathbb{C})$ such that all partial derivatives from zeroth up to first order, inclusively, are bounded. $B_{kl}(t,\cdot) \in C^1(\mathbb{R}^n, \mathbb{C})$ such that all partial derivatives from zeroth up to first order, inclusively, are bounded.
c) For all $j, m \in \{1, \dots, n\}$, $k, l \in \{1, \dots, p\}$: A^{j}_{kl}, $A^{j}_{kl,m}$ are continuously partially differentiable with respect to time with derivative defining a bounded function on $I \times \mathbb{R}^n$.

11

The Quasi-Linear Evolution Equation

In this chapter, it will be proved the well-posedness (local in time) of the initial value problem for quasi-linear evolution equations of the form

$$u'(t) = -A(t, u(t))u(t) + f(t, u(t))$$

under certain assumptions. See also [82, 109, 111, 112, 119, 168, 188, 209]. For the solution of the analogous second order equation, see [207], and for abstract quasilinear integrodifferential equations of hyperbolic type, see [163, 164]. See also [150]. Here t is from some non-empty open subinterval I of $\mathbb{R}$, $'$ denotes the ordinary derivative of functions assuming values in a Banach space X; $A(t, \xi)$, $(t, \xi) \in I \times W$ is a family of infinitesimal generators of strongly continuous semigroups on X where W is a subset of X; $f : I \times W \to X$ and $u : I \to X$ is such that $u(t) \in D(A(t)) \cap W$ for every $t \in I$. The proof proceeds by iteration using Theorem 9.0.6 for non-autonomous linear evolution equations and Theorem 11.0.5 which is a variation of Banach's fixed point theorem. The result will be strong enough to allow for variable domains of the operators occurring in the equation.

Theorem 11.0.4. (Banach fixed point theorem) Let E be a closed subset of a Banach space $(X, \| \|)$, and let $F : E \to E$ be a contraction, i.e., let there exist $\alpha \in [0, 1)$ such that

$$\|F\xi - F\eta\| \leqslant \alpha \cdot \|\xi - \eta\| \tag{11.0.1}$$

for all $\xi, \eta \in E$. Then F has a uniquely determined fixed point, i.e., a uniquely determined $\xi_* \in E$ such that

$$F\xi_* = \xi_* \, .$$

Further,

$$\|\xi - \xi_*\| \leqslant \frac{\|\xi - F\xi\|}{1 - \alpha} \tag{11.0.2}$$

and

$$\lim_{\nu \to \infty} F^\nu \xi = \xi_* \tag{11.0.3}$$

for every $\xi \in E$ where F^ν for $\nu \in \mathbb{N}$ is inductively defined by $F^0 := \mathrm{id}_E$ and $F^{k+1} := F \circ F^k$, for $k \in \mathbb{N}$.

Proof. Note that (11.0.1) implies that F is continuous. Further, define $f : E \to \mathbb{R}$ by

$$f(\xi) := \|\xi - F\xi\|$$

for all $\xi \in E$. Now let $\xi \in X$. Then it follows

$$\|F^{\nu+\mu'}\xi - F^{\nu+\mu}\xi\| \leqslant \sum_{k=\mu}^{\mu'-1} \|F^{\nu+k+1}\xi - F^{\nu+k}\xi\| \leqslant \sum_{k=\mu}^{\mu'-1} \alpha^{\nu+k} f(\xi) \leqslant \frac{\alpha^{\nu}}{1-\alpha} f(\xi)$$

for all $\nu, \mu, \mu' \in \mathbb{N}$ such that $\mu' \geqslant \mu$. Hence $(F^{\nu}\xi)_{\nu\in\mathbb{N}}$ is a Cauchy-sequence and hence, by the completeness of $(X, \| \, \|)$ and the closedness of E, convergent to some $\xi_* \in E$. Further, it follows by the continuity of F that ξ_* is a fixed point of F. Further, if $\bar{\xi} \in E$ is some fixed point of F, then

$$\|\xi_* - \bar{\xi}\| = \|F\xi_* - F\bar{\xi}\| \leqslant \alpha \cdot \|\xi_* - \bar{\xi}\|$$

and hence $\bar{\xi} = \xi_*$ since the assumption $\bar{\xi} \neq \xi_*$ leads to the contradiction that $1 \leqslant \alpha$. Finally, let η be some element of E. Then

$$\begin{aligned} \|\eta - \xi_*\| &= \|\eta - F\xi_*\| = \|\eta - F\eta + F\eta - F\xi_*\| \\ &\leqslant \|\eta - F\eta\| + \|F\eta - F\xi_*\| \leqslant f(\eta) + \alpha \cdot \|\eta - \xi_*\| \end{aligned}$$

and hence (11.0.2). □

Theorem 11.0.5. Let $(X, \| \, \|_X)$, $(Z, \| \, \|_Z)$ be Banach spaces, Y a subspace of X, $\| \, \|_Y$ a norm on Y such that $(Y, \| \, \|_Y)$ is a Banach space and such that the inclusion $\iota_{Y\hookrightarrow X}$ of $(Y, \| \, \|_Y)$ into X is continuous. Further, let $S \in L((Y, \| \, \|_Y), Z)$ be such that

$$\|\xi\|_Y \leqslant c \cdot (\|\xi\|_X + \|S\xi\|_Z)$$

for all $\xi \in Y$ and some $c > 0$. Finally, let E be a nonempty bounded closed subset of $(Y, \| \, \|_Y)$, $F : E \to E$, D a closed subset of Z containing SE, $\bar{E}$ the closure of E in $(X, \| \, \|_X)$ and $G : \bar{E} \times D \times D \to D$ a continuous map such that

(i) F is a contraction with respect to $\| \, \|_X$, i.e, there is $\alpha \in [0, 1)$ such that

$$\|F\xi - F\eta\|_X \leqslant \alpha \cdot \|\xi - \eta\|_X$$

for all $\xi, \eta \in E$,

(ii)

$$SF\xi = G(\xi, S\xi, SF\xi)$$

for all $\xi \in E$,

(iii) there are $\beta, \gamma \geqslant 0$ such that $\beta + \gamma < 1$ and

$$\|G(\xi, \eta_1', \eta_2') - G(\xi, \eta_1, \eta_2)\|_Z \leqslant \beta \cdot \|\eta_1' - \eta_1\|_Z + \gamma \cdot \|\eta_2' - \eta_2\|_Z \tag{11.0.4}$$

for all $\xi \in \bar{E}$, $\eta_1, \eta_2, \eta_1', \eta_2' \in D$.

Then F has a uniquely determined fixed point, i.e., a unique determined $\xi_* \in E$ such that

$$F\xi_* = \xi_* \ .$$

Further, for every

$$\|\xi - \xi_*\|_X \leqslant \frac{\|\xi - F\xi\|_X}{1-\alpha}$$

and

$$\lim_{\nu \to \infty, Y} F^\nu \xi = \xi_*$$

for every $\xi \in E$ where F^ν for $\nu \in \mathbb{N}$ is inductively defined by $F^0 := \mathrm{id}_E$ and $F^{k+1} := F \circ F^k$, for $k \in \mathbb{N}$.

Proof. For this, let $\xi \in E$. Since F is a contraction with respect to $\| \ \|_X$, it follows that $(F^\nu \xi)_{\nu \in \mathbb{N}}$ is a Cauchy-sequence in X and hence, by the completeness of $(X, \| \ \|)$ and the closedness of $\bar{E}$, convergent to some $\xi_* \in \bar{E}$. Further, $G_{\xi_*} : D \to D$, defined by $G_{\xi_*}\eta := G(\xi_*, \eta, \eta)$ for all $\eta \in D$, is a contraction as a consequence of (11.0.4). Hence there is a uniquely determined $\eta_* \in D$ such that

$$\eta_* = G(\xi_*, \eta_*, \eta_*) \ .$$

Further,

$$\begin{aligned}
\|SF^{\nu+1}\xi - \eta_*\|_Z &= \|G(F^\nu\xi, SF^\nu\xi, SF^{\nu+1}\xi) - \eta_*\|_Z \\
&= \|G(F^\nu\xi, SF^\nu\xi, SF^{\nu+1}\xi) - G(\xi_*, \eta_*, \eta_*)\|_Z \\
&\leqslant \|G(F^\nu\xi, SF^\nu\xi, SF^{\nu+1}\xi) - G(F^\nu\xi, \eta_*, \eta_*)\|_Z \\
&\quad + \|G(F^\nu\xi, \eta_*, \eta_*) - G(\xi_*, \eta_*, \eta_*)\|_Z \\
&\leqslant \beta \cdot \|SF^\nu\xi - \eta_*\|_Z + \gamma \cdot \|SF^{\nu+1}\xi - \eta_*\|_Z \\
&\quad + \|G(F^\nu\xi, \eta_*, \eta_*) - G(\xi_*, \eta_*, \eta_*)\|_Z
\end{aligned}$$

and hence

$$a_{\nu+1} \leqslant \frac{\beta}{1-\gamma} a_\nu + b_\nu \tag{11.0.5}$$

for every $\nu \in \mathbb{N}$, where

$$a_\nu := \|SF^\nu\xi - \eta_*\|_Z \ , \ b_\nu := \frac{1}{1-\gamma} \|G(F^\nu\xi, \eta_*, \eta_*) - G(\xi_*, \eta_*, \eta_*)\|_Z$$

and $\lim_{\nu \to \infty} b_\nu = 0$, since G is continuous and $\lim_{\nu \to \infty} \|F^\nu \xi - \xi_*\|_X = 0$. In the following, we prove that $\lim_{\nu \to \infty} a_\nu = 0$, too. Note that

$$a_\nu = \|SF^\nu\xi - \eta_*\|_Z \leqslant \|SF^\nu\xi\|_Z + \|\eta_*\|_Z \leqslant \|S\|_{\mathrm{Op},Y,Z} \cdot \|F^\nu\xi\|_Y + \|\eta_*\|_Z$$

and hence that $a_0, a_1, \dots$ is bounded since E is a bounded subset of $(Y, \langle | \rangle_Y)$. Therefore, by the Bolzano-Weierstrass Theorem, there is a convergent subsequence of

$a_0, a_1, \ldots$. Further, it follows from (11.0.5) that any such sequence has 0 as its limit. (Note that $\beta/(1-\gamma) = (\beta+\gamma-\gamma)/(1-\gamma) < 1$.) Hence for given $\varepsilon > 0$, there is $\nu_0 \in \mathbb{N}$ such that $a_{\nu_0} \leqslant \varepsilon$ and at the same time such that $b_\nu \leqslant \varepsilon$ for all $\nu \in \mathbb{N}$ such that $\nu \geqslant \nu_0$. Hence it follows by (11.0.5) inductively that

$$a_{\nu_0+\mu} \leqslant \varepsilon \cdot \sum_{k=0}^{\mu} \left(\frac{\beta}{1-\gamma}\right)^k$$

for all $\mu \in \mathbb{N}$ and hence that

$$a_\nu \leqslant \frac{\varepsilon}{1 - \frac{\beta}{1-\gamma}}$$

for all $\nu \in \mathbb{N}$ such that $\nu \geqslant \nu_0$ and hence, finally, that $\lim_{\nu\to\infty} a_\nu = 0$, i.e., that

$$\lim_{\nu\to\infty} \|SF^\nu\xi - \eta_*\|_Z = 0 \, .$$

Therefore, since

$$\|F^\mu\xi - F^\nu\xi\|_Y \leqslant c \cdot (\|F^\mu\xi - F^\nu\xi\|_X + \|SF^\mu\xi - SF^\nu\xi\|_Z)$$

for all $\mu, \nu \in \mathbb{N}$, it follows that $(F^\nu\xi)_{\nu\in\mathbb{N}}$ is a Cauchy sequence in Y and hence convergent to some element of E. Hence, since the inclusion $\iota_{Y\hookrightarrow X}$ of $(Y, \| \|_Y)$ into X is continuous, it follows that $\xi_* \in E$,

$$\lim_{\nu\to\infty} \|F^\nu\xi - \xi_*\|_Y = 0$$

and by the continuity of $F : (E, \| \|_X) \to (E, \| \|_X)$ that $F\xi_* = \xi_*$. Further, if $\bar{\xi} \in E$ is some fixed point of F, then

$$\|\xi_* - \bar{\xi}\|_X = \|F\xi_* - F\bar{\xi}\|_X \leqslant \alpha \cdot \|\xi_* - \bar{\xi}\|_X$$

and hence $\bar{\xi} = \xi_*$ since the assumption $\bar{\xi} \neq \xi_*$ leads to the contradiction $1 \leqslant \alpha$. Finally, let η be some element of E. Then

$$\begin{aligned}\|\eta - \xi_*\|_X &= \|\eta - F\xi_*\|_X = \|\eta - F\eta + F\eta - F\xi_*\|_X \\ &\leqslant \|\eta - F\eta\|_X + \|F\eta - F\xi_*\|_X \leqslant \|\eta - F\eta\|_X + \alpha \cdot \|\eta - \xi_*\|_X\end{aligned}$$

and hence

$$\|\eta - \xi_*\|_X \leqslant \frac{\|\eta - F\xi\|_X}{1-\alpha} \, .$$

□

Lemma 11.0.6. Let $(X, \| \|_X)$, $(Y, \| \|_Y)$ and $(Z, \| \|_Z)$ be normed vector spaces. Further, let $K \subset X$ be non-empty and compact, $A \subset Y$ be closed, $U \subset Y$ and $V \subset Z$ be non-empty open sets, $v \in C(K, Y)$, $w \in C(K, Z)$, $(v_\nu)_{\nu\in\mathbb{N}} \in (C(K, Y))^{\mathbb{N}}$ uniformly convergent to v and such that $\text{Ran}(v_\nu) \subset A$ for all $\nu \in \mathbb{N}$ and $(w_\nu)_{\nu\in\mathbb{N}} \in (C(K, Z))^{\mathbb{N}}$ be uniformly convergent to w.

(i) In addition, let $g \in C(K \times A \times Z, Z)$. Then there is $C \geqslant 0$ such that

$$\|g(s, v_\nu(s), w_\nu(s))\|_Z \leqslant C$$

for all $s \in K$.

(ii) In addition, let $B \in C_*(K \times A \times Z, L(Z, Z))$ and $(\bar{w}_\nu)_{\nu \in \mathbb{N}} \in (C(K, Z))^{\mathbb{N}}$ be uniformly convergent to $\bar{w} \in C(K, Z)$. Then there is $C' \geqslant 0$ such that

$$\|B(s, v_\nu(s), w_\nu(s))\bar{w}_\nu(s)\|_Z \leqslant C'$$

for all $s \in K$.

(iii) In addition, let $B \in C_*(K \times U \times V, L(Z, Z))$, $(Z, \| \|_Z)$ be complete and $K_1 \subset U$, $K_2 \subset V$ be a non-empty compact subsets of Y and Z, respectively. Then there are $\rho > 0$ and $C'' \geqslant 0$ such that

$$S_\rho := \bigcup_{(\xi,\eta) \in K_1 \times K_2} K \times B_{\rho,Y}(\xi) \times B_{\rho,Z}(\eta) \quad \subset \quad K \times U \times V$$

and

$$\|B(s, \xi', \eta')\|_{\mathrm{Op},Z} \leqslant C''$$

for all $(s, \xi', \eta') \in S$.

Proof. '(i)': The proof is indirect. Assume that there is no such C. Then for every $n \in \mathbb{N}$, there are $s_n \in K$ and $\nu_n \in \mathbb{N}$ such that

$$\|g(s_n, v_{\nu_n}(s_n), w_{\nu_n}(s_n))\|_Z \geqslant n\,.$$

Since $g(\cdot, v_\nu, w_\nu) : K \to Z$ is continuous and hence $\mathrm{Ran}\, g(\cdot, v_\nu, w_\nu)$ is compact, for every $\nu \in \mathbb{N}$ also the union of a finite number of $\mathrm{Ran}\, g(\cdot, v_\nu, w_\nu)$, $\nu \in \mathbb{N}$ is compact and hence bounded. Therefore, we can assume without restriction that $(\nu_n)_{\nu \in \mathbb{N}}$ is increasing. Further, because of the compactness of K, we can assume without restriction that $(s_n)_{n \in \mathbb{N}}$ is converging to some $s \in K$. Because of the uniform convergence of $(v_\nu)_{\nu \in \mathbb{N}} \in (C(K, Y))^{\mathbb{N}}$ to v, of $(w_\nu)_{\nu \in \mathbb{N}} \in (C(K, Z))^{\mathbb{N}}$ to w, the continuity of v,w and the closedness of A, it follows that

$$\lim_{n \to \infty} v_{\nu_n}(s_n) = v(s) \in A\,,\ \lim_{n \to \infty} w_{\nu_n}(s_n) = w(s)$$

and hence by the continuity of g that

$$\lim_{n \to \infty} g(s_n, v_{\nu_n}(s_n), w_{\nu_n}(s_n)) = g(s, v(s), w(s)) \ \lightning\,.$$

'(ii)': The proof is indirect. Assume that there is no such C'. We define the auxiliary function $h : K \times A \times Z \times Z \to Z$ by

$$h(s, \xi, \eta, \eta') := B(s, \xi, \eta)\eta'$$

for every $(s, \xi, \eta, \eta') \in K \times A \times Z \times Z$. Then it follows by Lemma 8.1.1 (ii) that $h \in C(K \times A \times Z \times Z, Z)$ and by our assumption that for every $n \in \mathbb{N}$ there are $s_n \in K$ and $\nu_n \in \mathbb{N}$ such that

$$\|h(s_n, v_{\nu_n}(s_n), w_{\nu_n}(s_n), \bar{w}_{\nu_n}(s_n))\|_Z \geqslant n .$$

Since $h(\cdot, v_\nu, w_\nu, \bar{w}_\nu) : K \to Z$ is continuous and hence $\operatorname{Ran} h(\cdot, v_\nu, w_\nu, \bar{w}_\nu)$ is compact, for every $\nu \in \mathbb{N}$ also the union of a finite number of $\operatorname{Ran} h(\cdot, v_\nu, w_\nu, \bar{w}_\nu)$, $\nu \in \mathbb{N}$ is compact and hence bounded. Therefore, we can assume without restriction that $(\nu_n)_{\nu \in \mathbb{N}}$ is increasing. Further, because of the compactness of K, we can assume without restriction that $(s_n)_{n \in \mathbb{N}}$ is converging to some $s \in K$. Because of the uniform convergence of $(v_\nu)_{\nu \in \mathbb{N}} \in (C(K, Y))^{\mathbb{N}}$ to v, the uniform convergence of $(w_\nu)_{\nu \in \mathbb{N}} \in (C(K, Z))^{\mathbb{N}}$ to w and the uniform convergence of $(\bar{w}_\nu)_{\nu \in \mathbb{N}} \in (C(K, Z))^{\mathbb{N}}$ to $\bar{w}$, the continuity of v,w,$\bar{w}$ and the closedness of A, it follows that

$$\lim_{n \to \infty} v_{\nu_n}(s_n) = v(s) \in A \ , \ \lim_{n \to \infty} w_{\nu_n}(s_n) = w(s) \ , \ \lim_{n \to \infty} \bar{w}_{\nu_n}(s_n) = \bar{w}(s)$$

and hence by the continuity of h that

$$\lim_{n \to \infty} h(s_n, v_{\nu_n}(s_n), w_{\nu_n}(s_n), \bar{w}_{\nu_n}(s_n)) = h(s, v(s), w(s), \bar{w}(s)) \ \lightning \ .$$

'(iii)': First, it follows the existence of $\delta > 0$ such that $K_1 + B_{\delta,Y}(0_Y) \subset U$ and $K_2 + B_{\delta,Z}(0_Z) \subset V$. Hence it follows $S_\delta \subset K \times U \times V$. Now, assume that B is unbounded on every S_ρ for every $\rho > 0$ such that $\rho \leqslant \delta$. Further, let $(n_\nu)_{\nu \in \mathbb{N}^*}$ be some unbounded sequence in $\mathbb{N}^*$. Then for every $\nu \in \mathbb{N}^*$ there are $s_\nu \in K$, $\xi_\nu \in K_1$, $\eta_\nu \in K_2$ and $\xi'_\nu \in B_{\delta/\nu,Y}(\xi_\nu)$, $\eta'_\nu \in B_{\delta/\nu,Z}(\eta_\nu)$ such that

$$\|B(s_\nu, \xi'_\nu, \eta'_\nu)\|_{\mathrm{Op},Z} \geqslant n_\nu + 1$$

Hence there is $\eta''_\nu \in B_{1,Z}(0_Z)$ such that

$$\|B(s_\nu, \xi'_\nu, \eta'_\nu)\eta''_\nu\|_Z \geqslant n_\nu$$

for all $\nu \in \mathbb{N}^*$. Because of the compactness of K, K_1 and K_2, we can assume without restriction (replacing $(n_\nu)_{\nu \in \mathbb{N}^*}$ by some unbounded subsequence) that $(s_\nu)_{\nu \in \mathbb{N}^*}$, $(\xi_\nu)_{\nu \in \mathbb{N}^*}$ and $(\eta_\nu)_{\nu \in \mathbb{N}^*}$ are converging to some $s \in K$, $\xi \in K_1$ and $\eta \in K_2$, respectively. As a consequence, also $\lim_{\nu \to \infty} \xi'_\nu = \xi$ and $\lim_{\nu \to \infty} \eta'_\nu = \eta$. It follows from the strong continuity of B that

$$\lim_{\nu \to \infty} B(s_\nu, \xi'_\nu, \eta'_\nu)\eta'' = B(s, \xi, \eta)\eta''$$

for every $\eta'' \in Z$ and hence by the principle of uniform boundedness[1] also the boundedness of the sequence of operator norms of

$$(\, B(s_\nu, \xi'_\nu, \eta'_\nu)\,)_{\nu \in \mathbb{N}^*}$$

[1] See, e.g, Theorem III.9 in the first volume of [179].

and hence because of $\|\eta_\nu''\| \leqslant 1$ for $\nu \in \mathbb{N}^*$ also the boundedness of the sequence

$$(\|B(s_\nu, \xi_\nu', \eta_\nu')\eta_\nu'' \|_Z)_{\nu \in \mathbb{N}^*} \cdot \natural$$

$\square$

Theorem 11.0.7. Let $\mathbb{K} \in \{\mathbb{R}, \mathbb{C}\}$, $(X, \| \ \|)$, $(Z, \| \ \|_Z)$ $\mathbb{K}$-Banach spaces, $a, b \in \mathbb{R}$ be such that $a < b$, $I := [a, b]$, $\Phi : (L(X, X), +, ., \circ) \rightarrow (L(Z, Z), +, ., \circ)$ a strongly sequentially continuous nonexpansive homomorphism and S a closed linear map from some dense subspace Y of X into Z. Then $(Y, \| \ \|_Y)$, where $\| \ \|_Y := \| \ \|_S$, is a $\mathbb{K}$-Banach space and the inclusion $\iota_{Y \hookrightarrow X}$ of $(Y, \| \ \|_Y)$ into X is continuous. Further, let $\tilde{X}$ be a subspace of X containing Y and $\| \ \|_{\tilde{X}}$ be a norm on $\tilde{X}$ such that $(\tilde{X}, \| \ \|_{\tilde{X}})$ is a Banach space, $\|\xi\|_X \leqslant \|\xi\|_{\tilde{X}}$ for all $\xi \in \tilde{X}$, $\|\xi\|_{\tilde{X}} \leqslant \|\xi\|_Y$ for all $\xi \in Y$ and such that the identity on any bounded subset of $(Y, \| \ \|_Y)$, equipped in domain with the topology induced by $\| \ \|_X$ and in range with the topology induced by $\| \ \|_{\tilde{X}}$, is uniformly continuous. In addition, let W be a bounded open subset of $(Y, \| \ \|_Y)$ such that the closure $\tilde{W}$ of W in $(\tilde{X}, \| \ \|_{\tilde{X}})$ is contained in Y. If not said otherwise, in the following, we assume that every subset of Y, $\tilde{X}$ and Z_0 (see (iv) below) is equipped with the topology induced by $\| \ \|_Y, \| \ \|_{\tilde{X}}, \| \ \|_0$, respectively. Finally, let $(A(t, \xi))_{(t,\xi) \in I \times W}$ be a family of infinitesimal generators of strongly continuous semigroups $(T(t, \xi))_{(t,\xi) \in I \times W}$ on X, $B \in C_*(I \times \tilde{W} \times Z, L(Z, Z))$ be such that

$$\|B(t, \xi, S\xi)\|_{\mathrm{Op}, Z, Z} \leqslant \lambda_B \ , \quad \|B(t, \xi, \eta) - B(t, \xi, \eta')\|_{\mathrm{Op}, Z, Z} \leqslant \mu_B \|\eta - \eta'\|_Z$$

for all $(t, \xi) \in I \times \tilde{W}$, $\eta, \eta' \in Z$ and some $\mu_B \in [0, \infty)$ and $\lambda_A, \lambda_B, \lambda_f, \mu_A, \mu_g \in [0, \infty)$ such that

(i) **(Stability)** For every $u \in C(I, W) \cap \mathrm{Lip}(I, \tilde{X})$, the corresponding family $(A(t, u(t)))_{t \in I}$ is stable and there are common stability constants for all elements $C(I, W) \cap \mathrm{Lip}(I, \tilde{X})$ sharing a Lipschitz constant.

(ii) **(Continuity)** $Y \subset D(A(t, \xi))$, $A(t, \xi)|_Y \in L(Y, \tilde{X})$ for every $(t, \xi) \in I \times W$ and $A := (I \times W \rightarrow L(Y, \tilde{X}), (t, \xi) \mapsto A(t, \xi)|_Y) \in C(I \times W, L(Y, \tilde{X}))$. Moreover,

$$\|A(t, \xi)\|_{\mathrm{Op}, Y, \tilde{X}} \leqslant \lambda_A \ , \quad \|A(t, \xi') - A(t, \xi)\|_{\mathrm{Op}, Y, X} \leqslant \mu_A \|\xi' - \xi\|_X$$

for all $t \in I$, $\xi, \xi' \in W$.

(iii) **(Intertwining relation)** For every $(t, \xi) \in I \times W$

$$S \circ [T(t, \xi)](s) \supset [\hat{T}(t, \xi)](s) \circ S \tag{11.0.6}$$

for all $s \in [0, \infty)$ where $\hat{T}(t, \xi) : [0, \infty) \rightarrow L(Z, Z)$ is the strongly continuous semigroup generated by $\Phi(A(t, \xi)) + B(t, \xi, S\xi)$.

(iv) Further, let Z_0 be a dense subspace of Z contained in $D(\Phi(A(t, \xi)))$ for all $(t, \xi) \in I \times W$, $\| \ \|_0$ a norm on Z_0 such that $(Z_0, \| \ \|_0)$ is a $\mathbb{K}$-Banach space and such that the inclusion $\iota_{Z_0 \hookrightarrow Z}$ of $(Z_0, \| \ \|_0)$ into Z is continuous. Finally, let $\Phi(A(t, \xi))|_{Z_0} \in L(Z_0, Z)$ for all $(t, \xi) \in I \times W$ and $\Phi A := (I \times W \rightarrow L(Z_0, Z), (t, \xi) \mapsto \Phi(A(t, \xi))|_{Z_0}) \in C_*(I \times W, L(Z_0, Z))$.

(v) In addition, let $f \in C(I \times W, Y)$ be such that

$$\|f(t,\xi)\|_Y \leqslant \lambda_f \, , \; \|f(t,\xi) - f(t,\xi')\|_X \leqslant \mu_f \|\xi - \xi'\|_X$$

for all $t \in I$ and $\xi, \xi' \in W$.

(vi) Finally, let $g \in C(I \times \tilde{W} \times Z, Z)$ be such that

$$S f(t,\xi) = g(t,\xi, S\xi)$$

for all $(t,\xi) \in I \times W$ and

$$\|g(t,\xi,\eta) - g(t,\xi,\eta')\|_Z \leqslant \mu_g \|\eta - \eta'\|_Z$$

for all $(t,\xi) \in I \times \tilde{W}$ and $\eta, \eta' \in Z$.

Then for every compact subset K of W, there is $T \in (a,b]$ such that for every $\xi \in K$ there is $u \in C([a,T], W) \cap C^1([a,T], \tilde{X})$, where we define $C^1([a,t], \tilde{X})$ to consist of those functions that are differentiable on (a,t) and whose derivatives have continuous extensions to $[a,t]$, such that

$$u'(t) + A(t,u(t))u(t) = f(t,u(t))$$

for all $t \in (a,T)$ where differentiation is with respect to $\tilde{X}$ and $u(a) = \xi$. If $v \in C([a,T], W) \cap C^1((a,T), \tilde{X})$ is such that $v'(t) + A(t,v(t))v(t) = f(t,v(t))$ for all $t \in (a,T)$ and $v(a) = \xi$, then $v = u$.

Proof. In the following, we are going to apply Theorem 11.0.5 where $(X, \| \|_X)$, $(Z, \| \|_Z)$ are $(C(I',X), \| \|_{\infty,X})$, $(C(I',Y), \| \|_{\infty,Y})$ and $(C(I',Z), \| \|_{\infty,Z})$, respectively, I' is a non-empty closed subinterval of I and

$$\mathcal{S} := (C(I',Y) \to C(I',Z) \, , \; f \mapsto S \circ f) \, .$$

Then the inclusion

$$(C(I',Y) \to C(I',X) \, , \; f \mapsto \iota_{Y \hookrightarrow X} \circ f)$$

is continuous, because of $\|f(t)\|_X \leqslant \|f(t)\|_Y$ for all $t \in I'$ and hence $\|f\|_{\infty,X} \leqslant \|f\|_{\infty,Y}$ for all $f \in C(I',Y)$. Further, $\mathcal{S}$ is linear and, because of

$$\|S(f(t))\|_Z \leqslant \|f(t)\|_Y$$

for every $t \in I'$ and hence

$$\|\mathcal{S}f\|_{\infty,Z} \leqslant \|f\|_{\infty,Y}$$

for every $f \in C(I',Y)$, also continuous. In addition,

$$\|f(t)\|_Y = \left[\|f(t)\|_X^2 + \|S(f(t))\|_Z^2\right]^{1/2} \leqslant \|f(t)\|_X + \|S(f(t))\|_Z$$

for every $t \in I'$ and hence

$$\|f\|_{\infty,Y} \leqslant \|f\|_{\infty,X} + \|Sf\|_{\infty,Z}$$

for every $f \in C(I', Y)$. Finally, E, F, D and G will be defined below.

First, obviously, it follows from the assumptions that the inclusions $\iota_{\tilde{X}\hookrightarrow X}$, $\iota_{Y\hookrightarrow\tilde{X}}$ and $\iota_{L(Y,\tilde{X})\hookrightarrow L(Y,X)}$ are continuous. In the following, let $\xi \in W, \rho \in (0,\infty)$ be such that the closed ball $B_{Y,\rho}(\xi) := \{\eta \in W : \|\eta - \xi\|_Y \leqslant \rho\}$ is contained in W. If K is some compact subset of W and $\xi \in K$, ρ is chosen such that $B_\rho(\xi) \subset W$ for all $\xi \in K$.

Further, let $T' \in [0, b-a]$, $I' := [a, a+T']$, $L \in [0,\infty)$ and

$$E := \{v \in C(I', Y) \cap \mathrm{Lip}(I', \tilde{X}) : \mathrm{Ran}\, v \subset B_{Y,\rho}(\xi) \wedge \|v(t) - v(s)\|_{\tilde{X}} \leqslant L|t-s|,\\ t, s \in I'\} .$$

E is a non-empty subset of $C(I', Y)$ since it contains $(I' \to Y, t \mapsto \xi)$. E is bounded in $C(I', Y)$ since for every $t \in I'$ it also follows that

$$\|v(t)\|_Y \leqslant \|v(t) - \xi\|_Y + \|\xi\|_Y \leqslant \|\xi\|_Y + \rho$$

for all $t \in I'$. Finally, E is closed in $C(I', Y)$ since for every $(v_\nu)_{\nu\in\mathbb{N}} \in E^{\mathbb{N}}$, $v \in C(I', Y)$ such that

$$\lim_{\nu\to\infty} \|v_\nu - v\|_{\infty,Y} = 0 ,$$

it follows for every $t \in I'$ the convergence of $(v_\nu(t))_{\nu\in\mathbb{N}}$ in Y to $v(t)$ and hence $v(t) \in B_{Y,\rho}(\xi)$ since $B_{Y,\rho}(\xi)$ is a closed subset of Y. Further, since $\iota_{Y\hookrightarrow\tilde{X}}$ is continuous, it also follows for every $t \in I'$ the convergence of $(v_\nu(t))_{\nu\in\mathbb{N}}$ in $\tilde{X}$ to $v(t)$ and hence

$$\lim_{\nu\to\infty} \|v_\nu(t) - v_\nu(s)\|_{\tilde{X}} = \|v(t) - v(s)\|_{\tilde{X}} \leqslant L|t-s|$$

for all $t, s \in I'$. For each $v \in E$, we define

$$A_v(t) := A(t, v(t)) \ , \ B_v(t) := B(t, v(t), Sv(t)) \ , \ f_v(t) := f(t, v(t))$$

for every $t \in I'$. Note that, obviously, there is an extension $\hat{v}$ of v to an element of $C(I, W) \cap \mathrm{Lip}(I, \tilde{X})$ such that $\mathrm{Ran}\, \hat{v} \subset B_{Y,\rho}(\xi)$ and $\|\hat{v}(t) - \hat{v}(s)\|_{\tilde{X}} \leqslant L|t-s|$ for all $t, s \in I$. In addition, because of

$$\begin{aligned} A_v &:= \iota_{L(Y,\tilde{X})\hookrightarrow L(Y,X)} \circ A \circ (I' \to I \times W, t \mapsto (t, v(t))) \\ &= (I' \to L(Y,X), t \mapsto A(t, v(t))|_Y) \in C(I', L(Y,X)) \end{aligned}$$

and

$$B_v = B \circ (I' \to I \times \tilde{W} \times Z, t \mapsto (t, v(t), Sv(t))) \in C_*(I', L(Z,Z)) ,$$

note that $S : Y \to Z$ is continuous, it follows by Theorem 9.0.6 that there is a unique Y/X-evolution operator $U_v \in C_*(\triangle(I'), L(X,X))$ for A_v and, if $\mu_L \in \mathbb{R}$ and $c_L \in [1,\infty)$ are the stability constants common to the elements of E, that

$$\|U_v(t,r)\|_{\text{Op},X} \leqslant C_L := c_L\, e^{|\mu_L|\cdot(b-a)} \;,\; \|U_{vY}(t,r)\|_{\text{Op,Y}} \leqslant C'_L := c_L\, e^{|\mu_L + c\lambda_B|\cdot(b-a)} \tag{11.0.7}$$

for all $(t,r) \in \Delta(I')$. Since

$$f_v = f \circ (I' \to I \times W, t \mapsto (t, v(t))) \in C(I', Y) \,,$$

it follows by Corollary 9.0.7 and its proof that $u_v : I' \to X$ defined by

$$u_v(t) := U_v(t,a)\xi + \int_{a,Y}^{t} U_v(t,s) f_v(s)\, ds$$

for all $t \in I'$, where the integrand is a continuous Y-valued map and integration denotes weak Lebesgue integration with respect to $L(Y,\mathbb{K})$, satisfies $u_v \in C(I',Y)$, $u_v(a) = \xi$, $u_v|_{\mathring{I}'}$ is differentiable and in particular

$$u_v'(t) = -A_v(t)u_v(t) + f_v(t)$$

for all $t \in \mathring{I}'$. Note that also $f_v \in C(I', \tilde{X})$. By Theorem 3.2.4 (ii), it also follows that

$$u_v(t) = U_v(t,a)\xi + \int_{a,\tilde{X}}^{t} U_v(t,s) f_v(s)\, ds$$

for all $t \in I'$ where integration denotes weak Lebesgue integration with respect to $L(\tilde{X},\mathbb{K})$. Since for every $s \in I'$ and $h \in \mathbb{R}$ such that $s + h \in I'$,

$$\begin{aligned}
\|A_v(s+h)u_v(s+h) - A_v(s)u_v(s)\|_{\tilde{X}} &\leqslant \|\,(A_v(s+h) - A_v(s))\,u_v(s)\|_{\tilde{X}} \\
&\quad + \|\,(A_v(s+h) - A_v(s))\,(u_v(s+h) - u_v(s))\,\|_{\tilde{X}} \\
&\quad + \|A_v(s)\,(u_v(s+h) - u_v(s))\,\|_{\tilde{X}} \\
&\leqslant \|\,(A_v(s+h) - A_v(s))\,u_v(s)\|_{\tilde{X}} + 3\lambda_A \cdot \|u_v(s+h) - u_v(s)\|_Y \,,
\end{aligned}$$

this implies that $(I' \to \tilde{X}, s \mapsto A_v(s)u_v(s))$ is continuous. Finally, for every $(t,r) \in \mathrm{D} := \{(t,r) : r \in I' \wedge t \in [r,T']\}$ and all $h, h' \in \mathbb{R}$ such that $(t,r) + (h,h') \in \mathrm{D}$

$$\begin{aligned}
&\|A_v(t+h')U_v(t+h', r+h)f_v(r+h) - A_v(t)U_v(t,r)f_v(r)\|_{\tilde{X}} \\
&= \|A_v(t+h')\,[U_v(t+h', r+h)f_v(r+h) - U_v(t,r)f_v(r)] \\
&\quad + (A_v(t+h') - A_v(t))\,U_v(t,r)f_v(r)\|_{\tilde{X}} \\
&\leqslant \lambda_A \cdot \|U_v(t+h', r+h)f_v(r+h) - U_v(t,r)f_v(r)\|_{\mathrm{Y}} \\
&\quad + \|\,(A_v(t+h') - A_v(t))\,U_v(t,r)f_v(r)\|_{\tilde{X}} \\
&= \lambda_A\, \|U_v(t+h', r+h)\,(f_v(r+h) - f_v(r)) \\
&\quad + (U_v(t+h', r+h) - U_v(t,r))f_v(r)\|_{\mathrm{Y}} \\
&\quad + \|\,(A_v(t+h') - A_v(t))\,U_v(t,r)f_v(r)\|_{\tilde{X}} \\
&\leqslant \lambda_A \cdot \|U_v\|_{\infty,\mathrm{Y,Y}} \cdot \|f_v(r+h) - f_v(r)\|_{\mathrm{Y}} \\
&\quad + \lambda_A \cdot \|U_v(t+h', r+h) - U_v(t,r))f_v(r)\|_{\mathrm{Y}} \\
&\quad + \|\,(A_v(t+h') - A_v(t))\,U_v(t,r)f_v(r)\|_{\tilde{X}}
\end{aligned}$$

and hence that $(D \to \tilde{X}, (t,r) \mapsto A_v(t)U_v(t,r)f_v(r))$ is continuous. As consequence, it follows by Theorem 3.2.4 (ii), Theorem 3.2.5, the Theorem of Fubini, and (9.0.14) that

$$\begin{aligned}
&\xi + \int_{a,\tilde{X}}^{t} (f_v(s) - A_v(s)u_v(s))\, ds \\
&= \xi + \int_{a,\tilde{X}}^{t} \left[f_v(s) - A_v(s) \left(U_v(s,a)\xi + \int_{a,Y}^{s} U_v(s,s')f_v(s')\, ds' \right) \right] ds \\
&= \xi + \int_{a,\tilde{X}}^{t} \left[f_v(s) - A_v(s)U_v(s,a)\xi - \int_{a,\tilde{X}}^{s} A_v(s)U_v(s,s')f_v(s')\, ds' \right] ds \\
&= \xi + \int_{a,\tilde{X}}^{t} f_v(s)\, ds - \int_{a}^{t} A_v(s)U_v(s,a)\xi\, ds \\
&\quad - \int_{a,\tilde{X}}^{t} \left(\int_{a,\tilde{X}}^{s} A_v(s)U_v(s,s')f_v(s')\, ds' \right) ds \\
&= \xi + \int_{a,\tilde{X}}^{t} f_v(s)\, ds - (\xi - U_v(t,a)\xi) - \int_{a,\tilde{X}}^{t} \left(\int_{s'}^{t} A_v(s)U_v(s,s')f_v(s')\, ds \right) ds' \\
&= U_v(t,a)\xi + \int_{a,\tilde{X}}^{t} f_v(s)\, ds - \int_{a,\tilde{X}}^{t} (f_v(s') - U_v(t,s')f_v(s'))\, ds' \\
&= U_v(t,a)\xi + \int_{a,\tilde{X}}^{t} U_v(t,s)f_v(s)\, ds = u_v(t)
\end{aligned}$$

for all $t \in I'$. Hence, finally, it follows by the auxiliary result in the proof of Lemma 8.5.2 that $\iota_{Y \hookrightarrow \tilde{X}} \circ u_v|_{\mathring{I}'}$ is differentiable and in particular that

$$u_v'(t) = -A_v(t)u_v(t) + f_v(t) \tag{11.0.8}$$

for all $t \in \mathring{I}'$, where differentiation is with respect to $\tilde{X}$, and hence also that $u_v \in C(I', Y) \cap C^1(I', \tilde{X})$ where we define $C^1(I', \tilde{X})$ to consist of those functions that are differentiable on $\mathring{I}'$ and whose derivatives have continuous extensions to I'. Further, it follows by Lemma 9.0.8 for every $\varepsilon > 0$ the existence of some $R \in [0, \infty)$ such that

$$\| (U_v(t,r) - \mathrm{id}_X)\, \xi \|_Y \leqslant \varepsilon + R \cdot (t - r) \tag{11.0.9}$$

for all $(t,r) \in \triangle(I')$ and hence by Theorem 3.2.5 that

$$\|u_v(t) - \xi\|_Y \leqslant \|U_v(t,a)\xi - \xi\|_Y + \left\| \int_{a,Y}^{t} U_v(t,s)f_v(s)\, ds \right\|_Y \leqslant \varepsilon + (R + \lambda_f C_L') \cdot T'$$

and

$$\begin{aligned}
\|u_v'(t)\|_{\tilde{X}} &\leqslant \|A_v(t)u_v(t)\|_{\tilde{X}} + \|f_v(t)\|_{\tilde{X}} \leqslant \lambda_A \|u_v(t)\|_Y + \lambda_f \\
&\leqslant \lambda_A \|u_v(t) - \xi\|_Y + \lambda_A \|\xi\|_Y + \lambda_f \leqslant \lambda_A \left[\, \|\xi\|_Y + \varepsilon + (R + \lambda_f C_L') \cdot T' \,\right] + \lambda_f
\end{aligned}$$

for all $t \in I'$. If K is some compact subset of W and $\xi \in K$, we choose R such that (11.0.9) is true for all $\xi \in K$ which is possible according to Lemma 9.0.8.

In the following, we choose $\varepsilon := \rho/2$,

$$L := \lambda_A \cdot (\rho + \|\xi\|_Y) + \lambda_f$$

or, if K is some compact subset of W and $\xi \in K$,

$$L := \lambda_A \cdot (\rho + \max\{\|\xi\|_Y : \xi \in K\}) + \lambda_f$$

and T' such that

$$(R + \lambda_f C'_L) \cdot T' \leqslant \frac{\rho}{2} \ .$$

Then

$$\operatorname{Ran} u_v \subset B_{Y,\rho}(\xi) \ , \ \|u'_v(t)\|_{\tilde{X}} \leqslant L$$

for all $t \in I'$ and hence also $\|u_v(t) - u_v(s)\|_{\tilde{X}} \leqslant L|t - s|$ for all $t, s \in I'$ and, finally, $u_v \in E$. We define $Fv := u_v$ in this way defining a map $F : E \to E$. In the following, it will be shown that

$$\|Fv - Fw\|_{\infty,\mathrm{X}} \leqslant \alpha \, \|v - w\|_{\infty,\mathrm{X}} \tag{11.0.10}$$

for all $w, v \in E$ and some $\alpha \in [0, 1)$. By the proof of Theorem 9.0.6, it follows

$$U_v(t,r)\eta - U_w(t,r)\eta = \int_r^t U_w(t,s)\,(A_w(s) - A_v(s))\,U_v(s,r)\eta\,ds \ ,$$

for every $\eta \in Y$ and $(t, r) \in \triangle(I')$ where the integrand is a continuous X- valued map and integration is weak Lebesgue integration with respect to $L(X, \mathbb{K})$. From this follows by Theorem 3.2.5

$$\|U_v(t,r)\eta - U_w(t,r)\eta\| \leqslant \mu_A\, C_L\, C'_L\, \|v - w\|_{\infty,\mathrm{X}}\, \|\eta\|_{\mathrm{Y}}\, T' \tag{11.0.11}$$

for every $\eta \in Y$ and $(t, r) \in \triangle(I')$ and hence

$$\begin{aligned} &\|U_v(t,s)f_v(s) - U_w(t,s)f_w(s)\| \\ &\quad \leqslant \|\,(U_v(t,s) - U_w(t,s))\,f_v(s)\| + \|U_w(t,s)\,(f_v(s) - f_w(s))\,\| \\ &\quad \leqslant C_L\,(\mu_f + \mu_A\,\lambda_f\,C'_L\,T')\,\|v - w\|_{\infty,\mathrm{X}} \end{aligned}$$

$(t, s) \in \triangle(I')$ and finally

$$\begin{aligned} \|u_v(t) - u_w(t)\| &\leqslant \mu_A\, C_L\, C'_L\, \|v - w\|_{\infty,\mathrm{X}}\, \|\xi\|_{\mathrm{Y}}\, T' \\ &\quad + C_L\,(\mu_f + \mu_A\,\lambda_f\,C'_L\,T')\,\|v - w\|_{\infty,\mathrm{X}}\,T' \\ &= C_L\,[\mu_f + \mu_A\,C'_L\,(\lambda_f\,T' + \|\xi\|_{\mathrm{Y}})]\,T'\|v - w\|_{\infty,\mathrm{X}} \end{aligned}$$

and hence (11.0.10) for small enough T'. In the following, we derive an integral equation for $S\,u_v$ where $v \in E$. Since $S : Y \to Z$ is continuous, it follows by Theorem 3.2.4 (ii), the proof of Theorem 9.0.6 and Lemma 8.1.1 (ii), (iii)c) that

$$\begin{aligned}
S\,u_\nu(t) &= S\,U_\nu(t,a)\xi + \int_{a,Z}^t S\,U_\nu(t,s) f_\nu(s)\,ds \\
&= \hat{U}_\nu(t,a)S\xi + \int_{a,Z}^t \hat{U}_\nu(t,s)S\,f_\nu(s)\,ds \\
&= \Phi(U_\nu(t,a))S\xi + \int_{a,Z}^t (\Phi\circ U_\nu)(t,s)S\,f_\nu(s)\,ds - [(\Phi\circ U_\nu)B_\nu\hat{U}_\nu](t,a)S\xi \\
&\quad - \int_{a,Z}^t [(\Phi\circ U_\nu)B_\nu\hat{U}_\nu](t,s)S\,f_\nu(s)\,ds
\end{aligned}$$

for every $t \in I'$ where

$$\hat{U}_\nu(t,r) := \text{volt}(\Phi\circ U_\nu, -B_\nu)$$

for all $(t,r) \in \triangle(I')$. Further, it follows by Lemma 8.1.1 (ii) the continuity of $[(\Phi \circ U_\nu)(t,\cdot)\circ pr_1](B_\nu \circ pr_1)\hat{U}_\nu\,[(S\,\circ f_\nu)\circ pr_2]$ where $pr_1 := (I'^2 \to \mathbb{R}, (s,s') \mapsto s)$ and $pr_2 := (I'^2 \to \mathbb{R}, (s,s') \mapsto s')$. Hence it follows by the proof of Lemma 8.1.1 (iii)b), the Theorem of Fubini, the change of variable formula and Theorem 3.2.4 (ii) that

$$\begin{aligned}
&\int_{a,Z}^t [(\Phi\circ U_\nu)B_\nu\hat{U}_\nu](t,s)S\,f_\nu(s)\,ds \\
&\quad = \int_{a,Z}^t \left(\int_{s,Z}^t \Phi(U_\nu(t,s'))B_\nu(s')\hat{U}_\nu(s',s)S\,f_\nu(s)\,ds' \right) ds \\
&\quad = \int_{a,Z}^t \left(\int_{a,Z}^{s'} \Phi(U_\nu(t,s'))B_\nu(s')\hat{U}_\nu(s',s)S\,f_\nu(s)\,ds \right) ds' \\
&\quad = \int_{a,Z}^t \left(\Phi(U_\nu(t,s'))B_\nu(s')S \int_{a,Y}^{s'} U_\nu(s',s) f_\nu(s)\,ds \right) ds' \\
&\quad = \int_{a,Z}^t \Phi(U_\nu(t,s'))B_\nu(s')\,(S\,u_\nu(s') - S\,U_\nu(s',a)\xi)\,ds'
\end{aligned}$$

and hence, finally,

$$\begin{aligned}
S\,u_\nu(t) &= \Phi(U_\nu(t,a))S\xi + \int_{a,Z}^t (\Phi\circ U_\nu)(t,s)\,(S\,f_\nu(s) - B_\nu(s)S\,u_\nu(s))\,ds \\
&= \Phi(U_\nu(t,a))S\xi \\
&\quad + \int_{a,Z}^t (\Phi\circ U_\nu)(t,s)\,[g(s,v(s),S\,v(s)) - B_\nu(s)S\,u_\nu(s)]\,ds \qquad (11.0.12)
\end{aligned}$$

for every $t \in I'$. In the following, let $\bar{E}$ be the closure of E in $C(I',X)$. $\bar{E}$ coincides with the closure $\tilde{E}$ of E in $C(I',\tilde{X})$. For this, note that $C(I',\tilde{X}) \subset C(I',X)$ since $\iota_{\tilde{X}\hookrightarrow X}$ is continuous. Further, since $\iota_{Y\hookrightarrow\tilde{X}}$ is continuous, it follows that $E \subset C(I',\tilde{X})$. Also for every $(v_\nu)_{\nu\in\mathbb{N}} \in E^{\mathbb{N}}$, $v \in C(I',\tilde{X})$ such that

$$\lim_{\nu\to\infty} \|v_\nu - v\|_{\infty,\tilde{X}} = 0\,,$$

it follows because of $\|\xi\|_X \leqslant \|\xi\|_{\tilde{X}}$ for all $\xi \in \tilde{X}$ also

$$\lim_{\nu\to\infty} \|v_\nu - v\|_{\infty,X} = 0$$

and hence $v \in \bar{E}$. Finally, let $(v_\nu)_{\nu\in\mathbb{N}} \in E^{\mathbb{N}}$, $v \in C(I', X)$ be such that

$$\lim_{\nu\to\infty} \|v_\nu - v\|_{\infty,X} = 0 \, .$$

Since the identity on $B_{Y,\rho}(\xi)$ equipped in domain with the topology induced by $\| \, \|_X$ and in range with the topology induced by $\| \, \|_{\tilde{X}}$ is uniformly continuous, it follows for given $\varepsilon > 0$ the existence of $\delta > 0$ such that for all $\eta, \eta' \in B_{Y,\rho}(\xi)$ satisfying $\|\eta - \eta'\|_X \leqslant \delta$ it follows $\|\eta - \eta'\|_{\tilde{X}} \leqslant \varepsilon$. In particular, if $N \in \mathbb{N}$ is such that for all $\mu, \nu \in \{N, N+1, \dots\}$

$$\|v_\mu - v_\nu\|_{\infty,X} \leqslant \delta \, ,$$

then it also follows that

$$\|v_\mu - v_\nu\|_{\infty,\tilde{X}} \leqslant \varepsilon$$

and hence that $(v_\nu)_{\nu\in\mathbb{N}}$ is a Cauchy sequence in $C(I', \tilde{X})$ and therefore convergent. Moreover, since $\|\xi\|_X \leqslant \|\xi\|_{\tilde{X}}$ for all $\xi \in \tilde{X}$, it follows that its limit is given by v and hence that $v \in \tilde{E}$. In the following, we define $U : E \to C_*(\Delta(I'), L(X, X))$ by $U(v) := U_v$ for all $v \in E$, $U\xi : E \to C(\Delta(I'), X)$ by $(U\xi)(v) := (\Delta(I') \to X, (t,r) \mapsto U_v(t,r)\xi)$ for every $\xi \in X$ and $\bar{U} : E \to C_*(\Delta(I'), L(Z, Z))$ by $U(v) := \Phi \circ U_v$ for all $v \in E$. By (11.0.11), (11.0.7), it follows for all $v, w \in E$, $(t,r) \in \Delta(I')$, $\xi \in X, \eta \in Y$ that

$$\begin{aligned} \|[(U\xi)(v)](t,r) - [(U\xi)(w)](t,r)\| &= \|U_v(t,r)\xi - U_w(t,r)\xi\| \qquad (11.0.13) \\ &\leqslant \|U_v(t,r)(\xi - \eta) - U_w(t,r)(\xi - \eta)\| + \|U_v(t,r)\eta - U_w(t,r)\eta\| \\ &\leqslant 2\,C_L\,\|\xi - \eta\| + \mu_A\,C_L\,C_L'\,\|v - w\|_{\infty,X}\,\|\eta\|_Y\,T' \, . \end{aligned}$$

Obviously, since Y is dense in X, from this it follows the continuity of $U\xi$ where E is equipped with the topology induced by $\| \, \|_{\infty,X}$. Further, let $v \in \tilde{E}$ and $(v_\nu)_{\nu\in\mathbb{N}} \in E^{\mathbb{N}}$ such that $\lim_{\nu\to\infty} \|v_\nu - v\|_{\infty,X}$ and $\varepsilon > 0$. Then $(v_\nu)_{\nu\in\mathbb{N}} \in E^{\mathbb{N}}$ is in particular a Cauchy sequence in $C(\Delta(I'), X)$. Since Y is dense in X, it follows from (11.0.13) that $((U\xi)(v_\nu))_{\nu\in\mathbb{N}} = ((\Delta(I') \to X, (t,r) \mapsto U_{v_\nu}(t,r)\xi))_{\nu\in\mathbb{N}}$ is a Cauchy-sequence in $C(\Delta(I'), X)$ and hence convergent to some $(\Delta(I') \to X, (t,r) \mapsto U_v(t,r)\xi) \in C(\Delta(I'), X)$. By (11.0.13) along with the denseness of Y in X, it also follows that this limit is independent of the approximating sequence $(v_\nu)_{\nu\in\mathbb{N}} \in E^{\mathbb{N}}$. Further, it follows by the linearity of $U_{v_\nu}(t,r)$ and the chain property of U_{v_ν} for every $\nu \in \mathbb{N}^*$, the linearity of $U_v(t,r)$ and the chain property of U_v and by (11.0.7)

$$\|U_v(t,r)\xi\|_X \leqslant C_L\,\|\xi\|_X$$

for every $\xi \in X$ and hence $U_v(t,r) \in L(X, X)$ and

$$\|U_v(t,r)\|_{\mathrm{Op},X} \leqslant C_L \qquad (11.0.14)$$

for all $(t,r) \in \triangle(I')$. In this way, we arrive at an extension $\tilde{U} : (\tilde{E} \to C_*(\triangle(I'), L(X, X))$ of U. Further, by a simple argument using 'diagonal sequences', it follows that $\tilde{U}\xi : \tilde{E} \to C(\triangle(I'), X)$ defined by $(\tilde{U}\xi)(v) := (\triangle(I') \to X, (t,r) \mapsto U_v(t,r)\xi)$ for every $v \in \tilde{E}$, $\xi \in X$ is continuous and hence that $\tilde{U}$ is strongly continuous. Therefore, since Φ is a strongly sequentially continuous nonexpansive homomorphism, the extension $\tilde{\bar{U}} := (\tilde{E} \to C_*(\triangle(I'), L(Z,Z)), v \mapsto \Phi \circ \tilde{U}(v))$ of $\bar{U}$ is strongly continuous, too. In particular, it follows by (11.0.14) that

$$\|\tilde{\bar{U}}_v(t,r)\|_{\text{Op},Z} \leqslant C_L \tag{11.0.15}$$

for all $(t,r) \in \triangle(I')$.

In the following, we define for every $(v, w, \bar{w}) \in \tilde{E} \times C(I', Z) \times C(I', Z)$ a corresponding $G(v, w, \bar{w}) \in C(I', Z)$ by

$$\begin{aligned}[G(v,w,\bar{w})](t) &:= \tilde{\bar{U}}_v(t,a)S\xi \\ &+ \int_{a,Z}^{t} \tilde{\bar{U}}_v(t,s)\,[g(s,v(s),w(s)) - B(s,v(s),w(s))\bar{w}(s)]\,ds\end{aligned}$$

for all $t \in I'$. Note that for every $v \in E$, it follows that $\text{Ran}\, v \subset B_{Y,\rho}(\xi) \subset W \subset \tilde{W}$ and hence also that $\text{Ran}\, v \subset \tilde{W}$ for all $v \in \tilde{E}$. In addition, for every $(v,w) \in \tilde{E} \times C(I', Z)$ the corresponding inclusion $\iota_{(v,w)} : I' \to I' \times \tilde{W} \times Z$, defined by $\iota_{(v,w)}(s) := (s, v(s), w(s))$ for every $s \in I'$, is continuous. Therefore, because of $g \in C(I \times \tilde{W} \times Z, Z)$, $B \in C_*(I \times \tilde{W} \times Z, L(Z,Z))$, it follows $g \circ \iota_{(v,w)} \in C(I', Z)$, $B \circ \iota_{(v,w)} \in C_*(I', L(Z,Z))$ and hence by Lemma 8.1.1 (ii) $\left(B \circ \iota_{(v,w)}\right)\bar{w} \in C(I', Z)$ for every $\bar{w} \in C(I', Z)$ and $\tilde{\bar{U}}_v(t,\cdot)\left[g \circ \iota_{(v,w)} - \left(B \circ \iota_{(v,w)}\right)\bar{w}\right] \in C([a,t], Z)$. In the following, let $k := g \circ \iota_{(v,w)} - \left(B \circ \iota_{(v,w)}\right)\bar{w} \in C(I', Z)$. In particular, it follows by Theorem 3.2.4 (ii) and Theorem 3.2.5 for every $t \in I'$ and $h \geqslant 0$ such that $t + h \in I'$

$$\begin{aligned}&\left\| \int_{a,Z}^{t+h} \tilde{\bar{U}}_v(t+h,s)k(s)\,ds - \int_{a,Z}^{t} \tilde{\bar{U}}_v(t,s)k(s)\,ds \right\|_Z \\ &\leqslant \left\| \int_{a,Z}^{t} \left(\tilde{\bar{U}}_v(t+h,s)k(s) - \tilde{\bar{U}}_v(t,s)k(s)\right) ds \right\|_Z + \left\| \int_{t,Z}^{t+h} \tilde{\bar{U}}_v(t+h,s)k(s)\,ds \right\|_Z \\ &\leqslant \left\| \left(\tilde{\bar{U}}_v(t+h,t) - \tilde{\bar{U}}_v(t,t)\right) \int_{a,Z}^{t} \tilde{\bar{U}}_v(t,s)k(s)\,ds \right\|_Z \\ &\quad + \|\tilde{\bar{U}}_v\|_{\infty,Z,Z} \cdot \int_{t,Z}^{t+h} \|k(s)\|_Z\,ds\,,\end{aligned}$$

for $h \leqslant 0$ such that $t + h \in I'$

$$\begin{aligned}&\left\| \int_{a,Z}^{t+h} \tilde{\bar{U}}_v(t+h,s)k(s)\,ds - \int_{a,Z}^{t} \tilde{\bar{U}}_v(t,s)k(s)\,ds \right\|_Z \\ &\leqslant \left\| \int_{a,Z}^{t+h} \left(\tilde{\bar{U}}_v(t,s)k(s) - \tilde{\bar{U}}_v(t+h,s)k(s)\right) ds \right\|_Z + \left\| \int_{t+h,Z}^{t} \tilde{\bar{U}}_v(t,s)k(s)\,ds \right\|_Z\end{aligned}$$

$$\leqslant \int_{a,Z}^{t+h} \left\| \tilde{\tilde{U}}_v(t,s)k(s) - \tilde{\tilde{U}}_v(t+h,s)k(s) \right\|_Z ds + \|\tilde{\tilde{U}}_v\|_{\infty,Z,Z} \cdot \int_{t+h,Z}^{t} \|k(s)\|_Z \, ds$$

and hence by $\tilde{\tilde{U}}_v \in C_*(\Delta(I'), L(Z,Z))$ and Lebesgue's dominated convergence theorem that $G(v,w,\bar{w}) \in C(I',Z)$.

In the following, it will be proved that G is continuous where $\tilde{E}$ is equipped with the induced topology by the uniform topology of $C(I',X)$. For this let $v \in \tilde{E}$, $w,\bar{w}$ be elements of $C(I',Z)$ and $(v_\mu)_{\mu\in\mathbb{N}} \in \tilde{E}^{\mathbb{N}}$ uniformly convergent to v and $(w_\mu)_{\mu\in\mathbb{N}}$, $(\bar{w}_\mu)_{\mu\in\mathbb{N}}$ sequences in $(C(I',Z)$ which are uniformly convergent to w and $\bar{w}$, respectively. Then it follows

$$\begin{aligned}
&\|\tilde{\tilde{U}}_{v_\mu}(t,s)\,[g(s,v_\mu(s),w_\mu(s)) - B(s,v_\mu(s),w_\mu(s))\bar{w}_\mu(s)] \\
&\quad - \tilde{\tilde{U}}_v(t,s)\,[g(s,v(s),w(s)) - B(s,v(s),w(s))\bar{w}(s)]\,\|_Z \\
&\leqslant \|\tilde{\tilde{U}}_{v_\mu}(t,s)\,[g(s,v_\mu(s),w_\mu(s)) - g(s,v(s),w(s)) \\
&\quad -B(s,v_\mu(s),w_\mu(s))\bar{w}_\mu(s) + B(s,v(s),w(s))\bar{w}(s)]\,\|_Z \\
&\quad + \left\| \left(\tilde{\tilde{U}}_{v_\mu}(t,s) - \tilde{\tilde{U}}_v(t,s)\right)[g(s,v(s),w(s)) - B(s,v(s),w(s))\bar{w}(s)] \right\|_Z \\
&\leqslant C_L\,\|g(s,v_\mu(s),w_\mu(s)) - g(s,v(s),w(s))\| \\
&\quad + C_L\,\|B(s,v_\mu(s),w_\mu(s))\bar{w}_\mu(s) - B(s,v(s),w(s))\bar{w}(s)\|_Z \\
&\quad + \left\| \left(\tilde{\tilde{U}}_{v_\mu}(t,s) - \tilde{\tilde{U}}_v(t,s)\right)[g(s,v(s),w(s)) - B(s,v(s),w(s))\bar{w}(s)] \right\|_Z
\end{aligned}$$

for every $(t,s) \in \Delta(I')$ and $v \in \mathbb{N}$ and hence for every $t \in I'$

$$\begin{aligned}
&\int_a^t \|\tilde{\tilde{U}}_{v_\mu}(t,s)\,[g(s,v_\mu(s),w_\mu(s)) - B(s,v_\mu(s),w_\mu(s))\bar{w}_\mu(s)] \\
&\quad - \tilde{\tilde{U}}_v(t,s)\,[g(s,v(s),w(s)) - B(s,v(s),w(s))\bar{w}(s)]\,\|_Z \, ds \\
&\leqslant C_L \int_{I'} \|g(s,v_\mu(s),w_\mu(s)) - g(s,v(s),w(s))\|_Z \, ds \\
&\quad + C_L \int_{I'} \|B(s,v_\mu(s),w_\mu(s))\bar{w}_\mu(s) - B(s,v(s),w(s))\bar{w}(s)\|_Z \, ds \\
&\quad + \int_a^t \left\| \left(\tilde{\tilde{U}}_{v_\mu}(t,s) - \tilde{\tilde{U}}_v(t,s)\right)[g(s,v(s),w(s)) - B(s,v(s),w(s))\bar{w}(s)] \right\|_Z ds\,.
\end{aligned} \tag{11.0.16}$$

Further, since $\tilde{\tilde{U}}$ is strongly continuous for every $\eta \in Z$ and given $\varepsilon > 0$, there is $\mu_0 \in \mathbb{N}$ such that

$$\|\tilde{\tilde{U}}_{v_\mu}\eta - \tilde{\tilde{U}}_v\eta\|_{\infty,Z} \leqslant \varepsilon$$

for all $v \in \mathbb{N}$ such that $v \geqslant v_0$. That such an $\mu_0 \in \mathbb{N}$ can be found uniformly for all elements of some compact subset $K \subset Z$ can be seen as follows. For this, let $\delta > 0$ be such that $4C_L \leqslant \varepsilon$, $n \in \mathbb{N}^*$ and $\eta_1,\ldots,\eta_n \in K$ be such that the union of $U_{\delta,Z}(\eta_1),\ldots,U_{\delta,Z}(\eta_n)$ is covering K. Then there is $\mu_0 \in \mathbb{N}$ such that

$$\|\tilde{\tilde{U}}_{v_\mu}\eta_k - \tilde{\tilde{U}}_v\eta_k\|_{\infty,Z} \leqslant \frac{\varepsilon}{2}$$

for all $k \in \{1, \dots, n\}$ and all $\nu \in \mathbb{N}$ such that $\nu \geqslant \nu_0$. For $\eta \in K$, there is $j \in \{1, \dots, n\}$ such that $\eta \in U_{\delta, Z}(\eta_j)$ and hence

$$\|\tilde{\bar{U}}_{\nu_\mu}\eta - \tilde{\bar{U}}_\nu \eta\|_{\infty,Z} \leqslant \|\tilde{\bar{U}}_{\nu_\mu}\eta - \tilde{\bar{U}}_{\nu_\mu}\eta_j\|_{\infty,Z} + \|\tilde{\bar{U}}_{\nu_\mu}\eta_j - \tilde{\bar{U}}_\nu \eta_j\|_{\infty,Z}$$
$$+ \|\tilde{\bar{U}}_\nu \eta_j - \tilde{\bar{U}}_\nu \eta\|_{\infty,Z} \leqslant 2C_L\,\delta + \frac{\varepsilon}{2} \leqslant \varepsilon \,.$$

Applying this to the compact range of $g \circ \iota_{(v,w)} - \left(B \circ \iota_{(v,w)}\right) \bar{w}$, it follows from (11.0.16) by Lemma 11.0.6, (11.0.15) and Lebesgue's dominated convergence for given ε the existence of $\mu_0 \in \mathbb{N}$ such that

$$\int_a^t \|\tilde{\bar{U}}_{\nu_\mu}(t,s)\,[g(s, v_\mu(s), w_\mu(s)) - B(s, v_\mu(s), w_\mu(s))\bar{w}_\mu(s)]$$
$$- \tilde{\bar{U}}_\nu(t,s)\,[g(s, v(s), w(s)) - B(s, v(s), w(s))\bar{w}(s)]\,\|_Z\, ds \leqslant \varepsilon$$

for all $\nu \in \mathbb{N}$ such that $\nu \geqslant \nu_0$. and hence finally by Theorem 3.2.5 and the strong continuity of $\tilde{\bar{U}}$ that

$$\lim_{\mu \to \infty} \|G(v_\mu, w_\mu, \bar{w}_\mu) - G(v, w, \bar{w})\|_{\infty,Z} = 0\,.$$

Further, it follows by Theorem 3.2.5 for $v \in \tilde{E}$, $w_1, \bar{w}_1, w_2, \bar{w}_2 \in C(I', Z)$ and $t \in I'$

$$\begin{aligned}
&\|[G(v, w_1, \bar{w}_1)](t) - [G(v, w_2, \bar{w}_2)](t)\|_Z \\
&\leqslant \int_{a,Z}^t \|\tilde{\bar{U}}_\nu(t,s)\,[g(s, v(s), w_1(s)) - g(s, v(s), w_2(s))]\,\|_Z\, ds \\
&\quad + \int_{a,Z}^t \|\tilde{\bar{U}}_\nu(t,s)\,[B(s, v(s), w_1(s))\bar{w}_1(s) - B(s, v(s), w_2(s))\bar{w}_2(s)]\,\|_Z\, ds \\
&\leqslant \mu_g\, C_L\, T'\, \|w_1 - w_2\|_{\infty,Z} \\
&\quad + C_L \int_{a,Z}^t \|B(s, v(s), w_1(s))\bar{w}_1(s) - B(s, v(s), w_2(s))\bar{w}_2(s)\|_Z\, ds \\
&\leqslant \mu_g\, C_L\, T'\, \|w_1 - w_2\|_{\infty,Z} \\
&\quad + C_L \int_{a,Z}^t \|\,[B(s, v(s), w_1(s)) - B(s, v(s), w_2(s))]\,\bar{w}_1(s)\|_Z\, ds \\
&\quad + C_L \int_{a,Z}^t \|B(s, v(s), w_2(s))(\bar{w}_2(s) - \bar{w}_1(s))\|_Z\, ds \\
&\leqslant C_L\, T'\, (\mu_g + \mu_B\, \|\bar{w}_1\|_{\infty,Z})\, \|w_1 - w_2\|_{\infty,Z} \\
&\quad + C_L \int_{a,Z}^t \|B(s, v(s), w_2(s))(\bar{w}_2(s) - \bar{w}_1(s))\|_Z\, ds \\
&\leqslant C_L\, T'\, (\mu_g + \mu_B\, \|\bar{w}_1\|_{\infty,Z})\, \|w_1 - w_2\|_{\infty,Z} \\
&\quad + C_L \int_{a,Z}^t \|\,(B(s, v(s), w_2(s)) - B(s, \xi, S\xi))\,(\bar{w}_2(s) - \bar{w}_1(s))\|_Z\, ds
\end{aligned}$$

$$+ C_L \int_{a,Z}^{t} \|B(s,\xi,S\xi)(\bar{w}_2(s) - \bar{w}_1(s))\|_Z \, ds$$
$$\leqslant C_L T' (\mu_g + \mu_B \|\bar{w}_1\|_{\infty,Z}) \|w_1 - w_2\|_{\infty,Z} + C_L \lambda_B T' \|\bar{w}_1 - \bar{w}_2\|_{\infty,Z}$$
$$+ C_L \int_{a,Z}^{t} \| (B(s,v(s),w_2(s)) - B(s,\xi,S\xi)) (\bar{w}_2(s) - \bar{w}_1(s))\|_Z \, ds \, .$$

Further, for every $s \in I'$

$$\begin{aligned} &\| (B(s,v(s),w_2(s)) - B(s,\xi,S\xi)) (\bar{w}_2(s) - \bar{w}_1(s))\|_Z \\ &\quad \leqslant \| (B(s,v(s),w_2(s)) - B(s,v(s),S\xi)) (\bar{w}_2(s) - \bar{w}_1(s))\|_Z \\ &\qquad + \| (B(s,v(s),S\xi) - B(s,\xi,S\xi)) (\bar{w}_2(s) - \bar{w}_1(s))\|_Z \\ &\quad \leqslant \mu_B \cdot (\|w_2\|_{\infty,Z} + \|S\xi\|_Z) \cdot \|\bar{w}_1 - \bar{w}_2\|_{\infty,Z} \\ &\qquad + \| (B(s,v(s),S\xi) - B(s,\xi,S\xi)) (\bar{w}_2(s) - \bar{w}_1(s))\|_Z \end{aligned}$$

By Lemma 11.0.6 (iii), there are $\rho_0 > 0$ and $C'' \geqslant 0$ such that if $\rho \leqslant \rho_0$, which is assumed to be the case in the following,

$$\|B(s,\xi',S\xi)\|_{\mathrm{Op},Z} \leqslant C'' \tag{11.0.17}$$

for all $\xi' \in B_{\tilde{X},\rho}(\xi)$ ($\supset B_{Y,\rho}(\xi)$) and $s \in I'$. Note that according to Lemma 11.0.6 (iii) ρ_0, C'' exist such that (11.0.17) is true for all $\xi' \in B_{\tilde{X},\rho}(\xi)$ ($\supset B_{Y,\rho}(\xi)$) and $s \in I'$ and all $\xi \in K$ if K is some compact subset of W. Hence

$$\begin{aligned} &\| (B(s,v(s),w_2(s)) - B(s,\xi,S\xi)) (\bar{w}_2(s) - \bar{w}_1(s))\|_Z \\ &\quad \leqslant \mu_B \cdot (\|w_2\|_{\infty,Z} + \|S\xi\|_Z) \cdot \|\bar{w}_1 - \bar{w}_2\|_{\infty,Z} + (\lambda_B + C'') \cdot \|\bar{w}_2 - \bar{w}_1\|_{\infty,Z} \end{aligned}$$

for all $s \in I'$ and, finally,

$$\begin{aligned} &\|[G(v,w_1,\bar{w}_1)] - [G(v,w_2,\bar{w}_2)]\|_{\infty,Z} \\ &\quad \leqslant C_L T' (\mu_g + \mu_B \|\bar{w}_1\|_{\infty,Z}) \|w_1 - w_2\|_{\infty,Z} \\ &\qquad + C_L T' \left[\mu_B \cdot (\|w_2\|_{\infty,Z} + \|S\xi\|_Z) + 2\lambda_B + C''\right] \cdot \|\bar{w}_1 - \bar{w}_2\|_{\infty,Z} \, . \end{aligned} \tag{11.0.18}$$

for all $t \in I'$. Further, it follows by Theorem 3.2.5 for $v \in \tilde{E}$, $w, \bar{w} \in C(I', Z)$ and $t \in I'$

$$\begin{aligned} &\|[G(v,w,\bar{w})](t) - \tilde{U}_v(t,a)S\xi\|_Z \\ &\quad \leqslant C_L \int_{a,Z}^{t} \|g(s,v(s),w(s))\|_Z \, ds + C_L \int_{a,Z}^{t} \|B(s,v(s),w(s))\bar{w}(s)\|_Z \, ds \\ &\quad \leqslant C_L \int_{a,Z}^{t} \|g(s,v(s),w(s)) - g(s,v(s),0_Z)\|_Z \, ds + C_L \int_{a,Z}^{t} \|g(s,v(s),0_Z)\|_Z \, ds \\ &\qquad + C_L \int_{a,Z}^{t} \|B(s,v(s),w(s))\bar{w}(s) - B(s,v(s),0_Z)\bar{w}(s)\|_Z \, ds \end{aligned}$$

$$+ C_L \int_{a,Z}^{t} \|B(s,v(s),0_Z)\bar{w}(s)\|_Z \, ds \leqslant C_L T' \cdot (\mu_g + \mu_B \|\bar{w}\|_{\infty,Z}) \cdot \|w\|_{\infty,Z}$$
$$+ C_L \int_{a,Z}^{t} \|g(s,v(s),0_Z)\|_Z \, ds + C_L \int_{a,Z}^{t} \|B(s,v(s),0_Z)\bar{w}(s)\|_Z \, ds \, .$$

By Lemma 11.0.6 (iii), there are $\rho_0' \in (0,\rho_0]$ and $C''', C^{IV} \geqslant 0$ such that if $\rho \leqslant \rho_0'$, which is assumed to be the case in the following,

$$\|g(s,\xi',0_Z)\|_Z \leqslant C''' \, , \; \|B(s,\xi',0_Z)\|_{\text{Op},Z} \leqslant C^{IV} \tag{11.0.19}$$

for all $\xi' \in B_{\tilde{X},\rho}(\xi) \, (\supset B_{Y,\rho}(\xi))$ and $s \in I'$. Note that according to Lemma 11.0.6 (iii) ρ_0', C''', C^{IV} exist such that (11.0.19) is true for all for all $\xi' \in B_{\tilde{X},\rho}(\xi) \, (\supset B_{Y,\rho}(\xi))$ and $s \in I'$ and $\xi \in K$ if K is some compact subset of W. Hence

$$\|[G(v,w,\bar{w})](t) - \tilde{\bar{U}}_v(t,a)S\xi\|_Z \leqslant C_L T' \cdot (\mu_g + \mu_B \|\bar{w}\|_{\infty,Z}) \cdot \|w\|_{\infty,Z}$$
$$+ C_L T' \cdot (C''' + C^{IV}\|\bar{w}\|_{\infty,Z})$$

and

$$\|[G(v,w,\bar{w})]\|_{\infty,Z} \leqslant C_L \|S\xi\|_Z + C_L T' \cdot [(\mu_g + \mu_B \|\bar{w}\|_{\infty,Z}) \cdot \|w\|_{\infty,Z} \tag{11.0.20}$$
$$+ C''' + C^{IV}\|\bar{w}\|_{\infty,Z} \cdot]$$

We define

$$D := B_{C_L\|S\xi\|_Z+\rho}(0_{C(I',Z)}) \subset C(I',Z) \, .$$

If K is some compact subset of W and $\xi \in K$, we define

$$D := B_{C_L \max\{\|S\xi\|_Z : \xi \in K\}+\rho}(0_{C(I',Z)}) \subset C(I',Z) \, .$$

Then it follows by (11.0.18) and (11.0.20) that for small enough $T' > 0$ that G maps $\tilde{E} \times D \times D$ into D and the existence of $\beta, \gamma \geqslant 0$ such that $\beta + \gamma < 1$ and

$$\|G(v,w_1,\bar{w}_1) - G(v,w_2,\bar{w}_2)\|_{\infty,Z} \leqslant \beta \cdot \|w_1 - w_2\|_{\infty,Z} + \gamma \cdot \|\bar{w}_1 - \bar{w}_2\|_{\infty,Z}$$

for all $v \in \tilde{E}$ and $w_1, \bar{w}_1, w_2, \bar{w}_2 \in D$. In particular, it follows for every $v \in E$

$$\|Sv(t)\|_Z \leqslant \|Sv(t) - S\xi\|_Z + \|S\xi\|_Z \leqslant \|v(t) - \xi\|_Y + \|S\xi\|_Z \leqslant C_L \|S\xi\|_Z + \rho$$

for every $t \in I'$ and hence that $\mathcal{S}E \subset D$. By (11.0.12) also

$$\mathcal{S}Fv = G(v, \mathcal{S}v, \mathcal{S}Fv)$$

for all $v \in E$. Hence, finally, it follows by Theorem 11.0.5 the existence of a unique $u \in E$ such that $Fu = u$. This implies (see (11.0.8)) that $u(a) = \xi$, that $\iota_{Y \hookrightarrow \tilde{X}} \circ u|_{\dot{I}'}$ is differentiable and in particular

$$u'(t) = -A(t,u(t))u(t) + f(t,u(t))$$

for all $t \in \mathring{I}'$, where differentiation is with respect to $\tilde{X}$, and hence also that $u \in C(I', Y) \cap C^1(I', \tilde{X})$.

If $v \in C(I', W) \cap C^1(I', \tilde{X})$ is such that

$$v'(t) + A(t, v(t))v(t) = f(t, v(t))$$

for all $t \in \mathring{I}'$ and $v(a) = \xi$. Then

$$\begin{aligned}\|v'(t)\|_{\tilde{X}} &\leqslant \|A(t, v(t))v(t)\|_{\tilde{X}} + \|f(t, v(t))\|_{\tilde{X}} \leqslant \lambda_A \|v(t)\|_Y + \lambda_f \\ &\leqslant \lambda_A \|v(t) - \xi\|_Y + \lambda_A \|\xi\|_Y + \lambda_f\end{aligned}$$

for all $t \in I'$. Hence there is T'' such that $0 \leqslant T'' \leqslant T'$ and

$$\operatorname{Ran} v \subset B_{Y,\rho}(\xi) \ , \ \|v'(t)\|_{\tilde{X}} \leqslant L$$

for all $t \in I'' := [a, a + T'']$. Then it follows by Corollary 9.0.7 (ii) that $u_v = v$ on I'' and hence that the restrictions of u and v to I'' are both fixed points of the contraction F associated to I'' and hence equal. Hence it follows the equality of u and v on a neighbourhood of a. Now assume that

$$\{t \in I' : v(t) \neq u(t)\}$$

is non-empty. Then

$$t_0 := \inf\{t \in I' : v(t) \neq u(t)\}$$

is smaller than b. Then applying the same reasoning to the interval $[t_0, b]$, where among others ξ is replaced by $v(t_0) = u(t_0)$, it follows the contradiction that $v = u$ on a neighbourhood of t_0.↯ Hence $v = u$. □

12

Examples of Quasi-Linear Evolution Equations

This chapter gives examples of the application of Theorem 11.0.7 to quasi-linear evolution equations. Here the main stress is on quasi-linear Hermitian hyperbolic systems. The treatment of a generalized inviscid Burgers' equation is included mainly in order to display the steps needed for the application of Theorem 11.0.7 to quasi-linear Hermitian hyperbolic systems in a technically simpler situation. Note that in all the examples of these notes, we consider solutions whose component functions are complex-valued because this generalization simplifies the application of spectral methods. In general, in the case of quasi-linear systems of partial differential equations taken from applications, this forces to extend coefficient functions, that are usually defined only in the real domain, into the complex domain. In general, this is easy to do in such cases and solutions corresponding to data with real-valued component functions lead to solutions with this property which don't depend on the extension. The main sources for this section are the papers [102, 109, 110]. Here, these results are 'adapted' to the late Kato's most recent approach to abstract quasi-linear evolution equations [114] given in this second part of the notes. For additional material on initial boundary value problems for quasi-linear symmetric hyperbolic systems see, for instance, [195, 196]. For examples of applications of Kato's older results [102, 109, 110] to problems in General Relativity and Astrophysics, see, for instance, [5, 17, 29, 33, 37, 44, 45, 61, 70–77, 117, 135, 147, 166, 215].

12.1 A Generalized Inviscid Burgers' Equation

In the following, we consider the solutions of the 1 + 1 equation

$$\frac{\partial u}{\partial t}(t,x) + a(u(t,x))\,\frac{\partial u}{\partial x}(t,x) = b(u(t,x))\,u(t,x),$$

for complex-valued u where $t, x \in \mathbb{R}$ and $a, b \in C^2(\mathbb{R}^2, \mathbb{C})$.[1]

[1] Note that $\mathbb{C} = \mathbb{R}^2$.

Lemma 12.1.1. (Elementary inequalities) Let $n \in \mathbb{N}^*$, $m \in \mathbb{N}$ and $k = (k_1, \dots, k_n) \in \mathbb{R}^n$. Then

$$\frac{1}{(n+1)^{m/2}} \sum_{\alpha \in \mathbb{N}^n, |\alpha| \leqslant m} |k^\alpha| \leqslant (1+|k|^2)^{m/2} \leqslant 2^m \sum_{\alpha \in \mathbb{N}^n, |\alpha| \leqslant m} |k^\alpha|.$$

Proof. First,

$$(1+|k|^2)^{m/2} \leqslant \left(1 + \sum_{j=1}^{n} |k_j|\right)^m \leqslant 2^m \sum_{\alpha \in \mathbb{N}^n, |\alpha| \leqslant m} |k^\alpha|$$

where the last inequality follows by induction over $m \in \mathbb{N}^*$. Further, it follows by the Cauchy-Schwartz inequality for the Euclidean scalar product on $\mathbb{R}^{n+1}$ that

$$1 + \sum_{j=1}^{n} |k_j| \leqslant (n+1)^{1/2} \left(1 + \sum_{j=1}^{n} |k_j|^2\right)^{1/2}$$

and hence that

$$(1+|k|^2)^{m/2} \geqslant (n+1)^{-m/2} \left(1 + \sum_{j=1}^{n} |k_j|\right)^m \geqslant (n+1)^{-m/2} \sum_{\alpha \in \mathbb{N}^n, |\alpha| \leqslant m} |k^\alpha|.$$

□

Lemma 12.1.2. (Sobolev inequalities) Let $n \in \mathbb{N}^*$, $k \in \{1, \dots, n\}$ and m, m' such that $m > n/2, m' > (n/2) + 1$. Further, let $e_1, \dots, e_n$ be the canonical basis of $\mathbb{R}^n$. Then

$$W^m_{\mathbb{C}}(\mathbb{R}^n), \partial^{e_k} W^{m'}_{\mathbb{C}}(\mathbb{R}^n) \subset C_\infty(\mathbb{R}^n, \mathbb{C})$$

and

$$\|h\|_\infty \leqslant C_{nm1} \|\|h\|\|_m, \|\partial^{e_k} h\|_\infty \leqslant C_{nm'2} \|\|h\|\|_{m'},$$

for all $h \in W^2_{\mathbb{C}}(\mathbb{R}^n)$ where $\|\|\ \|\|_l$ denotes the Sobolev norm on $W^l_{\mathbb{C}}(\mathbb{R}^n)$ for every $l \in \mathbb{N}$ and

$$C_{nm1} := 2^m \,(2\pi)^{-n/2} \,\|(1 + |\ |^2)^{-m/2}\|_2,$$

$$C_{nm'2} := 2^{m'} \,(2\pi)^{-n/2} \,\|(1 + |\ |^2)^{-(m'-1)/2}\|_2.$$

Proof. For this, let $F_1 : L^1_{\mathbb{C}}(\mathbb{R}^n) \to C_\infty(\mathbb{R}^n, \mathbb{C})$ and $F_2 : L^2_{\mathbb{C}}(\mathbb{R}^n) \to L^2_{\mathbb{C}}(\mathbb{R}^n)$ be the continuous Fourier transformations defined by

$$(F_1 f)(x) := \int_{\mathbb{R}^n} e^{-ixy} f(y)\, dy$$

for all $x \in \mathbb{R}^n$ and determined by $F_2 f := (2\pi)^{-n/2} F_1 f$ for every $f \in C_0^\infty(\mathbb{R}^n, \mathbb{C})$, respectively. Since $F_2 W_{\mathbb{C}}^m(\mathbb{R}^n)$ coincides with the the domain of the maximal multiplication operator in $L^2_{\mathbb{C}}(\mathbb{R}^n)$, by $(1+|\ |^2)^{m/2}$ and $m > n/2$, it follows for $h \in W_{\mathbb{C}}^m(\mathbb{R}^n)$

$$F_2 h = (1+|\ |^2)^{-m/2}\,(1+|\ |^2)^{m/2}\,F_2 h \in L^1_{\mathbb{C}}(\mathbb{R}^n) \cap L^2_{\mathbb{C}}(\mathbb{R}^n).$$

Hence

$$h = F_2^{-1} F_2 h = (2\pi)^{-n/2}\,[(F_1 F_2 h)\circ(-\mathrm{id}_{\mathbb{R}^n})] \in C_\infty(\mathbb{R}^n, \mathbb{C})$$

and

$$\begin{aligned}
\|h\|_\infty &\leqslant (2\pi)^{-n/2}\,\|F_2 h\|_1 = (2\pi)^{-n/2}\,\|(1+|\ |^2)^{-m/2}\,(1+|\ |^2)^{m/2} F_2 h\|_1 \\
&\leqslant (2\pi)^{-n/2}\,\|(1+|\ |^2)^{-m/2}\|_2\,\|(1+|\ |^2)^{m/2} F_2 h\|_2 \\
&\leqslant (2\pi)^{-n/2}\,2^m\,\|(1+|\ |^2)^{-m/2}\|_2\| \sum_{\alpha\in\mathbb{N}^m, |\alpha|\leqslant m} |k^\alpha|\,F_2 h\|_2 \\
&= 2^m\,(2\pi)^{-n/2}\,\|(1+|\ |^2)^{-m/2}\|_2\,|||h|||_m .
\end{aligned}$$

Further, since $F_2 W_{\mathbb{C}}^{m'}(\mathbb{R}^n)$ coincides with the domain of the maximal multiplication operator in $L^2_{\mathbb{C}}(\mathbb{R}^n)$ by $(1+|\ |^2)^{m'/2}$, it follows for $h \in W_{\mathbb{C}}^{m'}(\mathbb{R}^n)$ that

$$F_2\,\partial^{e_k} h = (1+|\ |^2)^{-(m'-1)/2}\,(1+|\ |^2)^{(m'-1)/2}\,F_2\,\partial^{e_k} h \in L^1_{\mathbb{C}}(\mathbb{R}^n) \cap L^2_{\mathbb{C}}(\mathbb{R}^n).$$

Hence

$$\partial^{e_k} h = F_2^{-1} F_2\,\partial^{e_k} h = (2\pi)^{-n/2}\,[(F_1 F_2\,\partial^{e_k} h)\circ(-\mathrm{id}_{\mathbb{R}^n})] \in C_\infty(\mathbb{R}^n, \mathbb{C})$$

and

$$\begin{aligned}
\|\partial^{e_k} h\|_\infty &\leqslant (2\pi)^{-n/2}\,\|F_2\,\partial^{e_k} h\|_1 \\
&= (2\pi)^{-n/2}\,\|(1+|\ |^2)^{-(m'-1)/2}\,(1+|\ |^2)^{(m'-1)/2} F_2\,\partial^{e_k} h\|_1 \\
&\leqslant (2\pi)^{-n/2}\,\|(1+|\ |^2)^{-(m'-1)/2}\|_2\,\|(1+|\ |^2)^{(m'-1)/2} F_2\,\partial^{e_k} h\|_2 \\
&= (2\pi)^{-n/2}\,\|(1+|\ |^2)^{-(m'-1)/2}\|_2\,\|\mathrm{pr}_k\,(1+|\ |^2)^{(m'-1)/2} F_2 h\|_2 \\
&\leqslant (2\pi)^{-n/2}\,\|(1+|\ |^2)^{-(m'-1)/2}\|_2\,\|(1+|\ |^2)^{m'/2} F_2 h\|_2 \\
&\leqslant 2^{m'}\,(2\pi)^{-n/2}\,\|(1+|\ |^2)^{-(m'-1)/2}\|_2\,|||h|||_{m'}
\end{aligned}$$

where pr_k denotes the k-th coordinate projection of $\mathbb{R}^n$. □

Corollary 12.1.3. Let $n \in \mathbb{N}^*$, $k \in \mathbb{N}$, $m \in \mathbb{N}$ such that $m > (n/2) + k$, $\alpha \in \mathbb{N}^n$ such that $|\alpha| \leqslant k$. Then

$$\partial^\alpha W_{\mathbb{C}}^m(\mathbb{R}^n) \subset C_\infty(\mathbb{R}^n, \mathbb{C})$$

and

$$\|\partial^\alpha f\|_\infty \leqslant C_{n(m-k)1}\,|||f|||_m$$

for all $f \in W_{\mathbb{C}}^m(\mathbb{R}^n)$.

Proof. Since $m > (n/2) + k$, it follows by Lemma 12.1.2 that

$$\partial^\alpha W_{\mathbb{C}}^m(\mathbb{R}^n) \subset W_{\mathbb{C}}^{m-|\alpha|}(\mathbb{R}^n) \subset W_{\mathbb{C}}^{m-k}(\mathbb{R}^n) \subset C_\infty(\mathbb{R}^n, \mathbb{C})$$

and

$$\|\partial^\alpha f\|_\infty \leqslant C_{n(m-k)1} \|\!|\partial^\alpha f|\!\|_{m-k} \leqslant C_{n(m-k)1} \|\!|f|\!\|_m$$

for all $f \in W_{\mathbb{C}}^m(\mathbb{R}^n)$. □

Theorem 12.1.4. Let $a, b \in C^2(\mathbb{R}^2, \mathbb{C})$ and

$$\begin{aligned}
&\alpha_0(R) := \|a|_{B_R(0)}\|_\infty, \alpha_1(R) := \|\,|\nabla a|\,|_{B_R(0)}\,\|_\infty, \beta_0(R) := \|b|_{B_R(0)}\|_\infty,\\
&\beta_1(R) := \|\,|\nabla b|\,|_{B_R(0)}\,\|_\infty, \alpha_2(R) := \max\{\|a_{,lm}|_{B_R(0)}\|_\infty : l, m \in \{1,2\}\},\\
&\beta_2(R) := \max\{\|\beta_{,lm}|_{B_R(0)}\|_\infty : l, m \in \{1,2\}\}
\end{aligned}$$

for $k \in \{0, 1\}$ and $R \in [0, \infty)$.[2] In addition, let $X := L_{\mathbb{C}}^2(\mathbb{R})$, $Y := W_{\mathbb{C}}^2(\mathbb{R})$ $(\subset C^1(\mathbb{R}, \mathbb{C}))$, $S := (Y \to X, h \mapsto (-D^{2*} + 1)h)$, $R > 0$,

$$W := \{h \in Y : \|\!|h|\!\|_2 < R\}$$

where $\|\!|\ |\!\|_2$ denotes the Sobolev norm for $W_{\mathbb{C}}^2(\mathbb{R})$ and $I = [c, d]$. Here $c, d \in \mathbb{R}$ are such that $c \leqslant d$.

(i) S is a densely-defined, linear self-adjoint and bijective operator in X. In particular,

$$\frac{\sqrt{3}}{2} \|\!|\ |\!\|_2 \leqslant \|\ \|_s \leqslant \sqrt{3} \|\!|\ |\!\|_2.$$

(ii) For every $(t, h) \in \mathbb{R} \times W_{\mathbb{C}}^2(\mathbb{R})$, by

$$A_0(t, h)k := -(a \circ h)D^{1*}k$$

for every $k \in W_{\mathbb{C}}^1(\mathbb{R})$, there is defined a closable linear operator $A_0(t, h)$ in X, whose closure $A(t, h)$ is quasi-accretive with bound

$$-\frac{1}{2}\alpha_1(C_1\|\!|h|\!\|_2)\, C_2\|\!|h|\!\|_2 \qquad (12.1.1)$$

and is the infinitesimal generator of a strongly continuous group on X. Here $C_1 := C_{121}, C_2 := C_{122}$ where C_{121}, C_{122} are defined as in Lemma 12.1.2.

(iii) The family $(A(t, u(t)))_{t \in I}$ is stable with stability constants

$$RC_2\,\alpha_1(RC_1)/2, 1$$

for every $u : I \to W$.

[2] Note that $\mathbb{C} = \mathbb{R}^2$.

(iv) $Y \subset D(A(t,h))$, $A(t,h)|_Y \in L(Y,X)$ for every $(t,h) \in I \times W$ and

$$A := (I \times W \to L(Y,X), (t,h) \mapsto A(t,h)|_Y) \in C(I \times W, L(Y,X)).$$

Moreover,

$$\|A(t,h)\| \leqslant \lambda_A, \|A(t,h_1) - A(t,h_2)\| \leqslant \mu_A \|h_1 - h_2\|_Y$$

for all $t \in I$, $h_1, h_2 \in W$ where

$$\lambda_A := (2\sqrt{3}/3)\,\alpha_0(RC_1), \mu_A := (2\sqrt{3}/3)\,C_2\,\alpha_1(3RC_1).$$

(v) Let $\tilde{D} := W^4_{\mathbb{C}}(\mathbb{R})$. Then

$$SA(t,h)k = (A(t,h) + B(t,h,Sh))Sk \tag{12.1.2}$$

for every $t \in I$, $h \in W$ and $k \in \tilde{D}$ where $B : I \times W \times X \to L(X,X)$ is defined by

$$B(t,h,\bar{h})k := -\sum_{l=1}^{2}(a_{,l} \circ h)\left[(\bar{h}_l - 1)D^{1*}S^{-1}k + 2h'_l D^{2*}S^{-1}k\right] + \sum_{l,m=1}^{2}(a_{,lm} \circ h)h'_l h'_m D^{1*}S^{-1}k$$

for all $(t,h,\bar{h}) \in I \times W \times X$ and $k \in X$. Further, B is strongly continuous and such that

$$\|B(t,h,Sh)\| \leqslant \lambda_B, \|B(t,h,\bar{h}_1) - B(t,h,\bar{h}_2)\| \leqslant \mu_B \|\bar{h}_1 - \bar{h}_2\|_2$$

for all $(t,h) \in I \times W$, $\bar{h}_1, \bar{h}_2 \in X$ where

$$\lambda_B := \alpha_1(R)[1 + 4C_2(2+\sqrt{2})R] + 2C_2^2\,\alpha_2(R)R^2,\ \mu_B := (4/3)\sqrt{6}\,C_2\,\alpha_1(R).$$

(vi) Let $(t,h) \in I \times W$ and $T(t,h) : [0,\infty) \to L(X,X)$, $\hat{T}(t,h) : [0,\infty) \to L(X,X)$ be the strongly continuous semigroups generated by $A(t,h)$ and $\hat{A}(t,h) := A(t,h) + B(t,h,Sh)$, respectively. Then

$$ST(t,h)(s) \supset \hat{T}(t,h)(s)S$$

for all $s \geqslant 0$.

(vii) By

$$f(t,h) := (b \circ h)h$$

for all $(t,h) \in I \times W$, there is defined an element $f \in C(I \times W, Y)$. In particular,

$$\|f(t,h)\|_Y \leqslant \lambda_f, \|f(t,h_1) - f(t,h_2)\|_2 \leqslant \mu_f \|h_1 - h_2\|_2$$

for all $t \in I$ and $h_1, h_2 \in W$ where

$$\lambda_f := \sqrt{3}R\,[3\beta_0(RC_1) + 2R\beta_1(RC_1)(C_1 + 3C_2) + 4R^2C_2^2\beta_2(RC_1)],$$
$$\mu_f := \beta_0(RC_1) + RC_1\beta_1(3RC_1).$$

(viii) By

$$g(t,h,k) := (b \circ h)k + \sum_{l=1}^{2} (b_{,l} \circ h)\, h\, k_l - \sum_{l=1}^{2} (b_{,l} \circ h)\, h\, h_l$$
$$- \sum_{l,m=1}^{2} (b_{,lm} \circ h)\, h_l' h_m' h - 2 \sum_{l=1}^{2} (b_{,l} \circ h)\, h_l' h'$$

for all $(t,h,k) \in I \times W \times X$, there is defined an element g of $C(I \times W \times X, X)$. In particular,

$$S f(t,h) = g(t,h,Sh)$$

for all $(t,h) \in I \times W$ and

$$\|g(t,h,k_1) - g(t,h,k_2)\|_2 \leqslant \mu_g \|k_1 - k_2\|_2$$

for all $(t,h) \in I \times W$ and $k_1, k_2 \in X$ where

$$\mu_g := \beta_0(RC_1) + 2RC_1\beta_1(RC_1).$$

(ix) Then for every compact subset K of W, there is $T \in (c,d]$ such that for every $h \in K$ there is $u \in C([c,T],W) \cap C^1([c,T],X)$ such that

$$u'(t) + A(t,u(t))u(t) = f(t,u(t))$$

for all $t \in (c,T)$ where differentiation is with respect to X and $u(c) = h$. If $v \in C([c,T],W) \cap C^1((c,T),X)$ is such that $v'(t) + A(t,v(t))v(t) = f(t,v(t))$ for all $t \in (c,T)$ and $v(c) = h$, then $v = u$.

Proof. '(i)': For this, let the Fourier transformation $F_2 : L^2_{\mathbb{C}}(\mathbb{R}) \to L^2_{\mathbb{C}}(\mathbb{R})$ be defined as in the proof of Lemma 12.1.2. Obviously, S is a densely-defined linear operator in X. Note that $F_2 W^m_{\mathbb{C}}(\mathbb{R})$ coincides with the domain of the maximal multiplication operator in $L^2_{\mathbb{C}}(\mathbb{R})$ by $(1 + |\ |^2)^{m/2}$ for every $m \in \mathbb{N}$. Let $h \in W^2_{\mathbb{C}}(\mathbb{R})$, then

$$\|h\|_s^2 = \|h\|_2^2 + \|(-D^{2*} + 1)h\|_2^2 \leqslant 3\|h\|_2^2 + 2\|D^{2*}h\|_2^2 \leqslant 3|\!|\!|h|\!|\!|_2^2$$

and

$$\|h\|_s^2 = \|h\|_2^2 + \|(-D^{2*} + 1)h\|_2^2 = \|h\|_2^2 + \|F_2(-D^{2*} + 1)h\|_2^2$$
$$= \|h\|_2^2 + \|(\mathrm{id}_{\mathbb{R}}^2 + 1)F_2 h\|_2^2 \geqslant \|h\|_2^2 + \frac{3}{4}\,\|\mathrm{id}_{\mathbb{R}}^2 F_2 h\|_2^2 + \|\mathrm{id}_{\mathbb{R}} F_2 h\|_2^2 \geqslant \frac{3}{4}\,|\!|\!|h|\!|\!|_2^2$$

and finally

$$\frac{\sqrt{3}}{2}\,|\!|\!|h|\!|\!|_2 \leqslant \|h\|_s \leqslant \sqrt{3}\,|\!|\!|h|\!|\!|_2.$$

Hence $(D(S), \|\ \|_S$ is complete and therefore S is closed. In addition, since

$$-D^{2*} + 1 = F_2^{-1} \circ T_{\mathrm{id}_{\mathbb{R}}^2 + 1} \circ F_2,$$

S is self-adjoint and bijective.

'(ii)': For this, let $(t, h) \in \mathbb{R} \times W^2_{\mathbb{C}}(\mathbb{R})$. Then $h \in C^1(\mathbb{R}, \mathbb{C})$, $a \circ h \in C^1(\mathbb{R}, \mathbb{C})$, $(a \circ h)' = (a' \circ h)\, h'$. Then it follows by Lemma 12.1.2 that $a \circ h, (a \circ h)' \in B(\mathbb{R}, \mathbb{C})$ and

$$\|a \circ h\|_\infty \leqslant \alpha_0(C_1 |||h|||_2),\ \|(a \circ h)'\|_\infty \leqslant \alpha_1(C_1 |||h|||_2)\, C_2 |||h|||_2.$$

Hence it follows by Corollary 5.5.3 that by

$$A_0(t, h)k := -(a \circ h) D^{1*} k$$

for every $k \in W^1_{\mathbb{C}}(\mathbb{R})$ there is defined a closable linear operator $A_0(t, h)$ in X whose closure $A(t, h)$ is quasi-accretive with bound given by (12.1.1) and is the infinitesimal generator of a strongly continuous group on X.
'(iii)': The statement is a simple consequence of (ii) along with Lemma 8.6.4.
'(iv)': First, $D(A(t, h)) \supset W^1_{\mathbb{C}}(\mathbb{R}) \supset Y$ and $A(t, h)|_Y \in L(Y, X)$ since the restriction of D^{1*} to Y defines an element of $L(Y, X)$ and $a \circ h \in B(\mathbb{R}, \mathbb{C})$ for all $(t, h) \in I \times W$. Further, it follows for $(t_1, h_1), (t_2, h_2) \in I \times W$ and $x \in \mathbb{R}$ that

$$|a(h_1(x)) - a(h_2(x))| \leqslant \left[\max_{\lambda \in [0,1]} |(\nabla a)(h_1(x) + \lambda(h_2(x) - h_1(x)))| \right] \cdot |h_1(x) - h_2(x)|$$

and hence because of

$$\|h_1 + \lambda(h_2 - h_1)\|_\infty \leqslant C_1 |||h_1 + \lambda(h_2 - h_1)|||_2 \leqslant 3RC_1$$

for every $\lambda \in [0, 1]$ that

$$|a \circ h_1(x) - a \circ h_2(x)| \leqslant \alpha_1(3RC_1)\, |h_1(x) - h_2(x)|.$$

Using this, it follows for every $k \in Y$

$$\begin{aligned} \|(A(t_1, h_1) - A(t_2, h_2))k\|_2 &\leqslant \alpha_1(3RC_1)\, \|h_1 - h_2\|_2\, \|D^{1*}k\|_\infty \\ &\leqslant C_2\, \alpha_1(3RC_1)\, \|h_1 - h_2\|_2\, |||k|||_2 \end{aligned}$$

and hence

$$\|A(t_1, h_1) - A(t_2, h_2)\| \leqslant C_2\, \alpha_1(3RC_1)\, \|h_1 - h_2\|_2 \leqslant C_2\, \alpha_1(3RC_1)\, |||h_1 - h_2|||_2.$$

Further, since

$$\|h_1\|_\infty \leqslant C_1 |||h_1|||_2 \leqslant RC_1,$$

it follows that

$$\|A(t_1, h_1)k\|_2 = \|(a \circ h_1) D^{1*}k\|_2 \leqslant \alpha_0(RC_1)\, \|D^{1*}k\|_2 \leqslant \alpha_0(RC_1) |||k|||_2$$

for every $k \in Y$ and hence

$$\|A(t_1, h_1)\| \leqslant \alpha_0(RC_1).$$

'(v)': First, it follows for $(t, h, \bar{h}) \in I \times W \times X$ and $k \in X$ by using the boundedness of $a_{,l} \circ h$, $a_{,lm} \circ h$, $D^{1*}S^{-1}k$ and h' that $B(t, h, \bar{h})k \in X$. In particular,

$$\|B(t,h,\bar{h})k\|_2 = \| - \sum_{l=1}^{2}(a_{,l} \circ h)\left[(\bar{h}_l - 1)D^{1*}S^{-1}k + 2h_l' D^{2*}S^{-1}k\right]$$

$$+ \sum_{l,m=1}^{2} (a_{,lm} \circ h)h_l' h_m' D^{1*}S^{-1}k\|_2$$

$$\leqslant \alpha_1(R)\, \|D^{1*}S^{-1}k\|_\infty \sum_{l=1}^{2} \|\bar{h}_l\|_2 + 4\,\alpha_1(R)C_2 R\, \|D^{2*}S^{-1}k\|_2$$

$$+ 2\left(\alpha_1(R) + 2C_2^2\, \alpha_2(R)R^2\right) \|D^{1*}S^{-1}k\|_2$$

$$\leqslant C_2\, \alpha_1(R) \| |S^{-1}k| \|_2 \sum_{l=1}^{2} \|\bar{h}_l\|_2 + [\alpha_1(R)(1 + 8C_2 R)$$

$$+ 2C_2^2\, \alpha_2(R)R^2]\, \|k\|_2$$

$$\leqslant (2/3)\,\sqrt{3}\, C_2\, \alpha_1(R) \|S^{-1}k\|_{\mathrm{S}} \sum_{l=1}^{2} \|\bar{h}_l\|_2 + [\alpha_1(R)(1 + 8C_2 R)$$

$$+ 2C_2^2\, \alpha_2(R)R^2]\, \|k\|_2$$

$$\leqslant \left[(2/3)\,\sqrt{6}\, C_2\, \alpha_1(R) \sum_{l=1}^{2} \|\bar{h}_l\|_2 + \alpha_1(R)(1 + 8C_2 R) + 2C_2^2\, \alpha_2(R)R^2\right] \|k\|_2 .$$

Since B is obviously linear, it follows that $B(t, h, \bar{h}) \in L(X, X)$ and since

$$\|S h\|_2 \leqslant \|h\|_{\mathrm{S}} \leqslant \sqrt{3}R$$

that

$$\|B(t, h, S h)\| \leqslant \lambda_{\mathrm{B}}$$

for every $(t, h) \in I \times W$. Further, it follows

$$\|B(t, h, \bar{h}_1)k - B(t, h, \bar{h}_2)k\|_2 \leqslant \sum_{l=1}^{2} \|(a_{,l} \circ h)(\bar{h}_{1l} - \bar{h}_{2l})D^{1*}S^{-1}k\|_2$$

$$\leqslant \mu_{\mathrm{B}} \|\bar{h}_1 - \bar{h}_2\|_2 \|k\|_2$$

for all $(t, h) \in I \times W$, $\bar{h}_1, \bar{h}_2 \in X$. Let $(t, h, \bar{h}) \in I \times W \times X$, $(t_1, h_1, \bar{h}_1), (t_2, h_2, \bar{h}_2), \ldots$ a sequence in $I \times W \times X$ which is convergent to $(t, h, \bar{h})$ and $k \in X$. As a consequence of

$$\|h_\nu - h\|_\infty \leqslant C_1 \| |h_\nu - h| \|_2, \|h_\nu' - h'\|_\infty \leqslant C_2 \| |h_\nu - h| \|_2,$$

and the continuity of $a_{,l}, a_{,lm}$ for all $l, m \in \{1,2\}$, it follows that

$$B(t_\nu, h_\nu, \bar{h})k = -\sum_{l=1}^{2}(a_{,l} \circ h_\nu)\left[(\bar{h}_l - 1)D^{1*}S^{-1}k + 2h'_{\nu l}D^{2*}S^{-1}k\right]$$
$$+ \sum_{l,m=1}^{2}(a_{,lm} \circ h_\nu)h'_{\nu l}h'_{\nu m}D^{1*}S^{-1}k$$

is a sequence in X which is everywhere pointwise convergent to $B(t, h, \bar{h})$ and which is majorized by the square summable function

$$2\,\alpha_1(R)\,[\,(\sqrt{6}/3)C_2\|k\|_2|\bar{h}_l| + (1+2C_2^2R^2)|D^{1*}S^{-1}k| + C_2R|D^{2*}S^{-1}k|\,].$$

Hence it follows by an application of Lebesgue's dominated convergence theorem that

$$\lim_{\nu\to\infty}\|B(t_\nu, h_\nu, \bar{h})k - B(t, h, \bar{h})k\|_2 = 0$$

and hence because of

$$\|B(t_\nu, h_\nu, \bar{h}_\nu)k - B(t, h, \bar{h})k\|_2 \leqslant \|B(t_\nu, h_\nu, \bar{h}_\nu)k - B(t_\nu, h_\nu, \bar{h})k\|_2$$
$$+\|B(t_\nu, h_\nu, \bar{h})k - B(t, h, \bar{h})k\|_2 \leqslant \mu_B\|\bar{h}_\nu - \bar{h}\|_2\|k\|_2$$
$$+\|B(t_\nu, h_\nu, \bar{h})k - B(t, h, \bar{h})k\|_2$$

also that

$$\lim_{\nu\to\infty}\|B(t_\nu, h_\nu, \bar{h}_\nu)k - B(t, h, \bar{h})k\|_2 = 0$$

and finally the strong continuity of B. Further, let $t \in I$, $h \in W \cap W^3_{\mathbb{C}}(\mathbb{R}) \subset C^2(\mathbb{R},\mathbb{C})$ and $k \in \tilde{D} \subset C^3(\mathbb{R},\mathbb{C})$. Then $a \circ h \in C^2(\mathbb{R},\mathbb{C}) \cap BC(\mathbb{R},\mathbb{C})$ and

$$(a \circ h)' = \sum_{l=1}^{2}(a_{,l} \circ h)\,h'_l \in B(\mathbb{R},\mathbb{C}),$$
$$(a \circ h)'' = \sum_{l=1}^{2}(a_{,l} \circ h)\,h''_l + \sum_{l,m=1}^{2}(a_{,lm} \circ h)\,h'_lh'_m \in B(\mathbb{R},\mathbb{C})$$

and hence

$$A(t,h)k = -(a \circ h)D^{1*}k = (a \circ h)\,k' \in Y \cap C^2(\mathbb{R},\mathbb{C}).$$

In particular,

$$-D^{2*}A(t,h)k = -[(a \circ h)\,k']'' = -\left[(a \circ h)\,k'' + \sum_{l=1}^{2}(a_{,l} \circ h)\,h'_l\,k'\right]'$$
$$= -(a \circ h)\,k''' - \sum_{l=1}^{2}(a_{,l} \circ h)(h''_l\,k' + 2h'_lk'') - \sum_{l,m=1}^{2}(a_{,lm} \circ h)h'_lh'_mk'$$

$$= -A(t,h)D^{2*}k - \sum_{l=1}^{2}(a_{,l} \circ h)[(-D^{2*}h)_l\, D^{1*}k + 2h_l' D^{2*}k]$$

$$+ \sum_{l,m=1}^{2}(a_{,lm} \circ h)h_l' h_m' D^{1*}k = -A(t,h)D^{2*}k + B(t,h,Sh)Sk$$

and hence

$$SA(t,h)k = [A(t,h) + B(t,h,Sh)]Sk.$$

Since $R > 0$ is otherwise arbitrary and $W^3_{\mathbb{C}}(\mathbb{R})$ is dense in $W^2_{\mathbb{C}}(\mathbb{R})$, it follows, by the continuity, strong continuity of A and B, respectively, the continuity of $S : (Y, \| \|_S) \to X$, the equivalence of the norms $\| \|_S, \||\ \||_2$ as well as the closedness of S, also that

$$SA(t,h)k = [A(t,h) + B(t,h,Sh)]Sk.$$

for all $(t,h) \in I \times W$, $k \in \tilde{D}$.
'(vi)': For this, let $(t,h) \in I \times W$. Then $S\tilde{D} = W^2_{\mathbb{C}}(\mathbb{R})$ is a core for $\hat{A}(t,h)$ and hence the statement follows from (12.1.2) by Theorem 6.2.2 (ii).
'(vii)': In a first step, we show that for $h \in W^2_{\mathbb{C}}(\mathbb{R})$ it follows that

$$(b \circ h)h \in W^2_{\mathbb{C}}(\mathbb{R}) \tag{12.1.3}$$

(note that $(b \circ h)h \in L^2_{\mathbb{C}}(\mathbb{R})$) and

$$D^{1*}(b \circ h)h = H_1 := (b \circ h)h' + \sum_{l=1}^{2}(b_{,l} \circ h)\, h_l' h \left(\in L^2_{\mathbb{C}}(\mathbb{R})\right)$$

$$D^{2*}(b \circ h)h = H_2 := \sum_{l=1}^{2}(b_{,l} \circ h)\, h\, (D^{2*}h)_l + \sum_{l,m=1}^{2}(b_{,lm} \circ h)\, h_l' h_m' h$$

$$+ 2\sum_{l=1}^{2}(b_{,l} \circ h)\, h_l' h' + (b \circ h)D^{2*}h \left(\in L^2_{\mathbb{C}}(\mathbb{R})\right). \tag{12.1.4}$$

For this, let $h_1, h_2, \ldots$ be a sequence in $W^3_{\mathbb{C}}(\mathbb{R}) \subset C^2(\mathbb{R}, \mathbb{C})$ such that

$$\lim_{\nu \to \infty} \||h_\nu - h\||_2 = 0.$$

Note that this implies that

$$\lim_{\nu \to \infty} \|h_\nu - h\|_\infty = \lim_{\nu \to \infty} \|h_\nu' - h'\|_\infty = 0$$

and the uniform boundedness of the sequences $h_1, h_2, \ldots, h_1', h_2', \ldots$. Then $b \circ h_\nu \in C^2(\mathbb{R}, \mathbb{C}) \cap BC(\mathbb{R}, \mathbb{C})$ and

$$(b \circ h_\nu)h_\nu \in L^2_{\mathbb{C}}(\mathbb{R})$$

$$[(b \circ h_\nu)h_\nu]' = (b \circ h_\nu)h_\nu' + \sum_{l=1}^{2}(b_{,l} \circ h_\nu)\, h_{\nu l}' h_\nu \in L^2_{\mathbb{C}}(\mathbb{R})$$

$$[(b\circ h_\nu)h_\nu]'' = \sum_{l=1}^{2}(b_{,l}\circ h_\nu)\,h_\nu\,h''_{\nu l} + \sum_{l,m=1}^{2}(b_{,lm}\circ h_\nu)\,h'_{\nu l}h'_{\nu m}h_\nu$$
$$+\,2\sum_{l=1}^{2}(b_{,l}\circ h_\nu)\,h'_{\nu l}h'_\nu + (b\circ h_\nu)h''_\nu \in L^2_{\mathbb{C}}(\mathbb{R}) \tag{12.1.5}$$

and hence $(b\circ h_\nu)h_\nu \in W^2_{\mathbb{C}}(\mathbb{R})$. We note that for any sequence $g_1, g_2, \dots \in BC(\mathbb{R},\mathbb{C})$ which is uniformly bounded by some $C > 0$ and converging almost everywhere on $\mathbb{R}$ pointwise to some $g \in BC(\mathbb{R},\mathbb{C})$ and any sequence $k_1, k_2, \dots \in L^2_{\mathbb{C}}(\mathbb{R})$ such that

$$\lim_{\nu\to\infty}\|k_\nu - k\|_2 = 0$$

for some $k \in L^2_{\mathbb{C}}(\mathbb{R})$, it follows that

$$\|g_\nu k_\nu - gk\|_2 \leqslant \|g_\nu(k_\nu - k)\|_2 + \|(g_\nu - g)k\|_2 \leqslant C\|k_\nu - k\|_2 + \|(g_\nu - g)k\|_2$$

and hence by an application of Lebesgue's dominated convergence theorem that

$$\lim_{\nu\to\infty}\|g_\nu k_\nu - gk\|_2 = 0.$$

Hence it follows that

$$\lim_{\nu\to\infty}\|b\circ h_\nu)h_\nu - b\circ h\|_2 = \lim_{\nu\to\infty}\|[(b\circ h_\nu)h_\nu]' - H_1\|_2$$
$$= \lim_{\nu\to\infty}\|[(b\circ h_\nu)h_\nu]'' - H_2\|_2 = 0$$

and therefore (12.1.3) and (12.1.4). Hence it follows that $f := (I\times W \to Y, (t,h)\mapsto (b\circ h)h)$ is well-defined and by the previous arguments also that $f \in C(I\times W, Y)$. In particular, it follows for $(t,h)\in I\times W$ that

$$\|f(t,h)\|_2 \leqslant R\beta_0(RC_1)$$
$$\|D^{1*}f(t,h))\|_2 \leqslant R\,[\beta_0(RC_1) + 2RC_2\beta_1(RC_1)\,]$$
$$\|D^{2*}f(t,h)\|_2 \leqslant R\,[\beta_0(RC_1) + 2R\beta_1(RC_1)(C_1 + 2C_2) + 4R^2C_2^2\beta_2(RC_1)\,]$$

and, finally,

$$\|f(t,h)\|_Y \leqslant \lambda_f.$$

Further, it follows for $t\in I, h_1, h_2 \in W$ that

$$|b\circ h_2)(x) - (b\circ h_1)(x)|^2 \leqslant \beta_1^2(3RC_1)\,|h_2(x) - h_1(x)|^2$$

for all $x\in\mathbb{R}$ and hence

$$\|f(t,h_2) - f(t,h_2)\|_2 = \|(b\circ h_2)h_2 - (b\circ h_1)h_1\|_2 \leqslant \|(b\circ h_2)(h_2 - h_1)\|_2$$
$$+\,\|((b\circ h_2) - (b\circ h_1))h_1\|_2 \leqslant \mu_f\,\|h_2 - h_1\|_2.$$

'(viii)': The well-definedness of g as a map is obvious. Also the continuity of g follows by an application of Lebesgue's dominated convergence theorem by using the remark after (12.1.5). Further, for $(t,h) \in I \times W$, it follows by (12.1.4) that

$$Sf(t,h) = (b \circ h)Sh + \sum_{l=1}^{2}(b_{,l} \circ h)\, h\, (Sh)_l - \sum_{l=1}^{2}(b_{,l} \circ h)\, h\, h_l$$
$$- \sum_{l,m=1}^{2}(b_{,lm} \circ h)\, h'_l h'_m h - 2\sum_{l=1}^{2}(b_{,l} \circ h)\, h'_l h' = g(t,h,Sh).$$

Finally, it follows for $(t,h) \in I \times W$ and $k_1, k_2 \in X$ that

$$\|g(t,h,k_1) - g(t,h,k_2)\|_2 \leqslant \|(b \circ h)k_1 - (b \circ h)k_2\|_2$$
$$+ \sum_{l=1}^{2} \|(b_{,l} \circ h)\, h(k_{1l} - k_{2l})\|_2 \leqslant \mu_g \|k_1 - k_2\|_2.$$

'(ix)' Is a consequence of (i) - (viii) and Theorem 11.0.7. □

12.2 Quasi-Linear Hermitian Hyperbolic Systems

In the following, we consider the solutions of the system

$$\frac{\partial u}{\partial t}(t,x) + \sum_{j=1}^{n} A^j(t,x,u(t,x)) \cdot \frac{\partial u}{\partial x^j}(t,x) = F(t,x,u(t,x))$$

for $t \in \mathbb{R}$, $x \in \mathbb{R}^n$, $\mathbb{C}^p$-valued u, $A^1, \dots, A^n$ assuming values in the set of Hermitian $p \times p$-matrices and $\mathbb{C}^p$-valued F where $p \in \mathbb{N}^*$, $\cdot$ denotes matrix multiplication and partial differentiation is to be interpreted component-wise.[3]

Lemma 12.2.1. (Commutator estimates) Let $n \in \mathbb{N}^*$.

(i) In addition, let $m, m', m'' \in \mathbb{N}$ such that $m' \geqslant m$, $m'' \geqslant m$ and $m' + m'' > m + (n/2)$ Then for all $f \in W^{m'}_{\mathbb{C}}(\mathbb{R}^n)$, $g \in W^{m''}_{\mathbb{C}}(\mathbb{R}^n)$

$$fg \in W^m_{\mathbb{C}}(\mathbb{R}^n)$$

and

$$|||fg|||_m \leqslant C_{nmm'm''} \, |||f|||_{m'} \, |||g|||_{m''}$$

where

$$C_{nmm'm''} := 2^{m'+m''+((n+1)/2)} \, (n+1)^{m/2} \, \pi^{n/2} \, \|(1 + |\ |^2)^{-(m'+m''-m)}\|_1^{1/2}$$

and $|||\ |||_l$ denotes the Sobolev norm for $W^l_{\mathbb{C}}(\mathbb{R}^n)$ for every $l \in \mathbb{N}$.

[3] The elements of $\mathbb{C}^p$ are as usual considered as column vectors.

(ii) Let $m \in \mathbb{N}$ such that $m > n/2$, $q \in \mathbb{N}^*$, $k_1, \dots, k_q \in \mathbb{N}$ such that $k_1 + \dots + k_q \leqslant m$. Then there is $C \geqslant 0$ such that

$$\|h_1 \dots h_q\|_{m-(k_1+\dots+k_q)} \leqslant C \, \|h_1\|_{m-k_1} \dots \|h_q\|_{m-k_q}$$

for all $h_1 \in W^{m-k_1}_{\mathbb{C}}(\mathbb{R}^n), \dots, h_q \in W^{m-k_q}_{\mathbb{C}}(\mathbb{R}^n)$.

(iii) Let $m \in \mathbb{N}$ such that $m > n/2$, $q \in \mathbb{N} \backslash \{0, 1\}$, $k_1, \dots, k_q \in \mathbb{N}$ such that $k_1 + \dots + k_q \leqslant m - 1$. Then

$$f_1 \dots f_q \in W^1_{\mathbb{C}}(\mathbb{R}^n)$$

and

$$\partial^{e_j}(f_1 \dots f_q) = (\partial^{e_j} f_1) \cdot f_2 \dots f_q + \dots + f_1 \dots f_{q-1} \cdot \partial^{e_j} f_q$$

for all $f_j \in W^{m-k_j}_{\mathbb{C}}(\mathbb{R}^n)$, $j \in \{1, \dots, q\}$ where $e_1, \dots, e_n$ denotes the canonical basis of $\mathbb{R}^n$.

(iv) Let $m \in \mathbb{N}^*$ such that $m > (n/2) + 1$ and $a \in C^1(\mathbb{R}^n, \mathbb{C}) \cap BC(\mathbb{R}^n, \mathbb{C})$ such that $a_{,j} \in W^{m-1}_{\mathbb{C}}(\mathbb{R}^n)$ $(\subset C_\infty(\mathbb{R}^n, \mathbb{C}))$ for all $j \in \{1, \dots, n\}$. Then

$$\begin{aligned}
&\|[T_a(-\Delta + 1)^{m/2} - (-\Delta + 1)^{m/2} T_a] [(-\Delta + 1)^{(m-1)/2}]^{-1} f\|_2 \\
&\leqslant D_{nm} \, \|f\|_2 \left(\sum_{j=1}^{n} \|a_{,j}\|^2_{m-1} \right)^{1/2}
\end{aligned}$$

for all $f \in W^1_{\mathbb{C}}(\mathbb{R}^n)$ where

$$D_{nm} := m \, [\, C^2_{n(m-1)1} + 2^{2(m-1)} \, m^{2n} \, C^2_{n(m-1)(m-1)(m-1)} \,]^{1/2}.$$

Here T_a denotes the maximal multiplication operator in $L^2_{\mathbb{C}}(\mathbb{R}^n)$ corresponding to a, $(-\Delta + 1)^{l/2} := [(-\Delta + 1)^{1/2}]^l$ for all $l \in \mathbb{N}^*$ and $C_{n(m-1)1} \in [0, \infty)$ is defined according to Lemma 12.1.2.

(v) Let $m \in \mathbb{N}^*$ such that $m > (n/2) + 1$, $a \in C^{m+1}(\mathbb{R}^n \times \mathbb{C}^p, \mathbb{C})$ such that

$$a_{,\alpha}|_{\mathbb{R}^n \times (U_R(0))^p}$$

is bounded for every $\alpha \in \mathbb{N}^{n+2p}$, $|\alpha| \leqslant m + 1$ and every $R \geqslant 0$. Finally, let $a(\cdot, 0, \dots, 0) = 0$.

a) In addition, let $h_1, \dots, h_p \in W^m_{\mathbb{C}}(\mathbb{R}^n)$ $(\subset C^1(\mathbb{R}^n, \mathbb{C}))$. Then

$$a(\cdot, h_1, \dots, h_p) \in W^m_{\mathbb{C}}(\mathbb{R}^n).$$

Hence according to (iv) the restriction of

$$\begin{aligned}
&[\, T_{a(\cdot, h_1, \dots, h_p)}(-\Delta + 1)^{m/2} - (-\Delta + 1)^{m/2} T_{a(\cdot, h_1, \dots, h_p)} \,] \\
&\cdot [(-\Delta + 1)^{(m-1)/2}]^{-1}
\end{aligned}$$

to $W^1_{\mathbb{C}}(\mathbb{R}^n)$ gives a densely-defined bounded linear operator in $L^2_{\mathbb{C}}(\mathbb{R}^n)$ whose unique extension to a bounded linear operator on $L^2_{\mathbb{C}}(\mathbb{R}^n)$ will be denoted by $b^1_a(h_1, \dots, h_p)$.

b) The map

$$b_a^1 := ((W^m_{\mathbb{C}}(\mathbb{R}^n))^p \to L(L^2_{\mathbb{C}}(\mathbb{R}^n), L^2_{\mathbb{C}}(\mathbb{R}^n)), (h_1, \dots, h_p) \\ \mapsto b_a^1(h_1, \dots, h_p))$$

is continuous as well as bounded on bounded subsets of $((W^m_{\mathbb{C}}(\mathbb{R}^n))^p$.

(vi) Let $m \in \mathbb{N}^*$, $a \in C^{m+1}(\mathbb{R}^n, \mathbb{C})$ such that $a_{,\alpha}$ is bounded for every $\alpha \in \mathbb{N}^n$, $|\alpha| \leqslant m+1$. By the restriction of

$$[\,T_a(-\Delta + 1)^{m/2} - (-\Delta + 1)^{m/2} T_a\,]\,[(-\Delta + 1)^{(m-1)/2}]^{-1}$$

to $W^1_{\mathbb{C}}(\mathbb{R}^n)$, there is given a densely-defined bounded linear operator in $L^2_{\mathbb{C}}(\mathbb{R}^n)$ whose unique extension to a bounded linear operator on $L^2_{\mathbb{C}}(\mathbb{R}^n)$ will be denoted by b_a^2. In particular, there is $C > 0$ such that for all such a

$$\|b_a^2\| \leqslant C \sum_{\alpha \in \mathbb{N}^{m+1}} \|a_{,\alpha}\|_\infty$$

Proof. '(i)': For this, let $F_1 : L^1_{\mathbb{C}}(\mathbb{R}^n) \to C_\infty(\mathbb{R}^n, \mathbb{C})$ and $F_2 : L^2_{\mathbb{C}}(\mathbb{R}^n) \to L^2_{\mathbb{C}}(\mathbb{R}^n)$ be the continuous Fourier transformations defined by

$$(F_1 f)(x) := \int_{\mathbb{R}^n} e^{-ixy} f(y)\, dy$$

for all $x \in \mathbb{R}^n$ and determined by $F_2 f := (2\pi)^{-n/2} F_1 f$ for every $f \in C_0^\infty(\mathbb{R}^n, \mathbb{C})$, respectively. In addition, let $f, g \in C_0^\infty(\mathbb{R}^n, \mathbb{C})$. Then

$$\begin{aligned}(1 + |k|^2)^{m/2} (F_2 fg)(k) &= (2\pi)^{-n/2}\, (1 + |k|^2)^{m/2}\, (F_1 fg)(k) \\ &= (2\pi)^{-n/2}\, (1 + |k|^2)^{m/2}\, (F_1 f * F_1 g)(k) \\ &= (2\pi)^{-n/2} \int_{\mathbb{R}^n} (1 + |k|^2)^{m/2} (F_1 f)(k - k')(F_1 g)(k')\, dk'\end{aligned}$$

for all $k \in \mathbb{R}^n$. Further, let $k, k' \in \mathbb{R}^n$. If

$$|k - k'| \leqslant |k'|,$$

then

$$\begin{aligned}(1 + |k|^2)^{m/2} &\leqslant [1 + (\,|k - k'| + |k'|\,)^2]^{m/2} \leqslant (1 + 4|k'|^2)^{m/2} \\ &\leqslant 2^m (1 + |k'|^2)^{m/2} = 2^m (1 + |k'|^2)^{m''/2}\, (1 + |k'|^2)^{(m-m'')/2} \\ &\leqslant 2^m (1 + |k'|^2)^{m''/2}\, (1 + |k - k'|^2)^{(m-m'')/2}.\end{aligned}$$

If

$$|k - k'| \geqslant |k'|,$$

then

$$\begin{aligned}(1+|k|^2)^{m/2} &\leqslant [1+(|k-k'|+|k'|)^2]^{m/2} \leqslant (1+4|k-k'|^2)^{m/2}\\ &\leqslant 2^m(1+|k-k'|^2)^{m/2} = 2^m(1+|k-k'|^2)^{m'/2}(1+|k-k'|^2)^{(m-m')/2}\\ &\leqslant 2^m(1+|k-k'|^2)^{m'/2}(1+|k'|^2)^{(m-m')/2}.\end{aligned}$$

Hence it follows for $k \in \mathbb{R}^n$ that

$$\begin{aligned}&(1+|k|^2)^{m/2}(F_2 fg)(k)\\ &= (2\pi)^{-n/2}\int_{|k-k'|\leqslant|k'|}(1+|k|^2)^{m/2}(F_1 f)(k-k')(F_1 g)(k')\,dk'\\ &+ (2\pi)^{-n/2}\int_{|k-k'|\geqslant|k'|}(1+|k|^2)^{m/2}(F_1 f)(k-k')(F_1 g)(k')\,dk'.\end{aligned}$$

Further,

$$\begin{aligned}&\left|\int_{|k-k'|\leqslant|k'|}(1+|k-k'|^2)^{(m-m'')/2}(F_1 f)(k-k')(1+|k'|^2)^{m''/2}(F_1 g)(k')\,dk'\right|^2\\ &\leqslant 2^{2m}\left|\int_{|k-k'|\leqslant|k'|}|(1+|k-k'|^2)^{(m-m'')/2}(F_1 f)(k-k')|\right.\\ &\qquad\left.\cdot|(1+|k'|^2)^{m''/2}(F_1 g)(k')|\,dk'\right|^2\\ &\leqslant 2^{2m}\left|\int_{\mathbb{R}^n}|(1+|k-k'|^2)^{m'/2}(F_1 f)(k-k')|\cdot(1+|k-k'|^2)^{-(m'+m''-m)/2}\right.\\ &\qquad\left.\cdot|(1+|k'|^2)^{m''/2}(F_1 g)(k')\,dk'\right|^2\\ &\leqslant \|(1+|\ |^2)^{m'/2}(F_1 f)\|_2^2\int_{\mathbb{R}^n}(1+|k-k'|^2)^{-(m'+m''-m)}\cdot|(1+|k'|^2)^{m''/2}\\ &\quad(F_1 g)(k')|^2\,dk'\end{aligned}$$

as well as

$$\begin{aligned}&\left|\int_{|k-k'|\geqslant|k'|}(1+|k-k'|^2)^{m'/2}(F_1 f)(k-k')(1+|k'|^2)^{(m-m')/2}(F_1 g)(k')\,dk'\right|^2\\ &\leqslant 2^{2m}\left|\int_{|k-k'|\geqslant|k'|}|(1+|k-k'|^2)^{m'/2}(F_1 f)(k-k')|\right.\\ &\qquad\left.\cdot|(1+|k'|^2)^{(m-m')/2}(F_1 g)(k')|\,dk'\right|^2\\ &\leqslant 2^{2m}\left|\int_{\mathbb{R}^n}|(1+|k-k'|^2)^{m'/2}(F_1 f)(k-k')|\cdot(1+|k-k'|^2)^{-(m'+m''-m)/2}\right.\\ &\qquad\left.\cdot|(1+|k'|^2)^{m''/2}(F_1 g)(k')\,dk'\right|^2\end{aligned}$$

$$\leqslant \|(1+|\ |^2)^{m'/2}(F_1 f)\|_2^2 \int_{\mathbb{R}^n} (1+|k-k'|^2)^{-(m'+m''-m)} |(1+|k'|^2)^{m''/2} \cdot (F_1 g)(k')|^2\, dk'.$$

Hence it follows by Fubini's theorem that

$$\|(1+|\ |^2)^{m/2}(F_2 fg)\|_2^2 \leqslant 2\,(2\pi)^n\,\|(1+|\ |^2)^{-(m'+m''-m)}\|_1 \cdot \|(1+|\ |^2)^{m'/2}(F_2 f)\|_2^2 \cdot \|(1+|\ |^2)^{m''/2}(F_2 g)\|_2^2.$$

From this, we conclude by Lemma 12.1.1

$$|||fg|||_m^2 \leqslant 2^{n+2m'+2m''+1}\,(n+1)^m\,\pi^n\,\|(1+|\ |^2)^{-(m'+m''-m)}\|_1 \cdot |||f|||_{m'}^2 \cdot |||g|||_{m''}^2$$

and therefore, since $C_0^\infty(\mathbb{R}^n, \mathbb{C})$ is dense in $W_{\mathbb{C}}^l(\mathbb{R}^n)$ for every $l \in \mathbb{N}$, the statement of the theorem.

'(ii)': The proof proceeds by induction over $q \in \mathbb{N}^*$. The statement is obviously true for $q = 1$. Let q be some element of $\mathbb{N}^*$ for which there is $C \geqslant 0$ such that

$$\|h_1 \dots h_q\|_{m-(k_1+\dots+k_q)} \leqslant C\,|||h_1|||_{m-k_1} \dots |||h_q|||_{m-k_q}$$

for all $h_1 \in W_{\mathbb{C}}^{m-k_1}(\mathbb{R}^n), \dots, h_q \in W_{\mathbb{C}}^{m-k_q}(\mathbb{R}^n)$. In addition, let $h_1 \in W_{\mathbb{C}}^{m-k_1}(\mathbb{R}^n), \dots, h_{q+1} \in W_{\mathbb{C}}^{m-k_{q+1}}(\mathbb{R}^n)$. Then by assumption

$$\|h_2 \dots h_{q+1}\|_{m-(k_2+\dots+k_{q+1})} \leqslant C\,|||h_2|||_{m-k_2} \dots |||h_{q+1}|||_{m-k_{q+1}}.$$

Further, we define

$$m' := m - k_1 \geqslant m - (k_1 + \dots + k_{q+1}),$$
$$m'' := m - (k_2 + \dots + k_{q+1}) \geqslant m - (k_1 + \dots + k_{q+1}).$$

Then

$$m' + m'' = m - k_1 + m - (k_2 + \dots + k_{q+1}) > (n/2) + m - (k_1 + \dots + k_{q+1}).$$

Hence it follows from (i) the existence of $C' \geqslant 0$ such that

$$|||h_1 \dots h_{q+1}|||_{m-(k_1+\dots+k_{q+1})} \leqslant C'\,|||h_1|||_{m-k_1}\,|||h_2 \dots h_{q+1}|||_{m-(k_2+\dots+k_{q+1})} \leqslant C\,C'\,|||h_1|||_{m-k_1} \dots |||h_{q+1}|||_{m-k_{q+1}}.$$

'(iii)': For this, let $j \in \{1, \dots, n\}$. By (ii), it follows that by

$$(W_{\mathbb{C}}^{m-k_1}(\mathbb{R}^n) \times \dots \times W_{\mathbb{C}}^{m-k_q}(\mathbb{R}^n) \to W_{\mathbb{C}}^{m-1-(k_1+\dots+k_q)}(\mathbb{R}^n), (f_1, \dots, f_q) \mapsto \partial^{e_j}(f_1 \dots f_q))$$

and

$$(W_{\mathbb{C}}^{m-k_1}(\mathbb{R}^n) \times \cdots \times W_{\mathbb{C}}^{m-k_q}(\mathbb{R}^n) \to W_{\mathbb{C}}^{m-1-(k_1+\cdots+k_q)}(\mathbb{R}^n), (f_1,\ldots,f_q)$$
$$\mapsto (\partial^{e_j} f_1)\, f_2 \ldots f_q + \cdots + f_1 \ldots f_{q-1}\, \partial^{e_j} f_q$$

there are defined continuous multilinear maps whose restrictions to $(C_0^\infty(\mathbb{R}^n, \mathbb{C}))^q$ coincide. From this follows (iii) by the denseness of $(C_0^\infty(\mathbb{R}^n, \mathbb{C}))^q$ in $W_{\mathbb{C}}^{m-k_1}(\mathbb{R}^n) \times \cdots \times W_{\mathbb{C}}^{m-k_q}(\mathbb{R}^n)$.

'(iv)': For every complex-valued function g which is a.e. defined on $\mathbb{R}^n$ and measurable, we denote by T_g the corresponding maximal multiplication operator in $L_{\mathbb{C}}^2(\mathbb{R}^n)$. Note that according to the proof of Theorem 12.2.2 (i), $(-\Delta + 1)^{l/2}$ is a linear, self-adjoint and bijective operator in $L_{\mathbb{C}}^2(\mathbb{R}^n)$ with domain $W_{\mathbb{C}}^l(\mathbb{R}^n)$ satisfying

$$F_2(-\Delta + 1)^{l/2} F_2^{-1} = T_{(|\ |^2+1)^{l/2}}.$$

Since $m > n/2 + 1$, $W_{\mathbb{C}}^m(\mathbb{R}^n)$ is closed under pointwise multiplication of its elements according to (i). In the first step, we assume in addition that $a \in L_{\mathbb{C}}^2(\mathbb{R}^n)$ which implies that $a \in W_{\mathbb{C}}^m(\mathbb{R}^n)$. Hence it follows that $W_{\mathbb{C}}^1(\mathbb{R}^n)$ is contained in the domain of

$$K_{am} := [\,T_a(-\Delta + 1)^{m/2} - (-\Delta + 1)^{m/2} T_a\,][(-\Delta + 1)^{(m-1)/2}]^{-1}.$$

In addition, since the restriction of T_a to $W_{\mathbb{C}}^m(\mathbb{R}^n)$ defines a continuous linear operator on $(W_{\mathbb{C}}^m(\mathbb{R}^n), \|\!|\ \|\!|_m)$, it follows that the restriction of K_{am} to $W_{\mathbb{C}}^1(\mathbb{R}^n)$ is a continuous linear map from $(W_{\mathbb{C}}^1(\mathbb{R}^n), \|\!|\ \|\!|_1)$ to $L_{\mathbb{C}}^2(\mathbb{R}^n)$. Further, it follows for $f \in F_2\, C_0^\infty(\mathbb{R}^n, \mathbb{C})$ and almost all $k \in \mathbb{R}^n$ that

$$
\begin{aligned}
&(F_2 K_{am} F_2^{-1} f)(k) \\
&\quad = \left(F_2\, a F_2^{-1}(|\ |^2 + 1)^{1/2} f - (|\ |^2 + 1)^{m/2} F_2\, a F_2^{-1}(|\ |^2 + 1)^{(1-m)/2} f\right)(k) \\
&\quad = (2\pi)^{n/2} \Big((F_2 a) * (|\ |^2 + 1)^{1/2} f \\
&\quad\ -(|\ |^2 + 1)^{m/2}(F_2 a) * (|\ |^2 + 1)^{(1-m)/2} f\Big)(k) \\
&\quad = \int_{\mathbb{R}^n} (2\pi)^{n/2} (F_2 a)(k - k')\,[\,(|k'|^2 + 1)^{m/2} - (|k|^2 + 1)^{m/2}\,]\,(|k'|^2 + 1)^{(1-m)/2} \\
&\quad\ f(k')\, dk'.
\end{aligned}
$$

Since

$$
\begin{aligned}
&|(|k|^2 + 1)^{m/2} - (|k'|^2 + 1)^{m/2}| \leqslant m \left[\max_{\lambda \in [0,1]} (|k + \lambda(k' - k)|^2 + 1)^{(m-1)/2}\right] \\
&\cdot |k - k'| \leqslant m \left[(|k|^2 + 1)^{(m-1)/2} + (|k'|^2 + 1)^{(m-1)/2}\right] |k - k'|,
\end{aligned}
$$

where in the last estimate it has been used that $([0,\infty) \to \mathbb{R}, x \mapsto (1+x^2)^{(m-1)/2})$ is increasing and that the polynomial $([0,1] \to \mathbb{R}, \lambda \mapsto [k + \lambda(k'-k)]^2)$ assumes its maximum at one of the endpoints of its domain,

$$|K_{am} F_2^{-1} f| \leqslant m\,(2\pi)^{n/2}\,[T_{(|\,|^2+1)^{(m-1)/2}}\,(|F_2 \nabla a| * T_{(|\,|^2+1)^{-(m-1)/2}} f) + |F_2 \nabla a| * f],$$

a.e. on $\mathbb{R}^n$ where $F_2 \nabla a := (F_2 a_{,1}, \dots, F_2 a_{,n})$. Note that

$$|F_2 \nabla a| \in D(T_{(|\,|^2+1)^{(m-1)/2}})$$

and hence that

$$\begin{aligned} & F_2^{-1}|F_2\nabla a|, (F_2^{-1}|F_2\nabla a|\,) \cdot F_2^{-1} f, (F_2^{-1}|F_2\nabla a|\,) \cdot F_2^{-1} T_{(|\,|^2+1)^{-(m-1)/2}} f \\ & \in W^{m-1}_{\mathbb{C}}(\mathbb{R}^n) \subset BC(\mathbb{R}^n, \mathbb{C}) \end{aligned}$$

where it has been used that $F_2^{-1} f, F_2^{-1} T_{(|\,|^2+1)^{-(m-1)/2}} f$ have bounded derivatives to any order as a rapidly decreasing functions. As a consequence,

$$\begin{aligned} & m\, F_2[(F_2^{-1}|F_2\nabla a|\,) \cdot F_2^{-1} f] = m\,(2\pi)^{n/2}\,|F_2\nabla a| * f \in D(T_{(|\,|^2+1)^{(m-1)/2}}), \\ & m\, F_2[(F_2^{-1}|F_2\nabla a|\,) \cdot F_2^{-1} T_{(|\,|^2+1)^{-(m-1)/2}} f] \\ & = m\,(2\pi)^{n/2}\,|F_2\nabla a| * T_{(|\,|^2+1)^{-(m-1)/2}} f \in D(T_{(|\,|^2+1)^{(m-1)/2}}) \end{aligned}$$

and therefore

$$\begin{aligned} & \|m\,(2\pi)^{n/2}\,|F_2\nabla a| * f\|_2^2 \\ & \leqslant m^2\, \|\, F_2^{-1}|F_2\nabla a|\,\|_\infty^2 \cdot \|f\|_2^2 \leqslant m^2\, C^2_{n(m-1)1}\, \|\, F_2^{-1}|F_2\nabla a|\,\|_{m-1}^2\, \|f\|_2^2 \\ & = m^2\, C^2_{n(m-1)1}\, \|f\|_2^2 \sum_{\alpha\in\mathbb{N}^n, |\alpha|\leqslant m-1} \|\,\partial^\alpha F_2^{-1}|F_2\nabla a|\,\|_2^2 \\ & = m^2\, C^2_{n(m-1)1}\, \|f\|_2^2 \sum_{j=1}^n \|\|a_{,j}\|\|_{m-1}^2 \end{aligned}$$

where $C_{n(m-1)1} \in [0,\infty)$ is defined according to Lemma 12.1.2. Further, it follows by Lemma 12.1.1 along with (i) that

$$\begin{aligned} & \|m\,(2\pi)^{n/2}\, T_{(|\,|^2+1)^{(m-1)/2}}\,(|F_2\nabla a| * T_{(|\,|^2+1)^{-(m-1)/2}} f\,)\|_2^2 \\ & \leqslant 2^{2(m-1)}\, m^2\, \|(2\pi)^{n/2} \sum_{\alpha\in\mathbb{N}^n, |\alpha|\leqslant m-1} |\,\mathrm{pr}^\alpha\,| \cdot \big(|F_2\nabla a| * T_{(|\,|^2+1)^{-(m-1)/2}} f\big)\,\|_2^2 \\ & \leqslant 2^{2(m-1)}\, m^{n+2} \sum_{\alpha\in\mathbb{N}^n, |\alpha|\leqslant m-1} \|(2\pi)^{n/2}\, \mathrm{pr}^\alpha \cdot \big(|F_2\nabla a| * T_{(|\,|^2+1)^{-(m-1)/2}} f\big)\,\|_2^2 \end{aligned}$$

$$
\begin{aligned}
&= 2^{2(m-1)}\, m^{n+2} \sum_{\alpha\in\mathbb{N}^n, |\alpha|\leqslant m-1} \|\partial^\alpha (2\pi)^{n/2} F_2^{-1} \left(|F_2\nabla a| * T_{(|\ |^2+1)^{-(m-1)/2}} f\right)\|_2^2 \\
&= 2^{2(m-1)}\, m^{n+2}\, \|(2\pi)^{n/2} F_2^{-1} \left(|F_2\nabla a| * T_{(|\ |^2+1)^{-(m-1)/2}} f\right)\|_{m-1}^2 \\
&= 2^{2(m-1)}\, m^{n+2}\, \|(F_2^{-1}|F_2\nabla a|\,)\cdot F_2^{-1} T_{(|\ |^2+1)^{-(m-1)/2}} f\|_{m-1}^2 \\
&\leqslant 2^{2(m-1)}\, m^{n+2}\, C^2_{n(m-1)(m-1)(m-1)} \|F_2^{-1}|F_2\nabla a|\,\|_{m-1}^2 \\
&\quad \cdot \|F_2^{-1} T_{(|\ |^2+1)^{-(m-1)/2}} f\|_{m-1}^2 \\
&\leqslant 2^{2(m-1)}\, m^{n+2}\, C^2_{n(m-1)(m-1)(m-1)} \|F_2^{-1} T_{(|\ |^2+1)^{-(m-1)/2}} f\|_{m-1}^2 \sum_{j=1}^n \|a_{,j}\|_{m-1}^2 \\
&= 2^{2(m-1)}\, m^{n+2}\, C^2_{n(m-1)(m-1)(m-1)} \sum_{\alpha\in\mathbb{N}^n, |\alpha|\leqslant m-1} \|\operatorname{pr}^\alpha T_{(|\ |^2+1)^{-(m-1)/2}} f\|_2^2 \\
&\quad \cdot \sum_{j=1}^n \|a_{,j}\|_{m-1}^2 \\
&\leqslant 2^{2(m-1)}\, m^{2(n+1)}\, C^2_{n(m-1)(m-1)(m-1)} \|f\|_2^2 \sum_{j=1}^n \|a_{,j}\|_{m-1}^2
\end{aligned}
$$

where

$$
\operatorname{pr}^\alpha := \prod_{j=1}^n \operatorname{pr}_j^{\alpha_j}
$$

for every $\alpha \in \mathbb{R}^n$ and pr_j denotes the coordinate projection of $\mathbb{R}^n$ onto the j-th coordinate, $j \in \{1, \dots, n\}$. Hence we arrive at

$$
\|K_{am} F_2^{-1} f\|_2 \leqslant D_{nm}\, \|F_2^{-1} f\|_2 \cdot \left(\sum_{j=1}^n \|a_{,j}\|_{m-1}^2 \right)^{1/2} .
$$

Since $C_0^\infty(\mathbb{R}^n, \mathbb{C})$ is dense in $(W^1_{\mathbb{C}}(\mathbb{R}^n), \|\ \|_1)$, the restriction of K_{am} to $W^1_{\mathbb{C}}(\mathbb{R}^n)$ is a continuous linear map from $(W^1_{\mathbb{C}}(\mathbb{R}^n), \|\ \|_1)$ to $L^2_{\mathbb{C}}(\mathbb{R}^n)$ and the inclusion of $(W^1_{\mathbb{C}}(\mathbb{R}^n), \|\ \|_1)$ into $L^2_{\mathbb{C}}(\mathbb{R}^n)$ is continuous, finally, it follows that

$$
\|K_{am} g\|_2 \leqslant D_{nm}\, \|g\|_2 \cdot \left(\sum_{j=1}^n \|a_{,j}\|_{m-1}^2 \right)^{1/2}
$$

for all $g \in W^1_{\mathbb{C}}(\mathbb{R}^n)$. In the final step, we drop the additional assumption that $a \in L^2_{\mathbb{C}}(\mathbb{R}^n)$ and define for $\nu \in \mathbb{N}^*$

$$
a_\nu := h_\nu a,\ h_\nu := e^{-|\ |^2/\nu^2} .
$$

Since a is bounded and h_ν and all its partial derivatives are bounded as well as square integrable, it follows that a_ν is satisfying the assumptions for a from the previous step. In particular, for $j \in \{1, \dots, n\}$ and $\beta \in \mathbb{N}^n$ such that $|\beta| \leqslant m - 1$,

$$\partial^\beta a_{\nu,j} = (\partial^\beta h_{\nu,j})\, a + \sum_{k=1}^{n} \sum_{\gamma \in \Gamma_k} \binom{\beta_1}{\gamma_1} \dots \binom{\beta_n}{\gamma_n} (\partial^{\beta-\gamma} h_{\nu,j})\, \partial^{\gamma - e_k} a_{,k}$$
$$+ \sum_{\gamma_1=0}^{\beta_1} \dots \sum_{\gamma_n=0}^{\beta_n} \binom{\beta_1}{\gamma_1} \dots \binom{\beta_n}{\gamma_n} (\partial^\gamma h_\nu)\, \partial^{\beta-\gamma} a_{,j}$$

where

$$\Gamma_k := \{\gamma \in \{0, \dots, \beta_1\} \times \dots \times \{0, \dots, \beta_n\} \backslash \{(0, \dots, 0)\} : \gamma_1 = \dots = \gamma_{k-1} = 0 \wedge \gamma_k \neq 0\}$$

for $k \in \{1, \dots, n\}$ and $e_1, \dots, e_n$ is the canonical basis in $\mathbb{R}^n$. Since for every $\gamma \in \mathbb{N}^n$ the corresponding $\partial^\gamma h_1, \partial^\gamma h_2, \dots$ converges in $L^2_{\mathbb{C}}(\mathbb{R}^n)$ to the zero function if $|\gamma| \neq 0$ and the constant function of value 1 if $|\gamma| = 0$ and at the same time is a bounded sequence in $BC(\mathbb{R}^n, \mathbb{C})$ converging everywhere pointwise on $\mathbb{R}^n$ to the zero function if $|\gamma| \neq 0$ and the constant function of value 1 if $|\gamma| = 0$, it follows by Lebesgue's dominated convergence theorem that

$$\lim_{\nu \to \infty} \|\partial^\beta a_{\nu,j} - \partial^\beta a_{,j}\|_2 = 0$$

and hence that

$$\lim_{\nu \to \infty} |||a_{\nu,j} - a_{,j}|||_{m-1} = 0. \tag{12.2.1}$$

In the following, we denote by $\bar{K}_{a_\nu}$ the unique extension of K_{a_ν} to a bounded linear operator on $L^2_{\mathbb{C}}(\mathbb{R}^n)$. Then

$$\|(K_{a_\mu m} - K_{a_\nu m}) f\|_2 =$$
$$\|[T_{a_\mu - a_\nu}(-\Delta + 1)^{m/2} - (-\Delta + 1)^{m/2} T_{a_\mu - a_\nu}][(-\Delta + 1)^{(m-1)/2}]^{-1} f\|_2$$
$$\leqslant D_{nm} \|f\|_2 \left(\sum_{j=1}^{n} |||a_{\mu,j} - a_{\nu,j}|||^2_{m-1} \right)^{1/2}$$

for all $f \in W^1_{\mathbb{C}}(\mathbb{R}^n)$ and hence

$$\|\bar{K}_{a_\mu m} - \bar{K}_{a_\nu m}\| \leqslant D_{nm} \left(\sum_{j=1}^{n} |||a_{\mu,j} - a_{\nu,j}|||^2_{m-1} \right)^{1/2} .$$

Hence it follows by (12.2.1) that $K_{a_1 m}, K_{a_2 m}, \dots$ is a Cauchy sequence in $L(L^2_{\mathbb{C}}(\mathbb{R}^n), L^2_{\mathbb{C}}(\mathbb{R}^n))$ and hence convergent to some $\bar{K}_{am} \in L(L^2_{\mathbb{C}}(\mathbb{R}^n), L^2_{\mathbb{C}}(\mathbb{R}^n))$ satisfying

$$\|\bar{K}_{am}\| \leqslant D_{nm} \left(\sum_{j=1}^{n} |||a_{,j}|||^2_{m-1} \right)^{1/2} .$$

Further, since $a_1, a_2, \ldots$ is a bounded sequence in $BC(\mathbb{R}^n, \mathbb{C})$ which is everywhere on $\mathbb{R}^n$ pointwise convergent to a, it follows for $f \in W^1_{\mathbb{C}}(\mathbb{R}^n)$ by Lebesgue's dominated convergence theorem that

$$\lim_{\nu\to\infty} \|T_{a_\nu}(-\Delta+1)^{1/2}f - T_a(-\Delta+1)^{1/2}f\|_2 = 0,$$
$$\lim_{\nu\to\infty} \|T_{a_\nu}(-\Delta+1)^{(1-m)/2}f - T_a(-\Delta+1)^{(1-m)/2}f\|_2 = 0$$

and therefore also the convergence of

$$(-\Delta+1)^{m/2}T_{a_1}\,[(-\Delta+1)^{(1-m)/2}]f, (-\Delta+1)^{m/2}T_{a_2}\,[(-\Delta+1)^{(1-m)/2}]f, \ldots$$

in $L^2_{\mathbb{C}}(\mathbb{R}^n)$. Since $(-\Delta+1)^{m/2}$ is closed, this implies that

$$T_a\,[(-\Delta+1)^{(1-m)/2}]f \in D((-\Delta+1)^{m/2})$$

and

$$\lim_{\nu\to\infty} \|(-\Delta+1)^{m/2}T_{a_\nu}(-\Delta+1)^{(1-m)/2}f - (-\Delta+1)^{m/2}T_a(-\Delta+1)^{(1-m)/2}f\|_2 = 0$$

as well as

$$\|[T_a(-\Delta+1)^{m/2} - (-\Delta+1)^{m/2}T_a]\,[(-\Delta+1)^{(m-1)/2}]^{-1}f\|_2 \leqslant D_{nm}\,\|f\|_2 \left(\sum_{j=1}^{n} \|a_{,j}\|^2_{m-1}\right)^{1/2}.$$

'(v)': For this, let $h_1, \ldots, h_p \in W^m_{\mathbb{C}}(\mathbb{R}^n)$ $(\subset (C^1(\mathbb{R}^n, \mathbb{C}))^p)$ and $\alpha \in \mathbb{N}^n \times \{0\}^{2p}$ such $|\alpha| \leqslant m$.[4] Then

$$a_{,\alpha}(\cdot, h_1, \ldots, h_p) \in L^2_{\mathbb{C}}(\mathbb{R}^n). \tag{12.2.2}$$

For the proof, we notice that $a_{,\alpha}(\cdot, h_1, \ldots, h_p) \in C(\mathbb{R}^n, \mathbb{C})$ and that $h_1, \ldots, h_p$ are bounded according to Lemma 12.1.2. Hence it follows by the mean value theorem that

$$|\,a_{,\alpha}(x, h_1(x), \ldots, h_p(x))\,|^2 = |\,a_{,\alpha}(x, h_1(x), \ldots, h_p(x)) - a_{,\alpha}(x, 0, \ldots, 0)\,|^2$$
$$\leqslant \left[\max_{\lambda\in[0,1]} |\,(\nabla a_{,\alpha})(x, \lambda h_1(x), \ldots, \lambda h_p(x)|\right]^2 \cdot |(h_1(x), \ldots, h_p(x))|^2$$
$$\leqslant C \cdot |(h_1(x), \ldots, h_p(x))|^2$$

for all $x \in \mathbb{R}^n$ and large enough $C \geqslant 0$. Hence it follows (12.2.2). Then it follows by (iii) and induction over $|\alpha|$ that

$$(a(\cdot, h_1, \ldots, h_p))_{,\alpha}$$

[4] Note that $\mathbb{C} = \mathbb{R}^2$. Also we identify $\mathbb{R}^n \times \mathbb{C}^p$ and $\mathbb{R}^{n+2p}$.

is a linear combination of

$$a_{,(\alpha_1,\ldots,\alpha_n,0,\ldots,0)}(\cdot, h_1, \ldots, h_p) \tag{12.2.3}$$

and, if $|\alpha| \neq 0$, of terms of the form

$$a_\beta(\cdot, h_1, \ldots, h_p)\,(\partial^{\gamma_{j_1}} h_{j_1 r_1}) \cdots \cdots (\partial^{\gamma_{j_q}} h_{j_q r_q}) \tag{12.2.4}$$

where $\beta \in \mathbb{N}^{n+2p}$ is such that $|\beta| \leqslant |\alpha|$, $q \in \mathbb{N}$ satisfying $1 \leqslant q \leqslant |\alpha|$, $j_1, \ldots, j_q \in \{1, \ldots, 2p\}$ satisfying $|\gamma_{j_1}|, \ldots, |\gamma_{j_q}| \geqslant 1$ as well as

$$q \leqslant |\gamma_{j_1}| + \cdots + |\gamma_{j_q}| \leqslant |\alpha|$$

and $r_1, \ldots, r_q \in \{1, 2\}$. Hence it follows by (12.2.2) and (ii) that

$$(a(\cdot, h_1, \ldots, h_p))_{,\alpha} \in L^2_{\mathbb{C}}(\mathbb{R}^n)$$

and therefore that

$$a(\cdot, h_1, \ldots, h_p) \in W^m_{\mathbb{C}}(\mathbb{R}^n).$$

Hence b^1_a is well-defined. By help of Lemma 12.1.2, (ii) and Lebesgue's dominated convergence theorem, it follows that all derivatives (12.2.3), (12.2.4) define continuous maps from $W^m_{\mathbb{C}}(\mathbb{R}^n)$ to $L^2_{\mathbb{C}}(\mathbb{R}^n)$ which are bounded on bounded subsets of $((W^m_{\mathbb{C}}(\mathbb{R}^n))^p$ and therefore by (iv) also that b^1_a is continuous as well as bounded on bounded subsets of $(W^m_{\mathbb{C}}(\mathbb{R}^n))^p$.
'(vi)': The statement is obvious for even m and a simple consequence of Lemma 10.2.1 (vii) if m is odd. □

Theorem 12.2.2. Let $n, p \in \mathbb{N}^*$, $m \in \mathbb{N}^*$ such that $m > (n/2) + 1$, $X := (L^2_{\mathbb{C}}(\mathbb{R}^n))^p$, $Y := (W^m_{\mathbb{C}}(\mathbb{R}^n))^p$ $(\subset (C^1(\mathbb{R}^n, \mathbb{C}))^p)$,

$$S := (Y \to X, u = (u_1, \ldots, u_p) \mapsto ((-\Delta + 1)^{m/2} u_1, \ldots, (-\Delta + 1)^{m/2} u_p))$$

where $(-\Delta + 1)^{m/2} := [\,(-\Delta + 1)^{1/2}\,]^m$, J be a non-empty open interval of $\mathbb{R}$, $c, d \in J$ such that $c \leqslant d$ and $I := [c, d]$. Further, let $A^1, \ldots, A^n : J \times \mathbb{R}^n \times \mathbb{C}^p \to M(p \times p, \mathbb{C})$, $F : J \times \mathbb{R}^n \times \mathbb{C}^p \to \mathbb{C}$ be such that $A^j(t, x, u)$ is Hermitian for all $(t, x, u) \in I \times \mathbb{R}^n \times \mathbb{C}^p$ and $j \in \{1, \ldots, n\}$ as well as such that[5] for every $j \in \{1, \ldots, n\}$, $k, l \in \{1, \ldots, p\}$ and $R > 0$

$$A^j_{kl}(t, \cdot, \cdot), F^j(t, \cdot, \cdot) \in C^{m+1}(\mathbb{R}^{n+1+2p}, \mathbb{C})$$

for every $t \in J$ and such that the restriction of $A^j_{kl,\alpha}, f$ to $I \times \mathbb{R}^n \times (U_R(0))^p$ is bounded for every $\alpha \in \{0\} \times \mathbb{N}^{n+2p}$ satisfying $|\alpha| \leqslant m + 1$. Also, for any such α, $A^j_{kl,\alpha}, F^j$ continuously is partially differentiable by time with a corresponding partial derivative whose restriction to $I \times \mathbb{R}^n \times (U_R(0))^p$ is bounded. In particular, we define

$$\alpha_0(R) := \max\{\| A^j_{kl}|_{I \times \mathbb{R}^n \times (U_R(0))^p} \|_\infty : j \in \{1, \ldots, n\}, k, l \in \{1, \ldots, p\}\},$$

[5] Note that $\mathbb{C} = \mathbb{R}^2$. The index '0' associated to time, i.e:, the coordinate projection of $\mathbb{R} \times \mathbb{R}^n \times (\mathbb{R}^2)^p$ onto the first coordinate. Finally, $\mathbb{R} \times \mathbb{R}^n \times \mathbb{C}^p$ and $\mathbb{R}^{n+1+2p}$ will be identified.

$$\alpha_1(R) := \max\{\| A^j_{kl,m}|_{I\times\mathbb{R}^n\times(U_R(0))^p}\|_\infty : j \in \{1,\dots,n\}, k,l \in \{1,\dots,p\},$$
$$m \in \{0,\dots,n+2p\}\}.$$

In addition, let F be such that

$$(I \to Y, t \mapsto (F^1(t,\cdot,0,\dots,0),\dots,F^p(t,\cdot,0,\dots,0))) \in C(I,Y).$$

Finally, let $R > 0$ and

$$W := \{h \in Y : \|h_1\|_m^2 + \dots \|h_p\|_m^2 < R^2\}.$$

Then

(i) S is a densely-defined linear, self-adjoint and bijective operator in X. In particular,

$$\frac{1}{n^{m/2}}\left(\sum_{k=1}^p |||u_k|||_m^2\right)^{1/2} \leqslant \|u\|_{\mathrm{Y}} \leqslant [1+(n+1)^m]^{1/2}\left(\sum_{k=1}^p |||u_k|||_m^2\right)^{1/2} \quad (12.2.5)$$

for all $u = (u_1,\dots,u_p) \in Y$ where $|||\ |||_m$ denotes the Sobolev norm for $W^m_{\mathbb{C}}(\mathbb{R}^n)$.

(ii) For every $t \in I, h \in W$, by

$$\Bigg((W^1_{\mathbb{C}}(\mathbb{R}^n))^p \to X, u \mapsto \sum_{j=1}^n A^j(t,\cdot,h)\partial^j u$$
$$:= \left(\sum_{j=1}^n\sum_{l=1}^p A^j_{1l}(t,\cdot,h)\partial^j u_l,\dots,\sum_{j=1}^n\sum_{l=1}^p A^j_{pl}(t,\cdot,h)\partial^j u_l)\right),$$

there is defined a densely-defined linear and quasi-accretive operator in X with bound

$$-np\,(1+2pRC_2)\,\alpha_1(RC_1)/2 \quad (12.2.6)$$

whose closure $A(t,h)$ is the infinitesimal generator of a strongly continuous group on X where $C_1 := C_{nm1}, C_2 := C_{nm2}$ and C_{nm1}, C_{nm2} are defined according to Lemma 12.1.2.

(iii) The family $(A(t,u(t)))_{t\in I}$ is stable with stability constants

$$np\,(1+2pRC_2)\,\alpha_1(RC_1)/2, 1$$

for every $u : I \to W$.

(iv) $Y \subset D(A(t,h)), A(t,h)|_Y \in L(Y,X)$ for every $(t,h) \in I \times W$ and

$$A := (I \times W \to L(Y,X), (t,h) \mapsto A(t,h)|_Y) \in C(I \times W, L(Y,X)).$$

Moreover,

$$\|A(t,h)\| \leqslant \lambda_{\mathrm{A}}, \|A(t,h_1) - A(t,h_2)\| \leqslant \mu_{\mathrm{A}}\|h_1 - h_2\|_{\mathrm{Y}}$$

for all $t \in I, h_1, h_2 \in W$ where

$$\lambda_{\mathrm{A}} := p\,(n+1+2p)^{1/2}\,n^{(m+2)/2}\,\alpha_0(RC_1),$$
$$\mu_{\mathrm{A}} := p\,(n+1+2p)^{1/2}\,n^{m+1}\,(2C_1)^{1/2}\,\alpha_1(3RC_1).$$

(v) Let $\tilde{D} := (W_{\mathbb{C}}^{m+1}(\mathbb{R}^n))^p$. Then

$$SA(t,h)k = (A(t,h) + B(t,h,Sh))Sk, \tag{12.2.7}$$

for every $t \in I$, $h \in W$ and $k \in \tilde{D}$ where $B : I \times W \times X \to L(X,X)$ is defined by

$$B(t,h,\bar{h})k := \left(\sum_{j=1}^{n} \sum_{l=1}^{p} b^1_{A^j_{1l}(t,\cdot,\cdot)-A^j_{1l}(t,\cdot,0)}(h)\,(-\Delta+1)^{-1/2}\partial^j k_l, \ldots, \right.$$
$$\left. \sum_{j=1}^{n} \sum_{l=1}^{p} b^1_{A^j_{pl}(t,\cdot,\cdot)-A^j_{pl}(t,\cdot,0)}(h)\,(-\Delta+1)^{-1/2}\partial^j k_l \right)$$
$$+ \left(\sum_{j=1}^{n} \sum_{l=1}^{p} b^2_{A^j_{1l}(t,\cdot,0)}\,(-\Delta+1)^{-1/2}\partial^j k_l, \ldots, \right.$$
$$\left. \sum_{j=1}^{n} \sum_{l=1}^{p} b^2_{A^j_{pl}(t,\cdot,0)}\,(-\Delta+1)^{-1/2}\partial^j k_l \right)$$

for all $(t,h,\bar{h}) \in I \times W \times X$ and $k \in X$. Here b^1, b^2 are defined according to Lemma 12.2.1. Further, B is continuous and there is $\lambda_B \geqslant 0$ such that

$$\|B(t,h,Sh)\| \leqslant \lambda_B, \|B(t,h,\bar{h}_1) - B(t,h,\bar{h}_2)\| = 0$$

for all $(t,h) \in I \times W, \bar{h}_1, \bar{h}_2 \in X$.

(vi) Let $(t,h) \in I \times W$ and $T(t,h) : [0,\infty) \to L(X,X)$, $\hat{T}(t,h) : [0,\infty) \to L(X,X)$ be the strongly continuous semigroups generated by $A(t,h)$ and $\hat{A}(t,h) := A(t,h) + B(t,h,Sh)$, respectively. Then

$$ST(t,h)(s) \supset \hat{T}(t,h)(s)S$$

for all $s \geqslant 0$.

(vii) By

$$f(t,h) := (F^1(t,\cdot,h), \ldots, F^p(t,\cdot,h))$$

for all $(t,h) \in I \times W$, there is defined an element $f \in C(I \times W, Y)$. In particular, there are $\lambda_f, \mu_f \geqslant 0$ such that

$$\|f(t,h)\|_Y \leqslant \lambda_f, \|f(t,h_1) - f(t,h_2)\|_2 \leqslant \mu_f \|h_1 - h_2\|_2$$

for all $t \in I$ and $h_1, h_2 \in W$.

(viii) By

$$g(t,h,k) := Sf(t,h)$$

for all $(t,h,k) \in I \times W \times X$, there is defined an element g of $C(I \times W \times X, X)$. In particular,

$$\|g(t,h,k_1) - g(t,h,k_2)\|_2 = 0$$

for all $(t,h) \in I \times W$ and $k_1, k_2 \in X$.

(ix) Then for every compact subset K of W, there is $T \in (c, d]$ such that for every $h \in K$ there is $u \in C([c, T], W) \cap C^1([c, T], X)$ such that

$$u'(t) + A(t, u(t))u(t) = f(t, u(t))$$

for all $t \in (c, T)$ where differentiation is with respect to X, and $u(c) = h$. If $v \in C([c, T], W) \cap C^1((c, T), X)$ is such that $v'(t) + A(t, v(t))v(t) = f(t, v(t))$ for all $t \in (c, T)$ and $v(c) = h$, then $v = u$.

Proof. '(i)': For this, let $\|| \, \||_k$ denote the Sobolev norm for $W^k_{\mathbb{C}}(\mathbb{R}^n)$ where $k \in \mathbb{N}$. Further, let $F_2 : L^2_{\mathbb{C}}(\mathbb{R}^n) \to L^2_{\mathbb{C}}(\mathbb{R}^n)$ be the unitary Fourier transformation determined by

$$(F_2 f)(x) := (2\pi)^{-n/2} \int_{\mathbb{R}^n} e^{-ixy} f(y)\, dy$$

for all $x \in \mathbb{R}^n$ and for every $f \in C_0^\infty(\mathbb{R}^n, \mathbb{C})$. Finally, for every complex-valued function g which is a.e. defined on $\mathbb{R}^n$ and measurable, we denote by T_g the corresponding maximal multiplication operator in $L^2_{\mathbb{C}}(\mathbb{R}^n)$. Then $(-\Delta + 1)^{m/2} := [(-\Delta + 1)^{1/2}]^m$ is a linear operator as composition of linear operators. Further,

$$F_2[(-\Delta + 1)^{1/2}]^m F_2^{-1} = [T_{(|\,|^2+1)^{1/2}}]^m \subset T_{(|\,|^2+1)^{m/2}}.$$

Further, for every $f \in D(T_{(|\,|^2+1)^{m/2}})$ and $k \in \{1, \dots, m\}$

$$[(|\,|^2 + 1)^{1/2}]^k f = \frac{1}{(|\,|^2 + 1)^{(m-k)/2}} (|\,|^2 + 1)^{m/2} f \in L^2_{\mathbb{C}}(\mathbb{R}^n)$$

and hence $f \in D([T_{(|\,|^2+1)^{1/2}}]^m)$. Therefore,

$$[T_{(|\,|^2+1)^{1/2}}]^m = T_{(|\,|^2+1)^{m/2}}$$

and hence $(-\Delta + 1)^{m/2}$ is self-adjoint and bijective with domain given by $W^m_{\mathbb{C}}(\mathbb{R}^n)$ since $F_2 W^m_{\mathbb{C}}(\mathbb{R}^n)$ coincides with the domain of $T_{(1+|\,|^2)^{m/2}}$. As a consequence, S is a densely-defined linear, self-adjoint and bijective operator in X. In the next step, it will be proved that

$$\frac{1}{\sqrt{n}} \||f\||_{k+1} \leqslant \||(-\Delta + 1)^{1/2} f\||_k \leqslant \sqrt{n+1}\, \||f\||_{k+1} \tag{12.2.8}$$

for all $f \in W^{k+1}_{\mathbb{C}}(\mathbb{R}^n)$ and $k \in \mathbb{N}$. For this, let $k \in \mathbb{N}$ and $f \in W^{k+1}_{\mathbb{C}}(\mathbb{R}^n)$. Then by (10.2.9), (10.2.3),

$$\begin{aligned}
\||(-\Delta + 1)^{1/2} f\||_k^2 &= \sum_{\alpha \in \mathbb{N}^n, |\alpha| \leqslant k} \|(-\Delta + 1)^{1/2} \partial^\alpha f\|_2^2 \\
&= \sum_{\alpha \in \mathbb{N}^n, |\alpha| \leqslant k} \|(-\Delta)^{1/2} \partial^\alpha f\|_2^2 + \sum_{\alpha \in \mathbb{N}^n, |\alpha| \leqslant k} \|\partial^\alpha f\|_2^2 \\
&\leqslant n \sum_{\alpha \in \mathbb{N}^n, |\alpha| \leqslant k} \sum_{j=1}^n \|\partial^j \partial^\alpha f\|_2^2 + \||f\||_k^2 \leqslant (n+1)\, \||f\||_{k+1}^2 .
\end{aligned}$$

Further, it follows by (10.2.9), (10.2.4)

$$\|(-\Delta + 1)^{1/2} f\|_2^2 \geqslant \|\partial^j f\|_2^2 + \|f\|_2^2$$

and hence

$$\|(-\Delta + 1)^{1/2} f\|_2^2 \geqslant \|f\|_2^2 + \frac{1}{n}\sum_{j=1}^{n} \|\partial^j f\|_2^2$$

for all $f \in W^1_{\mathbb{C}}(\mathbb{R}^n)$ and hence for $f \in W^{k+1}_{\mathbb{C}}(\mathbb{R}^n)$

$$\begin{aligned}
|||(-\Delta + 1)^{1/2} f|||_k^2 &= \sum_{\alpha\in\mathbb{N}^n, |\alpha|\leqslant k} \|(-\Delta + 1)^{1/2}\partial^\alpha f\|_2^2 \\
&\geqslant \sum_{\alpha\in\mathbb{N}^n, |\alpha|\leqslant k} \|\partial^\alpha f\|_2^2 + \frac{1}{n}\sum_{\alpha\in\mathbb{N}^n, |\alpha|\leqslant k}\sum_{j=1}^{n} \|\partial^j\partial^\alpha f\|_2^2 \\
&\geqslant \frac{1}{n}\,|||f|||_{k+1}^2 .
\end{aligned}$$

Inequality (12.2.8) implies that

$$n^{-m/2}\,|||f|||_m \leqslant \|(-\Delta + 1)^{m/2} f\|_2 \leqslant (n+1)^{m/2}\,|||f|||_m \tag{12.2.9}$$

for all $f \in W^m_{\mathbb{C}}(\mathbb{R}^n)$. Hence it follows

$$\|u\|_s^2 = \sum_{k=1}^{p}\left(\|u_k\|_2^2 + \|(-\Delta + 1)^{m/2} u_k\|_2^2\right) \leqslant [1 + (n+1)^m]\sum_{k=1}^{p} |||u_k|||_m^2$$

and

$$\|u\|_s^2 = \sum_{k=1}^{p}\left(\|u_k\|_2^2 + \|(-\Delta + 1)^{m/2} u_k\|_2^2\right) \geqslant n^{-m}\sum_{k=1}^{p} |||u_k|||_m^2$$

for all $u = (u_1, \dots, u_p) \in Y$.

'(ii)': For this, let $t \in I, h \in W$. By Theorem 10.2.2 (ii), it follows that by

$$\left((W^1_{\mathbb{C}}(\mathbb{R}^n))^p \to X, u \mapsto \left(\sum_{j=1}^{n}\sum_{l=1}^{p} A^j_{1l}(t,\cdot,h)\partial^j u_l, \dots, \sum_{j=1}^{n}\sum_{l=1}^{p} A^j_{pl}(t,\cdot,h)\partial^j u_l\right)\right)$$

there is defined a densely-defined linear and quasi-accretive operator in X with bound

$$-\frac{1}{2}\sum_{j=1}^{n}\left(\sum_{k,l=1}^{p} \|[A^j_{kl}(t,\cdot,h)]_{,j}\|_\infty^2\right)^{1/2}$$

whose closure $A(t,h)$ is the infinitesimal generator of a strongly continuous group on X. Further, it follows for every $j \in \{1, \dots, n\}$, $x \in \mathbb{R}^n$

$$\begin{aligned}
|[A^j_{kl}(t,\cdot,h)]_{,j}(x)| = {} & |A^j_{kl,j}(t,x,h(x)) \\
& + \sum_{r=1}^{p} [A^j_{kl,n+2r-1}(t,x,h(x))\, h_{r1,j}(x)
\end{aligned}$$

$$
\begin{aligned}
&+ A^j_{kl,n+2r}(t,x,h(x))\, h_{r2,j}(x)]| \\
&\leqslant (1+2pRC_2)\,\alpha_1(RC_1).
\end{aligned}
$$

'(iii)': The statement is a simple consequence of (ii) along with Lemma 8.6.4. '(iv)': First, $D(A(t,h)) \supset W^1_{\mathbb{C}}(\mathbb{R}^n) \supset Y$ and $A(t,h)|_Y \in L(Y,X)$ since the restrictions of $\partial^1,\dots,\partial^n$ to $W^m_{\mathbb{C}}(\mathbb{R}^n)$ define elements of $L(W^m_{\mathbb{C}}(\mathbb{R}^n), L^2_{\mathbb{C}}(\mathbb{R}^n))$ and $A^j_{kl}(t,\cdot,h) \in B(\mathbb{R}^n,\mathbb{C})$ for $j \in \{1,\dots,n\}$, $k,l \in \{1,\dots,p\}$ and $(t,h) \in I \times W$. Further, it follows for $(t_1,h_1),(t_2,h_2) \in I \times W$ and $j \in \{1,\dots,n\}$, $k,l \in \{1,\dots,p\}$ that

$$
\begin{aligned}
&|A^j_{kl}(t_1,x,h_1(x)) - A^j_{kl}(t_2,x,h_2(x))| \\
&\leqslant \left[\max_{\lambda\in[0,1]} |(\nabla A^j_{kl})(t_1+\lambda(t_2-t_1),x,h_1(x)+\lambda(h_2(x)-h_1(x)))|\right] \\
&\cdot |(t_1-t_2,0,h_1(x)-h_2(x))| \\
&\leqslant \left[\max_{\lambda\in[0,1]} |(\nabla A^j_{kl})(t_1+\lambda(t_2-t_1),x,h_1(x)+\lambda(h_2(x)-h_1(x)))|\right] \\
&\cdot \left[(t_1-t_2)^2 + \sum_{k=1}^{p}\sum_{r=1}^{2} |h_{1kr}(x)-h_{2kr}(x)|^2\right]^{1/2}
\end{aligned}
$$

and hence, because of

$$
\|h_{1k}+\lambda(h_{2k}-h_{1k})\|_\infty \leqslant C_1\|h_{1k}+\lambda(h_{2k}-h_{1k})\|_m \leqslant 3RC_1
$$

for every $\lambda \in [0,1]$, that

$$
\begin{aligned}
&\|A^j_{kl}(t_1,\cdot,h_1) - A^j_{kl}(t_2,\cdot,h_2)\|_\infty \\
&\leqslant (n+1+2p)^{1/2}\alpha_1(3RC_1)\left[(t_1-t_2)^2 + 2C_1\sum_{k=1}^{p} |||h_{1k}-h_{2k}|||_m^2\right]^{1/2}.
\end{aligned}
$$

This implies that

$$
\begin{aligned}
&\|A(t_1,h_1)u - A(t_2,h_2)u\|_X^2 \leqslant (n+1+2p)\,p^2n^2\,\alpha_1^2(3RC_1)\sum_{k=1}^{p} |||u_k|||_m^2 \\
&\cdot \left[(t_1-t_2)^2 + 2C_1\sum_{k=1}^{p} |||h_{1k}-h_{2k}|||_m^2\right]
\end{aligned}
$$

for every $u \in Y$ and hence that

$$
\begin{aligned}
&\|A(t_1,h_1)|_Y - A(t_2,h_2)|_Y\| \\
&\leqslant p\,(n+1+2p)^{1/2}n^{(m+2)/2}\alpha_1(3RC_1)\cdot\left[|t_1-t_2| + (2C_1n^m)^{1/2}\|h_1-h_2\|_Y\right].
\end{aligned}
$$

Hence A is continuous and

$$\|A(t,h_1) - A(t,h_2)\| \leqslant \mu_{\mathrm{A}} \|h_1 - h_2\|_{\mathrm{Y}}$$

for all $t \in I$, $h_1, h_2 \in W$. Analogously, it follows that

$$\|A(t,h)\| \leqslant \lambda_{\mathrm{A}}$$

for all $t \in I$, $h \in W$.

'(v)': For this, let $t \in I$, $h \in W$ and $k \in \tilde{D}$. Then it follows by Lemma 12.2.1 (v) that

$$\begin{aligned}
A(t,h)k &= \left(\sum_{j=1}^{n} \sum_{l=1}^{p} A^j_{1l}(t,\cdot,h)\partial^j k_l, \ldots, \sum_{j=1}^{n} \sum_{l=1}^{p} A^j_{pl}(t,\cdot,h)\partial^j k_l \right) \\
&= \left(\sum_{j=1}^{n} \sum_{l=1}^{p} [A^j_{1l}(t,\cdot,h) - A^j_{1l}(t,\cdot,0)]\partial^j k_l, \ldots, \right. \\
&\qquad \left. \sum_{j=1}^{n} \sum_{l=1}^{p} [A^j_{pl}(t,\cdot,h) - A^j_{pl}(t,\cdot,0)]\partial^j k_l \right) \\
&\qquad + \left(\sum_{j=1}^{n} \sum_{l=1}^{p} A^j_{1l}(t,\cdot,0)\partial^j k_l, \ldots, \sum_{j=1}^{n} \sum_{l=1}^{p} A^j_{pl}(t,\cdot,0)\partial^j k_l \right) \in Y
\end{aligned}$$

and hence that

$$\begin{aligned}
&SA(t,h)k - A(t,h)Sk \\
&= \left(\sum_{j=1}^{n} \sum_{l=1}^{p} \Big\{ (-\triangle + 1)^{m/2} [A^j_{1l}(t,\cdot,h) - A^j_{1l}(t,\cdot,0)] - [A^j_{1l}(t,\cdot,h) - A^j_{1l}(t,\cdot,0)] \right. \\
&\quad (-\triangle + 1)^{m/2} \Big\} (-\triangle + 1)^{(1-m)/2} (-\triangle + 1)^{-1/2} \partial^j (-\triangle + 1)^{m/2} k_l, \ldots, \\
&\quad \sum_{j=1}^{n} \sum_{l=1}^{p} \Big\{ (-\triangle + 1)^{m/2} [A^j_{pl}(t,\cdot,h) - A^j_{pl}(t,\cdot,0)] - [A^j_{pl}(t,\cdot,h) - A^j_{pl}(t,\cdot,0)] \\
&\qquad\qquad \left. (-\triangle + 1)^{m/2} \Big\} (-\triangle + 1)^{(1-m)/2} (-\triangle + 1)^{-1/2} \partial^j (-\triangle + 1)^{m/2} k_l \right) \\
&\quad + \left(\sum_{j=1}^{n} \sum_{l=1}^{p} \Big\{ (-\triangle + 1)^{m/2} A^j_{1l}(t,\cdot,0) - A^j_{1l}(t,\cdot,0)(-\triangle + 1)^{m/2} \Big\} \right. \\
&\qquad\qquad (-\triangle + 1)^{(1-m)/2} (-\triangle + 1)^{-1/2} \partial^j (-\triangle + 1)^{m/2} k_l, \ldots, \\
&\quad \sum_{j=1}^{n} \sum_{l=1}^{p} \Big\{ (-\triangle + 1)^{m/2} A^j_{pl}(t,\cdot,0) - A^j_{pl}(t,\cdot,0)(-\triangle + 1)^{m/2} \Big\} \\
&\qquad\qquad \left. (-\triangle + 1)^{(1-m)/2} (-\triangle + 1)^{-1/2} \partial^j (-\triangle + 1)^{m/2} k_l \right) \\
&= B(t,h,Sh)Sk.
\end{aligned}$$

The remaining statements are obvious, see the proofs of Lemma 12.2.1 (v)b), (vi).
'(vi)': For this, let $(t, h) \in I \times W$. Then $S\tilde{D} = (W^1_{\mathbb{C}}(\mathbb{R}^n))^p$ is a core for $\hat{A}(t, h)$ and hence the statement follows from (12.2.7) by Theorem 6.2.2 (ii).
'(vii)': The statements are obvious, see the proofs of Lemma 12.2.1 (v)b).
'(viii)': The statements are obvious consequences of the definition of g and (vii).
'(ix)': Is a consequence of (i) - (viii) and Theorem 11.0.7. □

13

Appendix

Lemma 13.0.3. **(Peetre's inequality)** Let $n \in \mathbb{N}^*$, $s \in \mathbb{R}$ and $k, k' \in \mathbb{R}^n$. Then

$$(1 + |k|^2)^{s/2} \leqslant 2^{|s|/2}(1 + |k - k'|^2)^{|s|/2}\,(1 + |k'|^2)^{s/2}. \tag{13.0.1}$$

Proof. If $s \geqslant 0$, it follows that

$$\begin{aligned}(1 + |k|^2)^{s/2} &\leqslant [1 + (\,|k - k'| + |k'|\,)^2]^{s/2} \leqslant [1 + 2|k - k'|^2 + 2|k'|^2]^{s/2} \\ &\leqslant [\,2(1 + |k - k'|^2)(1 + |k|^2)\,]^{s/2} \\ &= 2^{s/2}(1 + |k - k'|^2)^{s/2}\,(1 + |k'|^2)^{s/2}.\end{aligned}$$

If $s \leqslant 0$, it follows from the previous inequality by exchanging k and k' that

$$(1 + |k'|^2)^{|s|/2} \leqslant 2^{|s|/2}\,(1 + |k - k'|^2)^{|s|/2}\,(1 + |k|^2)^{|s|/2}$$

and hence that

$$2^{-|s|/2}(1 + |k - k'|^2)^{-|s|/2}\,(1 + |k'|^2)^{|s|/2} \leqslant (1 + |k|^2)^{|s|/2}$$

and, finally, that

$$2^{|s|/2}(1 + |k - k'|^2)^{|s|/2}\,(1 + |k'|^2)^{s/2} \geqslant (1 + |k|^2)^{s/2}.$$

□

Lemma 13.0.4. **(Sobolev's inequalities)** Let $n \in \mathbb{N}^*$, $p \in \mathbb{R}$ such that $1 \leqslant p < n$ and $m \in \mathbb{N}^*$ such that $n/m > p$. Then

$$\|f\|_{np/(n-mp)} \leqslant \left[\prod_{m'=1}^{m} \frac{p(n-1)}{n(n-m'p)}\right] \sum_{\alpha \in \mathbb{N}^n, |\alpha| = m} \left\| \frac{\partial^m f}{\partial x^\alpha} \right\|_p \tag{13.0.2}$$

for all $f \in C_0^\infty(\mathbb{R}^n, \mathbb{C})$.

Proof. For this, let $n \in \mathbb{N}^*$, $f \in C_0^1(\mathbb{R}^n, \mathbb{C})$ and $j \in \{1, \dots, n\}$. We define $f_j : \mathbb{R}^{n-1} \to \mathbb{R}$ by

$$f_j(x_1, \dots, x_{j-1}, x_{j+1}, \dots, x_n) := \int_{-\infty}^{\infty} |f_{,j}(x_1, \dots, x_n)| \, dx_j$$

for all $(x_1, \dots, x_{j-1}, x_{j+1}, \dots, x_n) \in \mathbb{R}^{n-1}$. Obviously, f_j is continuous with a compact support and

$$\begin{aligned} |f(x_1, \dots, x_n)| &= \left| \int_{-\infty}^{x_j} f_{,j}(x_1, \dots, x_{j-1}, t, x_{j+1}, x_n) \, dt \right| \\ &\leqslant f_j(x_1, \dots, x_{j-1}, x_{j+1}, \dots, x_n) \end{aligned}$$

for all $(x_1, \dots, x_n) \in \mathbb{R}^n$ and $j \in \{1, \dots, n\}$. Hence it follows that

$$|f(x_1, \dots, x_n)|^{n/(n-1)} \leqslant \prod_{j=1}^{n} [f_j(x_1, \dots, x_{j-1}, x_{j+1}, \dots, x_n)]^{1/(n-1)}$$

for all $(x_1, \dots, x_n) \in \mathbb{R}^n$, which by integration and application of the generalized Hölder inequality for $(n-1)$-factors leads to

$$\begin{aligned} &\int_{\mathbb{R}} |f(x_1, \dots, x_n)|^{n/(n-1)} \, dx_1 \\ &\leqslant [f_1(x_2, \dots, x_n)]^{1/(n-1)} \int_{\mathbb{R}} \prod_{j=2}^{n} [f_j(x_1, \dots, x_{j-1}, x_{j+1}, \dots, x_n)]^{1/(n-1)} \, dx_1 \\ &\leqslant [f_1(x_2, \dots, x_n)]^{1/(n-1)} \prod_{j=2}^{n} \left(\int_{\mathbb{R}} f_j(x_1, \dots, x_{j-1}, x_{j+1}, \dots, x_n) \, dx_1 \right)^{1/(n-1)} \end{aligned}$$

for all $(x_2, \dots, x_n) \in \mathbb{R}^n$. Note that only $n-1$ of these factors in the last expression depend on x_2. Hence it follows by integration, Fubini's theorem and application of the generalized Hölder inequality for $(n-1)$–factors that

$$\begin{aligned} &\int_{\mathbb{R}^2} |f(x_1, \dots, x_n)|^{n/(n-1)} \, dx_1 \, dx_2 \\ &\leqslant \left(\int_{\mathbb{R}} f_1(x_2, \dots, x_n) \, dx_2 \right)^{1/(n-1)} \cdot \left(\int_{\mathbb{R}} f_2(x_1, x_3, \dots, x_n) \, dx_1 \right)^{1/(n-1)} \\ &\quad \cdot \prod_{j=3}^{n} \left(\int_{\mathbb{R}} f_j(x_1, \dots, x_{j-1}, x_{j+1}, \dots, x_n) \, dx_1 dx_2 \right)^{1/(n-1)} \end{aligned}$$

for all $(x_3, \dots, x_n) \in \mathbb{R}^n$. Repeating these previous steps $(n-2)$–times leads to

$$\begin{aligned} \int_{\mathbb{R}^n} |f|^{n/(n-1)} \, dv^n &\leqslant \prod_{j=1}^{n} \left(\int_{\mathbb{R}^{n-1}} f_j \, dv^{n-1} \right)^{1/(n-1)} = \left(\prod_{j=1}^{n} \|f_{,j}\|_1 \right)^{1/(n-1)} \\ &\leqslant \left(\frac{1}{n} \sum_{j=1}^{n} \|f_{,j}\|_1 \right)^{n/(n-1)} \end{aligned}$$

and hence, finally, to

$$\|f\|_{n/(n-1)} \leqslant \frac{1}{n} \sum_{j=1}^{n} \|f_{,j}\|_1 . \tag{13.0.3}$$

Now let $p \in \mathbb{R}$ such that $1 < p < n$,

$$r := \frac{p(n-1)}{n-p} = 1 + \frac{(p-1)n}{n-p} > 1$$

and $f \in C_0^1(\mathbb{R}^n, \mathbb{C})$. Note that $|\ |^r \in C^1(\mathbb{R}^2, \mathbb{R})$. Obviously, since $|\ |^r = (|\ |^2)^{r/2}$, it follows that $|\ |^r|_{\mathbb{R}^2 \setminus \{(0,0)\}} \in C^1(\mathbb{R}^2 \setminus \{(0,0)\}, \mathbb{R})$. Further,

$$\lim_{h \to 0, h \neq 0} \frac{|(h,0)|^r}{h} = \lim_{h \to 0, h \neq 0} \frac{|(0,h)|^r}{h} = \lim_{h \to 0, h \neq 0} \frac{|h|}{h} |h|^{r-1} = 0.$$

Hence $|\ |^r$ is partially differentiable with continuous partial derivatives $(|\ |^r)_{,1}$, $(|\ |^r)_{,2} : \mathbb{R}^2 \to \mathbb{R}$ given by

$$(|\ |^r)_{,1}(x,y) = \begin{cases} \frac{x}{|(x,y)|} |(x,y)|^{r-1} & \text{if } (x,y) \neq (0,0) \\ 0 & \text{if } (x,y) = (0,0). \end{cases}$$

$$(|\ |^r)_{,2}(x,y) = \begin{cases} \frac{y}{|(x,y)|} |(x,y)|^{r-1} & \text{if } (x,y) \neq (0,0) \\ 0 & \text{if } (x,y) = (0,0). \end{cases}$$

for all $(x,y) \in \mathbb{R}^2$. As a consequence, $|f|^r \in C_0^1(\mathbb{R}^n, \mathbb{R})$ and by (13.0.3) and Hölder's inequality

$$\begin{aligned} \left(\|f\|_{np/(n-p)}\right)^r &= \|\,|f|^r\|_{n/(n-1)} \leqslant \frac{1}{n} \sum_{j=1}^{n} \|(|f|^r)_{,j}\|_1 \\ &= \frac{r}{n} \sum_{j=1}^{n} \|\,|f|^{r-2}\,|\mathrm{Re}(f^* f_{,j})|\,\|_1 \leqslant \frac{r}{n} \sum_{j=1}^{n} \|\,|f|^{r-1}\,|f_{,j}|\,\|_1 \\ &\leqslant \frac{r}{n} \sum_{j=1}^{n} \|\,|f|^{r-1}\|_{p/(p-1)} \, \|f_{,j}\|_p = \frac{r}{n} \left(\|f\|_{np/(n-p)}\right)^{r-1} \sum_{j=1}^{n} \|f_{,j}\|_p , \end{aligned}$$

where we use the conventions that the arguments of $\|\ \|_1$ are zero in all zeros of f, and hence

$$\|f\|_{np/(n-p)} \leqslant \frac{p(n-1)}{n(n-p)} \sum_{j=1}^{n} \|f_{,j}\|_p . \tag{13.0.4}$$

For the final step, let $p \in \mathbb{R}$ such that $1 \leqslant p < n$ and $m \in \mathbb{N}^*$ such that $n/m > p$. Then

$$\|f\|_{np/(n-mp)} \leqslant \frac{[p(n-1)/n]^m}{\prod_{m'=1}^{m}(n-m'p)} \sum_{\alpha \in \mathbb{N}^n, |\alpha| = m} \left\| \frac{\partial^m f}{\partial x^\alpha} \right\|_p \tag{13.0.5}$$

for every $f \in C_0^\infty(\mathbb{R}^n, \mathbb{C})$. The proof proceeds by finite induction over m. The case $m = 1$ is a special case of the previously proved. Now let $m \in \mathbb{N}^*$ be such that

$n/m > p$ and such that (13.0.5) is true for every $f \in C_0^\infty(\mathbb{R}^n, \mathbb{C})$. Then, if $n/(m+1) > p$, it follows for $f \in C_0^\infty(\mathbb{R}^n, \mathbb{C})$

$$\begin{aligned}
&\frac{p(n-1)}{n[n-(m+1)p]} \sum_{j=1}^{n} \|f_{,j}\|_{np/(n-mp)} \\
&\leqslant \frac{[p(n-1)/n]^{m+1}}{\prod_{m'=1}^{m+1}(n-m'p)} \sum_{j=1}^{n} \sum_{\alpha\in\mathbb{N}^n, |\alpha|=m} \left\| \frac{\partial}{\partial x^j} \frac{\partial^m f}{\partial x^\alpha} \right\|_p \\
&= \frac{[p(n-1)/n]^{m+1}}{\prod_{m'=1}^{m+1}(n-m'p)} \sum_{\alpha\in\mathbb{N}^n, |\alpha|=m+1} \left\| \frac{\partial^{m+1} f}{\partial x^\alpha} \right\|_p .
\end{aligned}$$

We define $\bar{p} := np/(n-mp)$. Then

$$\bar{p} = p + \frac{mp^2}{n-mp} \geqslant p \geqslant 1, \bar{p} = \frac{n}{\frac{n}{p} - m} < \frac{n}{(m+1)-m} = n.$$

Hence it follows by (13.0.3), (13.0.4)

$$\|f\|_{n\bar{p}/(n-\bar{p})} \leqslant \frac{\bar{p}(n-1)}{n(n-\bar{p})} \sum_{j=1}^{n} \|f_{,j}\|_{\bar{p}}$$

and therefore, since

$$\frac{\bar{p}(n-1)}{n(n-\bar{p})} = \frac{\frac{np}{n-mp}(n-1)}{n\left(n - \frac{np}{n-mp}\right)} = \frac{p(n-1)}{n[n-(m+1)p]}$$

$$\frac{n\bar{p}}{n-\bar{p}} = \frac{\frac{n^2 p}{n-mp}}{n - \frac{np}{n-mp}} = \frac{np}{n-(m+1)p},$$

it follows that

$$\|f\|_{np/[n-(m+1)p]} \leqslant \frac{[p(n-1)/n]^{m+1}}{\prod_{m'=1}^{m+1}(n-m'p)} \sum_{\alpha\in\mathbb{N}^n, |\alpha|=m+1} \left\| \frac{\partial^{m+1} f}{\partial x^\alpha} \right\|_p .$$

□

References

1. Abramowitz M, Stegun I A (Eds.) 1984, *Pocketbook of Mathematical Functions*, Harri Deutsch: Thun.
2. Adams R A, Fournier J J F 2003, *Sobolev spaces*, 2nd ed., Academic Press: New York.
3. Alpert B, Greengard L, Hagstrom Thomas 2000, *Rapid evaluation of nonreflecting boundary kernels for time-domain wave propagation*, SIAM J. Numer. Anal. **37**, 1138–1164.
4. Altman M 1986, *A unified theory of nonlinear operator and evolution equations with applications. A new approach to nonlinear partial differential equations*, Monographs and Textbooks in Pure and Applied Mathematics **103**, Marcel Dekker: New York.
5. Anderson M T, Chrusciel P T 2005, *Asymptotically simple solutions of the vacuum Einstein equations in even dimensions*, Commun. Math. Phys. **260**, 557–577.
6. Arendt W 1987, *Vector valued Laplace transforms and Cauchy problems*, Israel J. Math. **59**, 327–352.
7. Arendt W 1989, *Integrated solutions of Volterra integrodifferential equations and applications*, Proceedings of the Conference on Volterra integro-differential equations in Banach spaces and applications, Trento 1987, G. Da Prato and M. Iannelli (Eds.), Pitman Research Notes in Mathematics Series **190**, 21–51.
8. Arendt W, Batty C, Hieber M, Neubrander F 2001, *Vector-valued Laplace transforms and Cauchy problems*, Monographs in Mathematics **96**, Birkhäuser: Basel.
9. Bachelot A 2000, *Creation of fermions at the charged black-hole horizon*, CR Acad. Sci. I-Math. **330**, 29–34.
10. Bachelot A 2000, *Creation of fermions at the charged black-hole horizon*, Ann. Henri Poincare **1**, 1043–1095.
11. Bachelot A, Motet-Bachelot A 1993, *Les Resonances D'un Trou Noir De Schwarzschild*, Ann. Inst. Henri Poincare physique theorique **59**, 3–68.
12. Batkai A, Piazzera S 2005, *Semigroups for delay equations*, A K Peters: Wellesley.
13. Baez J C, Segal I E, Zhou Z 1992, *Introduction to algebraic and constructive quantum field theory*, Princeton University Press: Princeton.
14. Bambusi D, Galgani L 1993, *Some rigorous results on the Pauli-Fierz model of classical electrodynamics*, Ann. I. H. Poincare Phy. **58**, 155–171.
15. Barbu V 1976, *Nonlinear semigroups and differential equations in Banach spaces*, Noordhoff: Leyden.
16. Bayliss A, Turkel E 1980, *Radiation boundary conditions for wave-like equations*, Comm. Pure Appl. Math. **33**, 707–725.

17. Beig R, Schmidt B G 2003, *Relativistic elasticity*, Class. Quantum Gravity **20**, 889–904.
18. Belenkaya L, Friedlander S, Yudovich V 1999, *The unstable spectrum of oscillating shear flows*, Siam J. Appl. Math. **59**, 1701–1715.
19. Belleni-Morante A, McBride A C 1998, *Applied nonlinear semigroups: An introduction*, Wiley: New York.
20. Bers L, John F, Schechter M 1964, *Partial differential equations*, Wiley: New York.
21. Beyer H 1991, *Remarks on Fulling's quantization*, Class. Quantum Grav. **8**, 1091–1112.
22. Beyer H R 1995, *The spectrum of radial adiabatic stellar oscillations*, J. Math. Phys. **36**, 4815–4825.
23. Beyer H R 1995, *The spectrum of adiabatic stellar oscillations*, J. Math. Phys. **36**, 4792–4814.
24. Beyer H R, Schmidt B G 1995, *Newtonian stellar oscillations*, Astron. Astrophys. **296**, 722–726.
25. Beyer H R 1999, *On the Completeness of the Quasinormal modes of the Pöschl-Teller potential*, Commun. Math. Phys. **204**, 397–423.
26. Beyer H R, Kokkotas K D 1999, *On the r-mode spectrum of relativistic stars*, Mon. Not. R. Astron. Soc. **308**, 745–750.
27. Beyer H R 2001, *On the stability of the Kerr metric*, Commun. Math. Phys. **221**, 659–676.
28. Beyer H R 2002, *A framework for perturbations and stability of differentially rotating stars*, Proc. Roy. Soc. Lond. A **458**, 359–380.
29. Beyer H, Sarbach O 2004, *Well-posedness of the Baumgarte-Shapiro-Shibata-Nakamura formulation of Einstein's field equations*, Phys. Rev. D **70**, 104004.
30. Beyer H R 2006, *Results on the spectrum of R-Modes of slowly rotating, relativistic Stars*, Class. Quantum Gravity **23**, 2409–2425.
31. Bjorken J D, Drell S D 1964, *Relativistic quantum mechanics*, McGraw-Hill: New York.
32. Bjorken J D, Drell S D 1965, *Relativistic quantum fields*, McGraw-Hill: New York.
33. Brauer U 1992, *An existence theorem for perturbed Newtonian cosmological models*, J. Math. Phys. **33**, 1224–1233.
34. Braz e Silva B 2003, *Stability of plane Couette flow: The resolvent method*, Ph.D. thesis, The University of New Mexico, Albuquerque, 2003.
35. Brezis H 1983, *Analyse fonctionnelle: Thorie et applications*, Collection Mathmatiques Appliques pour la Matrise, Masson: Paris.
36. McBride A C 1987, *Semigroups of linear operators: An introduction*, Longman: Harlow.
37. Brill D, Reula O, Schmidt B 1987, *Local linearization stability*, J. Math. Phys. **28**, 1844–1847.
38. Buchholz D 2000, *Algebraic quantum field theory: A status report*, Plenary talk given at XIIIth International Congress on Mathematical Physics, London, http://xxx.lanl.gov/abs/math-ph/0011044.
39. Butzer P L, Berens H 1967, *Semi-groups of operators and approximation*, Springer: New York.
40. Calderon A P 1965, *Commutators of singular integral operators*, Proc. Nat. Acad. Sci. USA **53**, 1092–1099.
41. Cannarsa P, Da Prato G, Zolesio J.-P. 1990, *The damped wave equation in a moving domain*, J. Differential Equations **85**, 1–16.
42. Carroll R W 1969, *Abstract methods in partial differential equations*, Harper & Row: New York.
43. Chandrasekhar S 1961, *Hydrodynamic and hydromagnetic stability*, Oxford University Press: Oxford.

44. Claudel C M, Newman K P 1998, *The Cauchy problem for quasi-linear hyperbolic evolution problems with a singularity in the time*, Proc. Roy. Soc. A **454**, 1073–1107.
45. Cutler C, Wald R M 1989, *Existence of radiating Einstein-Maxwell solutions which are C^∞ on all of P+ and P−*, Class. Quantum Gravity **6**, 453–466.
46. Constantin A 2001, *The construction of an evolution system in the hyperbolic case and applications*, Math. Nachr. **224**, 49–73.
47. Davies E B 1980, *One-parameter semigroups*, Academic Press: London.
48. Davis E B, Pang M M H 1987, *The Cauchy problem and a generalization of the Hille-Yosida theorem*, Proc. London Math. Soc. **55**, 181–208.
49. Diestel J, Uhl J J 1977, *Vector measures*, AMS: Providence.
50. Dorroh J R 1975, *A simplified proof of a theorem of Kato on linear evolution equations*, J. Math. Soc. Japan **27**, 474–478.
51. Drazin P G, Reid W H 2004, *Hydrodynamic Stability*, 2nd ed., Cambridge University Press, Cambridge.
52. Dunford N, Schwartz J T 1957, *Linear operators, Part I: General theory*, Wiley: New York.
53. Dunford N, Schwartz J T 1963, *Linear operators, Part II: Spectral theory: Self adjoint operators in Hilbert space theory*, Wiley: New York.
54. Dyson J, Schutz B F 1979, *Perturbations and stability of rotating stars. I. Completeness of normal modes*, Proc. Roy. Soc. London Ser. A **368**, 389–410.
55. Eisner T, Zwart H 2006, *Continuous-time Kreiss resolvent condition on infinite-dimensional spaces*, Math. Comp. **75**, 1971–1985.
56. Engel K-J, Nagel R 1990, *Cauchy problems for polynomial operator matrices on abstract energy spaces*, Forum Math. **2**, 89–102.
57. Engel K-J, Nagel R 2000, *One-parameter semigroups for linear evolution equations*, Springer: New York.
58. Engquist B, Majda A 1977, *Absorbing boundary conditions for the numerical simulation of waves*, Math. Comp. **31**, 629–651.
59. Evans D E 1976, *Time dependent perturbations and scattering of strongly continuous groups on Banach spaces*, Math. Ann. **221**, 275–290.
60. Fattorini H O 1983, *The Cauchy problem*, Addison-Wesley: Reading, Massachusetts.
61. Fischer A E, Marsden J E: 1972, *The Einstein evolution equations as a first-order quasilinear symmetric hyperbolic system*, Commun. Math. Phys. **28**, 1–38.
62. Finster F, Kamran N, Smoller J, Yau S-T 2000, *Non-existence of time-periodic solutions of the Dirac equation in an axisymmetric black hole geometry*, Commun. Pure Appl. Math. **53**, 902–929.
63. Finster F, Kamran N, Smoller J, Yau S-T 2002, *Decay rates and probability estimates for massive Dirac particles in the Kerr-Newman black hole geometry*, Commun. Math. Phys. **230**, 201–244.
64. Finster F, Kamran N, Smoller J, Yau S-T 2005, *An integral spectral representation of the propagator for the wave equation in the Kerr geometry*, Commun. Math. Phys. **260**, 257–298.
65. Finster F, Kamran N, Smoller J, Yau S-T 2006, *Decay of solutions of the wave equation in the Kerr geometry*, Commun. Math. Phys. **264**, 465–503.
66. Flato M, Simon J, Taflin E 1994, *The Maxwell-Dirac equations - Asymptotic completeness and the infrared problem*, Rev. Math. Phys. **6**, 1071–1083.
67. Flato M, Simon J, Taflin E 1987, *On global-solutions of the Maxwell-Dirac equations*, Commun. Math. Phys. **1**, 21–49.
68. Folland G B 1999, *Real analysis*, 2nd ed., Wiley: New York.

69. Friedman J L, Morris, M S 2000, *Schwarzschild perturbations die in time*, J. Math. Phys. **41**, 7529–7534.
70. Friedrich H 1981, *The asymptotic characteristic initial value problem for Einstein's vacuum field equations as an initial value problem for a first-order quasilinear symmetric hyperbolic system*, Proc. Roy. Soc. London Ser. A **378**, 401–421.
71. Friedrich H 1983, *Cauchy problems for the conformal vacuum field equations in general relativity*, Commun. Math. Phys. **91**, 445–472.
72. Friedrich H 1985, *On the hyperbolicity of Einstein's and other gauge field equations*, Commun. Math. Phys. **100**, 25–543.
73. Friedrich H 1986, *On the existence of n-geodesically complete or future complete solutions of Einstein's field equations with smooth asymptotic structure*, Commun. Math. Phys. **107**, 587–609.
74. Friedrich H 1991, *On the global existence and the asymptotic behavior of solutions to the Einstein-Maxwell-Yang-Mills equations*, J. Differential Geom. **34**, 275–345.
75. Friedrich H 1998, *Gravitational fields near space-like and null infinity*, J. Geom. Phys. *24*, 83–163.
76. Friedrich H, Nagy, Gabriel 1999, *The initial boundary value problem for Einstein's vacuum field equation*, Commun. Math. Phys. **201**, 619–655.
77. Friedrich H 2005, *On the nonlinearity of subsidiary systems*, Classical Quant. Grav. **22**, L77–L82.
78. Friedrich H, Rendall A 2000, *The Cauchy problem for the Einstein equations*, in: Schmidt B ed., Einstein's field equations and their physical implications, Lecture Notes in Physics **540**, Springer: Berlin.
79. Friedrichs K O 1944, *The identity of weak and strong extensions of differential operators*, Trans. Amer. Math. Soc. **55**, 132–151.
80. Fring A, Kostrykin V, Schrader R 1997, *Ionization probabilities through ultra-intense fields in the extreme limit*, J. Phys. A **30**, 8599–8610.
81. Fulling S A 1989, *Aspects of quantum field theory in curved spacetime*, Cambridge University Press, Cambridge.
82. Gil M I 2001, *Solution estimates for nonlinear nonautonomous evolution equations in spaces with normalizing mappings*, Acta Appl. Math. **69**, 1–23.
83. Gill T L, Zachary W W 1987, *Time-ordered operators and Feynman-Dyson algebras*, J. Math. Phys. **28**, 1459–1470.
84. Gesztesy F, Mitter H 1981, *A note on quasi-periodic states*, J. Phys. A **14**, L79–L85.
85. Givoli D 1991, *Nonreflecting boundary conditions*, J. Comput. Phys. **94**, 1–29.
86. Goldberg S 1985, *Unbounded linear operators* Dover: New York.
87. Goldstein J A 1970, *Semigroups of operators and abstract Cauchy problems*, Tulane University: New Orleans.
88. Goldstein J A 1972, *Lectures on semigroups of nonlinear operators*, Tulane University: New Orleans.
89. Goldstein J A, Sandefur J T 1982, *Equipartition of energy for higher order abstract hyperbolic equations*, Comm. Partial Diff. Eqs. **7**, 1217–1251.
90. Goldstein J A 1985, *Semigroups of linear operators and applications*, Oxford University Press: New York.
91. Goldstein J A 1987, *An abstract d'Alembert formula*, SIAM J. Math. Anal. **18**, 842–856.
92. Hagstrom T, Hariharan S. I. A 1998, *Formulation of asymptotic and exact boundary conditions using local operators. Absorbing boundary conditions* Appl. Numer. Math. **27**, 403–416.

93. Hiroshima F 1999, *Weak coupling limit and removing an ultraviolet cutoff for a Hamiltonian of particles interacting with a quantized scalar field*, J. Math. Phys. **40**, 1215–1236.
94. Hiroshima F 2000, *Essential self-adjointness of translation-invariant quantum field models for arbitrary coupling constants*, Commun. Math. Phys. **211**, 585–613.
95. Haag R 1996, *Local quantum physics: Fields, particles, algebras*, Springer: New York.
96. Hawking S W, *Particle creation by black holes* 1975, Commun. Math. Phys. **43**, 199–220.
97. Heard M L 1984, *A quasilinear hyperbolic integrodifferential equation related to a nonlinear string*, Trans. Amer. Math. Soc. **285**, 805–823.
98. Hille E 1948, *Functional analysis and semi-groups*, AMS: New York.
99. Hille E, Phillips R S 1957, *Functional analysis and semi-groups*, Revised ed., AMS: Providence.
100. Hirschmann T, Schimming R 1986, *Generalized harmonic gauge conditions in General Relativity as field-equations for lapse and shift*, Astron Nachr **307**, 293–301.
101. Hirzebruch F, Scharlau W 1971, *Einführung in die Funktionalanalysis*, BI: Mannheim.
102. Hughes T J R, Kato T, Marsden J E 1977, *Well-posed quasi-linear second-order hyperbolic systems with applications to nonlinear elastodynamics and general relativity*, Arch. Rational Mech. Anal. **63**, 273–294.
103. Howland J S 1974, *Stationary scattering theory for time-dependent Hamiltonians*, Math. Ann. **207**, 315–335.
104. Ishii S 1982, *Linear evolution equations $du/dt + A(t)u = 0$: a case where $A(t)$ is strongly uniform-measurable*, J. Math. Soc. Japan **34**, 413–424.
105. Jörgens K 1962, *Spectral theory of second-order ordinary differential operators*, Lectures delivered at the University of Aarhus 1962/63.
106. Kato T 1966, Perturbation theory for linear operators, Springer: New York.
107. Kato T 1970, *Linear evolution equations of "hyperbolic" type*, J. Fac. Sci. Univ. Tokyo **17**, 241–258.
108. Kato T 1973, *Linear evolution equations of "hyperbolic" type, II*, J. Mat. Soc. Japan **25**, 648–666.
109. Kato T 1975, *Quasi-linear equations of evolution, with applications to partial differential equations*, Dold A, Eckmann B (Eds.), Spectral theory and differential equations, Proceedings of the symposium held at Dundee, Scotland, 1-19 July, 1974, Lecture Notes in Mathematics 448, Springer: Berlin.
110. Kato T 1975, *The Cauchy problem for quasi-linear symmetric hyperbolic systems*, Arch. Rat. Mech. Anal. **58**, 181–205.
111. Kato T 1976, *Linear and quasi-linear equations of evolution of hyperbolic type*, C.I.M.E, II Ciclo 1976, *Hyperbolicity*, Liguori Editore: Naples.
112. Kato T 1983, *Quasi-linear equations of evolution in nonreflexive Banach spaces*, Fujita H, Lax P D, Strang G (Eds.) 1983, *Nonlinear partial differential equations in applied science*; Proceedings of the U.S.-Japan Seminar, Tokyo, 1982, Lecture Notes in Numerical and Applied Analysis Vol. 5, North-Holland: Amsterdam.
113. Kato T, Ponce G 1988, *Commutator estimates and the Euler and Navier-Stokes equations*, Comm. Pure Appl. Math., **XLI**, 891–907.
114. Kato T 1993, *Abstract evolution equations, linear and quasilinear, revisited* in: Komatsu H (ed) 1993, *Functional analysis and related topics, 1991*, Proceedings of the International conference in memory of professor Kosaku Yośida held at RIMS, Kyoto University, Japan, July 29-Aug. 2, 1991, Lecture Notes in Mathematics 1540, Springer: Berlin.
115. Kato T, Yalima K 1991, *Dirac equations with moving nuclei*, Ann. I. H. Poincare Phy. **54**, 209–221.

116. Kay B S, Wald R M 1987, *Linear stability of Schwarzschild under perturbations which are nonvanishing on the bifurcation 2-sphere*, Classical Quant. Grav. **4**, 893–898.
117. Klainerman S, Nicolo F 1999, *On local and global aspects of the Cauchy problem in general relativity*, Classical Quant. Grav. **16**, R73–R157.
118. Kobayasi K 1979, *On a theorem for linear evolution equations of hyperbolic type*, J. Math. Soc. Japan **31**, 647–654.
119. Kobayasi K, Sanekata N 1989, *A method of iterations for quasi-linear evolution equations in nonreflexive Banach spaces*, Hiroshima Math J. **19**, 521–540.
120. Krein S G 1971, *Linear differential equations in Banach space*, AMS: Providence.
121. Krein S G, Khazan M I 1985, *Differential equations in a Banach space*, J. Soviet Math. **30**, 2154–2239.
122. Krein M G, Langer H 1978, *On some mathematical principles in the linear theory of damped oscillations of continua I*, Integral Eqs. Operator Th. **1**, 364–399.
123. Kreiss H O 1970, *Initial Boundary value problems for hyperbolic systems*, Comm. Pure Appl. Math., XXIII (1970), 277–298.
124. Gustafsson B, Kreiss H-O, Oliger J 1995,*Time dependent problems and difference methods*, Wiley: New York.
125. Kreiss H-O, Lorenz J 2000, *Resolvent estimates and quantification of nonlinear stability*, Acta Math. Sin. (Engl, Ser.) **16**, 1–20.
126. Krein M G, Langer H 1978, *On some mathematical principles in the linear theory of damped oscillations of continua I*, Integral Eqs. Operator Th. **1**, 539–566.
127. Ladas G E, Lakshmikantham V 1972, *Differential equations in abstract spaces*, Academic Press: New York.
128. Lang S 1969, *Real analysis*, Addison-Wesley: Reading, Mass.
129. DeLaubenfels R 1994, *Existence families, functional calculi and evolution equations*, Lecture Notes in Mathematics 1570, Springer: Berlin.
130. Lax P D, Phillips R S 1960, *Local boundary conditions for dissipative symmetric linear differential operators*, Comm. Pure Appl. Math. **13**, 427–455.
131. Leray J 1953, *Hyperbolic differential equations*, Institute for Advanced Study: Princeton.
132. Lions J L, Magenes E 1972, 1972, 1973, *Non-homogeneous boundary value problems and applications*, Vols. I - III, Springer: Berlin.
133. Lions J L 1957, *Une remarque sur les applications du theoreme de Hille-Yosida*, J. Math. Soc. Japan **9**, 62–70.
134. Mackey G W 2004, *Mathematical foundations of quantum mechanics*, Dover Publications: Dover.
135. Makino T, Ukai, S 1995, *Local smooth solutions of the relativistic Euler equation*, J. Math. Kyoto U. **35**, 105–114.
136. Marsden J E, Weinstein A 1982, *The Hamiltonian-Structure of the Maxwell-Vlasov equations*, Physica D **4**, 394–406.
137. Markus A S 1988, *Introduction to the spectral theory of operator pencils*, AMS: Providence.
138. Martin R H 1976, *Nonlinear operators and differential equations*, Wiley: New York.
139. Massey Frank J III 1972, *Abstract evolution equations and the mixed problem for symmetric hyperbolic systems*, Trans. Amer. Math. Soc. **168**, 165–188.
140. Meyer Y 1992, *Wavelets and operators*, Cambridge University Press: Cambridge.
141. Meyer Y, Coifman R 1997, *Wavelets: Calderon-Zygmund and multilinear operators*, Cambridge University Press: Cambridge.
142. Mikhlin S G 1970, *Mathematical physics, an advanced course*, North-Holland: Amsterdam.

143. Milani A 1982, em Local in time existence for the complete Maxwell equations with monotone characteristic in a bounded domain, Ann. Mat. Pur. Appl. **131**, 233–254.
144. Misiolek G 2002, *Classical solutions of the periodic Camassa-Holm equation*, Geom. Funct. Anal. **12**, 1080–1104.
145. Mitter H, Thaller B 1985, *Particles in spherical electromagnetic-radiation fields*, Phys. Rev. A **31**, 2030–2037.
146. Miyadera I 1992, *Nonlinear semigroups*, AMS: Providence.
147. Mizohata K 1994, *Global weak solutions for the equation of isothermal gas around a star*, J. Math. Kyoto U. **34**, 585–598.
148. Nagel R 1985, *Well-posedness and positivity for systems of linear evolution equations*, Confer. Sem. Mat. Univ. Bari No. 203.
149. Nagel R 1995, *Semigroup methods for non-autonomous Cauchy problems*, In: Ferreyra G, Goldstein G R, Neubrander F (eds.): Evolution Equations. Lect. Notes Pure Appl. Math. **168**, 301–316.
150. Nagel R, Sinestrari E 1996, *Nonlinear hyperbolic Volterra integrodifferential equations*, Nonlinear Analysis **27**, 167–186.
151. Nagel R, Nickel G 2002, *Wellposedness for nonautonomous abstract Cauchy problems*, in: Neumann W R ed., Evolution equations, semigroups and functional analysis (Milano, 2000), 279–293, Progr. Nonlinear Differential Equations Appl. **50**, Birkhuser: Basel.
152. Nagy G, Sarbach O 2006, *A minimization problem for the lapse and the initial-boundary value problem for Einstein's field*, http://arxiv.org/abs/gr-qc/0601124.
153. Neidhardt H 1981, *On abstract linear evolution-equations.1*, Math. Nachr. **103**, 283–298.
154. Neubrander F 1988, *Well-posedness of higher order abstract Cauchy problems*, Trans. Amer. Math. Soc. **295**, 257–290.
155. Neubrander F 1988, *Integrated semigroups and their applications to the abstract Cauchy problem*, Pacific J. Math. **135**, 111–155.
156. Neubrander F 1989, *Integrated semigroups and their application to complete second order Cauchy problems*, Semigroup Forum **38**, 233–251.
157. von Neumann J 1932, *Mathematische Grundlagen der Quantenmechanik*, Springer: Berlin.
158. Nishitani T, Takayama M 1996, *A characteristic initial-boundary value problem for a symmetric positive system*, Hokkaido Math. J. **25**, 167–182.
159. Nickel G, Schnaubelt R 1998, *An extension of Kato's stability condition for nonautonomous Cauchy problems*, Taiwan J. Math. **2**, 483–496.
160. Nickel G 2000, *Evolution semigroups and product formulas for nonautonomous Cauchy problems*, Math. Nachr. **212**, 101–116.
161. Nishitani T, Takayama M 1998, *Characteristic initial-boundary value problems for symmetric hyperbolic systems*, Osaka J. Math. *35*, 29–657.
162. Obrecht E 1991, *Phase space for an n-th order differential equation in Banach spaces*, in: Semigroup theory and evolution equations, eds. Clement P, Mitidieri E and de Pagter B, Lecture notes in pure and applied mathematics **135**, Marcel Dekker: New York, 391–399.
163. Oka H, Tanaka N 1997, *Abstract quasilinear integrodifferential equations of hyperbolic type*, Nonlinear Anal-Theor. **29**, 903–925.
164. Oka H 1997, *Abstract quasilinear Volterra integrodifferential equations*, Nonlinear Anal-Theor. **28**, 1019–1045.
165. Oka H, Tanaka N 2005, *Evolution operators generated by non-densely defined operators*, Math. Nachr. **278**, 1285–1296.

166. Pan R, Smoller J A 2006, *Blowup of smooth solutions for relativistic Euler equations*, Commun. Math. Phys. **262**, 729–755.
167. Pavel N H 1987, *Nonlinear evolution operators and semigroups*, Springer: Berlin.
168. Pazy A 1983, *Semigroups of linear operators and applications to partial differential equations*, Springer: New York.
169. Phillips R S, Sarason L 1966, *Singular symmetric positive first order differential operators*, J. Math. Mech., 235–272.
170. Da Prato G 1966, *Semigruppi regolarizzabili*, Ricerche Mat. **15**, 223–248.
171. Da Prato G, Grisvard, P 1975, *Sommes d'operateurs lineaires et equations differentielles operationnelles* J. Math. Pures Appl. **54**, 305–387.
172. Da Prato G, Iannelli M 1976, *On a method for studying abstract evolution equations in the hyperbolic case*, Comm. Partial Differential Equations **1**, 585–608.
173. Da Prato G, Grisvard P 1994, *The damped wave equation in a noncylindrical domain*, Differential Integral Equations **7**, 735–746.
174. Da Prato G, Zolesio J P 1990, *Existence and optimal-control for wave-equation in moving domain*, Lect. Notes Contr. Inf. **147**, 167–190.
175. Prugovecki E 1981, *Quantum Mechanics in Hilbert Space*, Academic Press: New York.
176. Rauch J 1985, *Symmetric positive systems with boundary characteristic of constant multiplicity*, Trans. Amer. Math. Soc. **291**, 167–187.
177. Rauch J, Massey Frank J III 1974, *Differentiability of solutions to hyperbolic initial-boundary value problems*, Trans. Amer. Math. Soc. **189**, 303–318.
178. Rauch J 1994, *Boundary value problems with nonuniformly characteristic boundary*, J. Math. Pures Appl. **73**, 347–353.
179. Reed M and Simon B, 1980, 1975, 1979, 1978, *Methods of modern mathematical physics*, Volume I, II, III, IV, Academic: New York.
180. Renardy M and Rogers R C 1993, *An introduction to partial differential equations*, Springer: New York.
181. Reula O, Sarbach O 2005, *A model problem for the initial-boundary value formulation of Einstein's field equations*, Journal of Hyperbolic Differential Equations, 397–435.
182. Riesz F and Sz-Nagy B 1955, *Functional analysis*, Unger: New York.
183. Rodman L 1989, *An Introduction to Operator Polynomials*, Birkäuser: Basel.
184. Romanov V A 1973, Stability of plane-parallel Couette flow, Funct. Anal. Appl. **7**, 137–146.
185. Rudin W 1962, *Fourier analysis on groups*, Interscience Publishers: New York.
186. Rudin W 1991, *Functional analysis*, 2nd ed., MacGraw-Hill: New York.
187. Sarbach O, Heusler H, Brodbeck O 2001, *Self-adjoint wave equations for dynamical perturbations of self-gravitating fields*, Phys. Rev. D **63**, 104015.
188. Sanekata N 1989, *Abstract quasi-linear equations of evolution in nonreflexive Banach spaces*, Hiroshima Math. J. **19**, 109–139.
189. Sandefur J 1977, *Higher order abstract Cauchy problems*, J. Math. Anal. Appl. **60**, 728–742.
190. Sandefur J 1984, *Convergence of solutions of a second-order Cauchy problem*, J. Math. Anal. Appl. **100**, 470–477.
191. Schroer B 2001, *Lectures on algebraic quantum field theory and operator algebras*, http://xxx.lanl.gov/abs/math-ph/0102018.
192. Schutz B F 2004, *The art and science of black hole mergers*, Proceedings of 'Growing Black Holes', Garching 21–25 June 2004, http://arxiv.org/abs/gr-qc/0410121.
193. Secchi P 1995, *Linear symmetric hyperbolic systems with characteristic boundary*, Math. Methods Appl. Sci. **18**, 855–870.

194. Secchi P 1996, *The initial boundary value problem for linear symmetric hyperbolic systems with characteristic boundary of constant multiplicity*, J. Diff. Int. Eq. **9**, 671–700.
195. Secchi P 1996, *Well-posedness of characteristic symmetric hyperbolic systems*, Arch. Rational Mech. Anal. **134**, 155–197.
196. Secchi P 1997, *Characteristic symmetric hyperbolic systems with dissipation. Global existence and asymptotics*, Math. Meth. Appl. Sci. 20, 583–597.
197. Secchi P 1998, *A symmetric positive system with nonuniformly characteristic boundary*, Differential Integral Equations **11**, 605–621.
198. Secchi P 2000, *Full regularity of solutions to a nonuniformly characteristic boundary value problem for symmetric positive systems*, Adv. Math. Sci. Appl. **10**, 39–55.
199. Showalter R E 1997, *Monotone operators in Banach space and nonlinear partial differential equations*, Mathematical Surveys and Monographs, 49. American Mathematical Society: Providence.
200. Stein E M 1970, *Singular integrals and differentiability properties of functions*, Princeton University Press: Princeton.
201. Stein E M 1993, *Harmonic analysis: Real-variable methods, orthogonality and oscillatory integrals*, Princeton University Press: Princeton.
202. Streater R F, Wightman A S 2000, *PCT, Spin and statistics, and all that*, Princeton University Press: Princeton.
203. Tanabe H 1979, *Equations of evolution*, Pitman: London.
204. Tanabe H 1997, *Functional analytic methods for partial differential equations*, Marcel Dekker: New York.
205. Tanaka N, Okazawa N 1990, *Local C-semigroups and local integrated semigroups*, Proc. London Math. Soc. **61**, 63–90.
206. Tanaka N 2000, *Generation of linear evolution operators*, Proc. Am. Math. Soc. **128**, 2007–2015.
207. Tanaka N 2000, *A class of abstract quasi-linear evolution equations of second order*, J. Lond. Math. Soc. **62**, 198–212.
208. Tanaka N 2001, *A characterization of evolution operators*, Stud. Math. **146**, 285–299.
209. Tanaka N 2004, *Abstract Cauchy problems for quasi-linear evolution equations in the sense of Hadamard*, Proc. Lond. Math. Soc. **89**, 123–160.
210. Tanaka N 2004, *Nonautonomous abstract Cauchy problems for strongly measurable families*, Math. Nachr. **274–275**, 130–153.
211. Tarfulea N 2005, *Constraint Preserving Boundary Conditions for Hyperbolic Formulations of Einstein's Equations*, http://arxiv.org/abs/gr-qc/0508014.
212. Wald R M 1979, *Note on the stability of the Schwarzschild metric*, J. Math. Phys. **20**, 1056–1058.
213. Wald R M 1980, *Erratum: "Note on the stability of the Schwarzschild metric" (J. Math. Phys. 20 (1979), 1056–1058)*, J. Math. Phys. **21**, 218.
214. Wald R M 1994, *Quantum field theory in curved spacetime and black hole thermodynamics*, University of Chicago Press: Chicago.
215. Walton R A 2005, *A symmetric hyperbolic structure for isentropic relativistic perfect fluids*, Houston J. Math. **31**, 145–160.
216. Weidmann J 1980, *Linear Operators in Hilbert spaces*, Springer: New York.
217. Wloka J 1987, *Partial differential equations*, Cambridge: Cambridge.
218. Xiao T, Liang J 1998, *The Cauchy problem for higher order abstract differential equations*, Lecture Notes in Mathematics **1701**, Springer: New York.
219. Yagi A 1980, *Remarks on proof of a theorem of Kato and Kobayasi on linear evolution equations*, Osaka J. Math. **17**, 233–243.

220. Yakubov S, Yakubov Y 2000, *Differential-operator equations: Ordinary and partial differential equations*, Chapman & Hall: London.
221. Yamada Y 1987, *Some nonlinear degenerate wave equations*, Nonlinear Anal. **11**, 1155–1168.
222. Yamazaki T 2000, *Bounded solutions of quasilinear hyperbolic equations of Kirchhoff type with dissipative and forcing terms*, Funkcial. Ekvac. **43**, 511–528.
223. Yosida K 1956, *An operator-theoretical integration of the wave equation*, J. Math. Soc. **8**, 79–92.
224. Yosida K 1957, *Lectures on semi-group theory and its application to Cauchy's problem in partial differential equations*, Tata Institute of Fundamental Research: Bombay.
225. Yosida K 1968, *Functional analysis*, 2nd ed., Springer: Berlin.
226. Zheng Q 1994, *Strongly continuous M, N-families of bounded operators*, Integral Eqs. Operator Th. **19**, 105–119.
227. Ziemer W P 1989, *Weakly differentiable functions*, Springer: New York.

Index of Notation

Index of Terminology

Lecture Notes in Mathematics

For information about earlier volumes please contact your bookseller or Springer LNM Online archive: springerlink.com

Vol. 1711: W. Ricker, Operator Algebras Generated by Commuting Projećtions: A Vector Measure Approach (1999)

Vol. 1712: N. Schwartz, J. J. Madden, Semi-algebraic Function Rings and Reflectors of Partially Ordered Rings (1999)

Vol. 1713: F. Bethuel, G. Huisken, S. Müller, K. Steffen, Calculus of Variations and Geometric Evolution Problems. Cetraro, 1996. Editors: S. Hildebrandt, M. Struwe (1999)

Vol. 1714: O. Diekmann, R. Durrett, K. P. Hadeler, P. K. Maini, H. L. Smith, Mathematics Inspired by Biology. Martina Franca, 1997. Editors: V. Capasso, O. Diekmann (1999)

Vol. 1715: N. V. Krylov, M. Röckner, J. Zabczyk, Stochastic PDE's and Kolmogorov Equations in Infinite Dimensions. Cetraro, 1998. Editor: G. Da Prato (1999)

Vol. 1716: J. Coates, R. Greenberg, K. A. Ribet, K. Rubin, Arithmetic Theory of Elliptic Curves. Cetraro, 1997. Editor: C. Viola (1999)

Vol. 1717: J. Bertoin, F. Martinelli, Y. Peres, Lectures on Probability Theory and Statistics. Saint-Flour, 1997. Editor: P. Bernard (1999)

Vol. 1718: A. Eberle, Uniqueness and Non-Uniqueness of Semigroups Generated by Singular Diffusion Operators (1999)

Vol. 1719: K. R. Meyer, Periodic Solutions of the N-Body Problem (1999)

Vol. 1720: D. Elworthy, Y. Le Jan, X-M. Li, On the Geometry of Diffusion Operators and Stochastic Flows (1999)

Vol. 1721: A. Iarrobino, V. Kanev, Power Sums, Gorenstein Algebras, and Determinantal Loci (1999)

Vol. 1722: R. McCutcheon, Elemental Methods in Ergodic Ramsey Theory (1999)

Vol. 1723: J. P. Croisille, C. Lebeau, Diffraction by an Immersed Elastic Wedge (1999)

Vol. 1724: V. N. Kolokoltsov, Semiclassical Analysis for Diffusions and Stochastic Processes (2000)

Vol. 1725: D. A. Wolf-Gladrow, Lattice-Gas Cellular Automata and Lattice Boltzmann Models (2000)

Vol. 1726: V. Marić, Regular Variation and Differential Equations (2000)

Vol. 1727: P. Kravanja M. Van Barel, Computing the Zeros of Analytic Functions (2000)

Vol. 1728: K. Gatermann Computer Algebra Methods for Equivariant Dynamical Systems (2000)

Vol. 1729: J. Azéma, M. Émery, M. Ledoux, M. Yor (Eds.) Séminaire de Probabilités XXXIV (2000)

Vol. 1730: S. Graf, H. Luschgy, Foundations of Quantization for Probability Distributions (2000)

Vol. 1731: T. Hsu, Quilts: Central Extensions, Braid Actions, and Finite Groups (2000)

Vol. 1732: K. Keller, Invariant Factors, Julia Equivalences and the (Abstract) Mandelbrot Set (2000)

Vol. 1733: K. Ritter, Average-Case Analysis of Numerical Problems (2000)

Vol. 1734: M. Espedal, A. Fasano, A. Mikelić, Filtration in Porous Media and Industrial Applications. Cetraro 1998. Editor: A. Fasano. 2000.

Vol. 1735: D. Yafaev, Scattering Theory: Some Old and New Problems (2000)

Vol. 1736: B. O. Turesson, Nonlinear Potential Theory and Weighted Sobolev Spaces (2000)

Vol. 1737: S. Wakabayashi, Classical Microlocal Analysis in the Space of Hyperfunctions (2000)

Vol. 1738: M. Émery, A. Nemirovski, D. Voiculescu, Lectures on Probability Theory and Statistics (2000)

Vol. 1739: R. Burkard, P. Deuflhard, A. Jameson, J.-L. Lions, G. Strang, Computational Mathematics Driven by Industrial Problems. Martina Franca, 1999. Editors: V. Capasso, H. Engl, J. Periaux (2000)

Vol. 1740: B. Kawohl, O. Pironneau, L. Tartar, J.-P. Zolesio, Optimal Shape Design. Tróia, Portugal 1999. Editors: A. Cellina, A. Ornelas (2000)

Vol. 1741: E. Lombardi, Oscillatory Integrals and Phenomena Beyond all Algebraic Orders (2000)

Vol. 1742: A. Unterberger, Quantization and Non-holomorphic Modular Forms (2000)

Vol. 1743: L. Habermann, Riemannian Metrics of Constant Mass and Moduli Spaces of Conformal Structures (2000)

Vol. 1744: M. Kunze, Non-Smooth Dynamical Systems (2000)

Vol. 1745: V. D. Milman, G. Schechtman (Eds.), Geometric Aspects of Functional Analysis. Israel Seminar 1999-2000 (2000)

Vol. 1746: A. Degtyarev, I. Itenberg, V. Kharlamov, Real Enriques Surfaces (2000)

Vol. 1747: L. W. Christensen, Gorenstein Dimensions (2000)

Vol. 1748: M. Ruzicka, Electrorheological Fluids: Modeling and Mathematical Theory (2001)

Vol. 1749: M. Fuchs, G. Seregin, Variational Methods for Problems from Plasticity Theory and for Generalized Newtonian Fluids (2001)

Vol. 1750: B. Conrad, Grothendieck Duality and Base Change (2001)

Vol. 1751: N. J. Cutland, Loeb Measures in Practice: Recent Advances (2001)

Vol. 1752: Y. V. Nesterenko, P. Philippon, Introduction to Algebraic Independence Theory (2001)

Vol. 1753: A. I. Bobenko, U. Eitner, Painlevé Equations in the Differential Geometry of Surfaces (2001)

Vol. 1754: W. Bertram, The Geometry of Jordan and Lie Structures (2001)

Vol. 1755: J. Azéma, M. Émery, M. Ledoux, M. Yor (Eds.), Séminaire de Probabilités XXXV (2001)

Vol. 1756: P. E. Zhidkov, Korteweg de Vries and Nonlinear Schrödinger Equations: Qualitative Theory (2001)

Vol. 1757: R. R. Phelps, Lectures on Choquet's Theorem (2001)
Vol. 1758: N. Monod, Continuous Bounded Cohomology of Locally Compact Groups (2001)
Vol. 1759: Y. Abe, K. Kopfermann, Toroidal Groups (2001)
Vol. 1760: D. Filipović, Consistency Problems for Heath-Jarrow-Morton Interest Rate Models (2001)
Vol. 1761: C. Adelmann, The Decomposition of Primes in Torsion Point Fields (2001)
Vol. 1762: S. Cerrai, Second Order PDE's in Finite and Infinite Dimension (2001)
Vol. 1763: J.-L. Loday, A. Frabetti, F. Chapoton, F. Goichot, Dialgebras and Related Operads (2001)
Vol. 1764: A. Cannas da Silva, Lectures on Symplectic Geometry (2001)
Vol. 1765: T. Kerler, V. V. Lyubashenko, Non-Semisimple Topological Quantum Field Theories for 3-Manifolds with Corners (2001)
Vol. 1766: H. Hennion, L. Hervé, Limit Theorems for Markov Chains and Stochastic Properties of Dynamical Systems by Quasi-Compactness (2001)
Vol. 1767: J. Xiao, Holomorphic Q Classes (2001)
Vol. 1768: M. J. Pflaum, Analytic and Geometric Study of Stratified Spaces (2001)
Vol. 1769: M. Alberich-Carramiñana, Geometry of the Plane Cremona Maps (2002)
Vol. 1770: H. Gluesing-Luerssen, Linear Delay-Differential Systems with Commensurate Delays: An Algebraic Approach (2002)
Vol. 1771: M. Émery, M. Yor (Eds.), Séminaire de Probabilités 1967-1980. A Selection in Martingale Theory (2002)
Vol. 1772: F. Burstall, D. Ferus, K. Leschke, F. Pedit, U. Pinkall, Conformal Geometry of Surfaces in S^4 (2002)
Vol. 1773: Z. Arad, M. Muzychuk, Standard Integral Table Algebras Generated by a Non-real Element of Small Degree (2002)
Vol. 1774: V. Runde, Lectures on Amenability (2002)
Vol. 1775: W. H. Meeks, A. Ros, H. Rosenberg, The Global Theory of Minimal Surfaces in Flat Spaces. Martina Franca 1999. Editor: G. P. Pirola (2002)
Vol. 1776: K. Behrend, C. Gomez, V. Tarasov, G. Tian, Quantum Comohology. Cetraro 1997. Editors: P. de Bartolomeis, B. Dubrovin, C. Reina (2002)
Vol. 1777: E. García-Río, D. N. Kupeli, R. Vázquez-Lorenzo, Osserman Manifolds in Semi-Riemannian Geometry (2002)
Vol. 1778: H. Kiechle, Theory of K-Loops (2002)
Vol. 1779: I. Chueshov, Monotone Random Systems (2002)
Vol. 1780: J. H. Bruinier, Borcherds Products on O(2,1) and Chern Classes of Heegner Divisors (2002)
Vol. 1781: E. Bolthausen, E. Perkins, A. van der Vaart, Lectures on Probability Theory and Statistics. Ecole d' Eté de Probabilités de Saint-Flour XXIX-1999. Editor: P. Bernard (2002)
Vol. 1782: C.-H. Chu, A. T.-M. Lau, Harmonic Functions on Groups and Fourier Algebras (2002)
Vol. 1783: L. Grüne, Asymptotic Behavior of Dynamical and Control Systems under Perturbation and Discretization (2002)
Vol. 1784: L. H. Eliasson, S. B. Kuksin, S. Marmi, J.-C. Yoccoz, Dynamical Systems and Small Divisors. Cetraro, Italy 1998. Editors: S. Marmi, J.-C. Yoccoz (2002)
Vol. 1785: J. Arias de Reyna, Pointwise Convergence of Fourier Series (2002)
Vol. 1786: S. D. Cutkosky, Monomialization of Morphisms from 3-Folds to Surfaces (2002)
Vol. 1787: S. Caenepeel, G. Militaru, S. Zhu, Frobenius and Separable Functors for Generalized Module Categories and Nonlinear Equations (2002)
Vol. 1788: A. Vasil'ev, Moduli of Families of Curves for Conformal and Quasiconformal Mappings (2002)
Vol. 1789: Y. Sommerhäuser, Yetter-Drinfel'd Hopf algebras over groups of prime order (2002)
Vol. 1790: X. Zhan, Matrix Inequalities (2002)
Vol. 1791: M. Knebusch, D. Zhang, Manis Valuations and Prüfer Extensions I: A new Chapter in Commutative Algebra (2002)
Vol. 1792: D. D. Ang, R. Gorenflo, V. K. Le, D. D. Trong, Moment Theory and Some Inverse Problems in Potential Theory and Heat Conduction (2002)
Vol. 1793: J. Cortés Monforte, Geometric, Control and Numerical Aspects of Nonholonomic Systems (2002)
Vol. 1794: N. Pytheas Fogg, Substitution in Dynamics, Arithmetics and Combinatorics. Editors: V. Berthé, S. Ferenczi, C. Mauduit, A. Siegel (2002)
Vol. 1795: H. Li, Filtered-Graded Transfer in Using Noncommutative Gröbner Bases (2002)
Vol. 1796: J.M. Melenk, hp-Finite Element Methods for Singular Perturbations (2002)
Vol. 1797: B. Schmidt, Characters and Cyclotomic Fields in Finite Geometry (2002)
Vol. 1798: W.M. Oliva, Geometric Mechanics (2002)
Vol. 1799: H. Pajot, Analytic Capacity, Rectifiability, Menger Curvature and the Cauchy Integral (2002)
Vol. 1800: O. Gabber, L. Ramero, Almost Ring Theory (2003)
Vol. 1801: J. Azéma, M. Émery, M. Ledoux, M. Yor (Eds.), Séminaire de Probabilités XXXVI (2003)
Vol. 1802: V. Capasso, E. Merzbach, B. G. Ivanoff, M. Dozzi, R. Dalang, T. Mountford, Topics in Spatial Stochastic Processes. Martina Franca, Italy 2001. Editor: E. Merzbach (2003)
Vol. 1803: G. Dolzmann, Variational Methods for Crystalline Microstructure – Analysis and Computation (2003)
Vol. 1804: I. Cherednik, Ya. Markov, R. Howe, G. Lusztig, Iwahori-Hecke Algebras and their Representation Theory. Martina Franca, Italy 1999. Editors: V. Baldoni, D. Barbasch (2003)
Vol. 1805: F. Cao, Geometric Curve Evolution and Image Processing (2003)
Vol. 1806: H. Broer, I. Hoveijn. G. Lunther, G. Vegter, Bifurcations in Hamiltonian Systems. Computing Singularities by Gröbner Bases (2003)
Vol. 1807: V. D. Milman, G. Schechtman (Eds.), Geometric Aspects of Functional Analysis. Israel Seminar 2000-2002 (2003)
Vol. 1808: W. Schindler, Measures with Symmetry Properties (2003)
Vol. 1809: O. Steinbach, Stability Estimates for Hybrid Coupled Domain Decomposition Methods (2003)
Vol. 1810: J. Wengenroth, Derived Functors in Functional Analysis (2003)
Vol. 1811: J. Stevens, Deformations of Singularities (2003)
Vol. 1812: L. Ambrosio, K. Deckelnick, G. Dziuk, M. Mimura, V. A. Solonnikov, H. M. Soner, Mathematical Aspects of Evolving Interfaces. Madeira, Funchal, Portugal 2000. Editors: P. Colli, J. F. Rodrigues (2003)
Vol. 1813: L. Ambrosio, L. A. Caffarelli, Y. Brenier, G. Buttazzo, C. Villani, Optimal Transportation and its

Applications. Martina Franca, Italy 2001. Editors: L. A. Caffarelli, S. Salsa (2003)

Vol. 1814: P. Bank, F. Baudoin, H. Föllmer, L.C.G. Rogers, M. Soner, N. Touzi, Paris-Princeton Lectures on Mathematical Finance 2002 (2003)

Vol. 1815: A. M. Vershik (Ed.), Asymptotic Combinatorics with Applications to Mathematical Physics. St. Petersburg, Russia 2001 (2003)

Vol. 1816: S. Albeverio, W. Schachermayer, M. Talagrand, Lectures on Probability Theory and Statistics. Ecole d'Eté de Probabilités de Saint-Flour XXX-2000. Editor: P. Bernard (2003)

Vol. 1817: E. Koelink, W. Van Assche (Eds.), Orthogonal Polynomials and Special Functions. Leuven 2002 (2003)

Vol. 1818: M. Bildhauer, Convex Variational Problems with Linear, nearly Linear and/or Anisotropic Growth Conditions (2003)

Vol. 1819: D. Masser, Yu. V. Nesterenko, H. P. Schlickewei, W. M. Schmidt, M. Waldschmidt, Diophantine Approximation. Cetraro, Italy 2000. Editors: F. Amoroso, U. Zannier (2003)

Vol. 1820: F. Hiai, H. Kosaki, Means of Hilbert Space Operators (2003)

Vol. 1821: S. Teufel, Adiabatic Perturbation Theory in Quantum Dynamics (2003)

Vol. 1822: S.-N. Chow, R. Conti, R. Johnson, J. Mallet-Paret, R. Nussbaum, Dynamical Systems. Cetraro, Italy 2000. Editors: J. W. Macki, P. Zecca (2003)

Vol. 1823: A. M. Anile, W. Allegretto, C. Ringhofer, Mathematical Problems in Semiconductor Physics. Cetraro, Italy 1998. Editor: A. M. Anile (2003)

Vol. 1824: J. A. Navarro González, J. B. Sancho de Salas, $\mathscr{C}^\infty$ – Differentiable Spaces (2003)

Vol. 1825: J. H. Bramble, A. Cohen, W. Dahmen, Multiscale Problems and Methods in Numerical Simulations, Martina Franca, Italy 2001. Editor: C. Canuto (2003)

Vol. 1826: K. Dohmen, Improved Bonferroni Inequalities via Abstract Tubes. Inequalities and Identities of Inclusion-Exclusion Type. VIII, 113 p, 2003.

Vol. 1827: K. M. Pilgrim, Combinations of Complex Dynamical Systems. IX, 118 p, 2003.

Vol. 1828: D. J. Green, Gröbner Bases and the Computation of Group Cohomology. XII, 138 p, 2003.

Vol. 1829: E. Altman, B. Gaujal, A. Hordijk, Discrete-Event Control of Stochastic Networks: Multimodularity and Regularity. XIV, 313 p, 2003.

Vol. 1830: M. I. Gil', Operator Functions and Localization of Spectra. XIV, 256 p, 2003.

Vol. 1831: A. Connes, J. Cuntz, E. Guentner, N. Higson, J. E. Kaminker, Noncommutative Geometry, Martina Franca, Italy 2002. Editors: S. Doplicher, L. Longo (2004)

Vol. 1832: J. Azéma, M. Émery, M. Ledoux, M. Yor (Eds.), Séminaire de Probabilités XXXVII (2003)

Vol. 1833: D.-Q. Jiang, M. Qian, M.-P. Qian, Mathematical Theory of Nonequilibrium Steady States. On the Frontier of Probability and Dynamical Systems. IX, 280 p, 2004.

Vol. 1834: Yo. Yomdin, G. Comte, Tame Geometry with Application in Smooth Analysis. VIII, 186 p, 2004.

Vol. 1835: O.T. Izhboldin, B. Kahn, N.A. Karpenko, A. Vishik, Geometric Methods in the Algebraic Theory of Quadratic Forms. Summer School, Lens, 2000. Editor: J.-P. Tignol (2004)

Vol. 1836: C. Năstăsescu, F. Van Oystaeyen, Methods of Graded Rings. XIII, 304 p, 2004.

Vol. 1837: S. Tavaré, O. Zeitouni, Lectures on Probability Theory and Statistics. Ecole d'Eté de Probabilités de Saint-Flour XXXI-2001. Editor: J. Picard (2004)

Vol. 1838: A.J. Ganesh, N.W. O'Connell, D.J. Wischik, Big Queues. XII, 254 p, 2004.

Vol. 1839: R. Gohm, Noncommutative Stationary Processes. VIII, 170 p, 2004.

Vol. 1840: B. Tsirelson, W. Werner, Lectures on Probability Theory and Statistics. Ecole d'Eté de Probabilités de Saint-Flour XXXII-2002. Editor: J. Picard (2004)

Vol. 1841: W. Reichel, Uniqueness Theorems for Variational Problems by the Method of Transformation Groups (2004)

Vol. 1842: T. Johnsen, A. L. Knutsen, K_3 Projective Models in Scrolls (2004)

Vol. 1843: B. Jefferies, Spectral Properties of Noncommuting Operators (2004)

Vol. 1844: K.F. Siburg, The Principle of Least Action in Geometry and Dynamics (2004)

Vol. 1845: Min Ho Lee, Mixed Automorphic Forms, Torus Bundles, and Jacobi Forms (2004)

Vol. 1846: H. Ammari, H. Kang, Reconstruction of Small Inhomogeneities from Boundary Measurements (2004)

Vol. 1847: T.R. Bielecki, T. Björk, M. Jeanblanc, M. Rutkowski, J.A. Scheinkman, W. Xiong, Paris-Princeton Lectures on Mathematical Finance 2003 (2004)

Vol. 1848: M. Abate, J. E. Fornaess, X. Huang, J. P. Rosay, A. Tumanov, Real Methods in Complex and CR Geometry, Martina Franca, Italy 2002. Editors: D. Zaitsev, G. Zampieri (2004)

Vol. 1849: Martin L. Brown, Heegner Modules and Elliptic Curves (2004)

Vol. 1850: V. D. Milman, G. Schechtman (Eds.), Geometric Aspects of Functional Analysis. Israel Seminar 2002-2003 (2004)

Vol. 1851: O. Catoni, Statistical Learning Theory and Stochastic Optimization (2004)

Vol. 1852: A.S. Kechris, B.D. Miller, Topics in Orbit Equivalence (2004)

Vol. 1853: Ch. Favre, M. Jonsson, The Valuative Tree (2004)

Vol. 1854: O. Saeki, Topology of Singular Fibers of Differential Maps (2004)

Vol. 1855: G. Da Prato, P.C. Kunstmann, I. Lasiecka, A. Lunardi, R. Schnaubelt, L. Weis, Functional Analytic Methods for Evolution Equations. Editors: M. Iannelli, R. Nagel, S. Piazzera (2004)

Vol. 1856: K. Back, T.R. Bielecki, C. Hipp, S. Peng, W. Schachermayer, Stochastic Methods in Finance, Bressanone/Brixen, Italy, 2003. Editors: M. Fritelli, W. Runggaldier (2004)

Vol. 1857: M. Émery, M. Ledoux, M. Yor (Eds.), Séminaire de Probabilités XXXVIII (2005)

Vol. 1858: A.S. Cherny, H.-J. Engelbert, Singular Stochastic Differential Equations (2005)

Vol. 1859: E. Letellier, Fourier Transforms of Invariant Functions on Finite Reductive Lie Algebras (2005)

Vol. 1860: A. Borisyuk, G.B. Ermentrout, A. Friedman, D. Terman, Tutorials in Mathematical Biosciences I. Mathematical Neurosciences (2005)

Vol. 1861: G. Benettin, J. Henrard, S. Kuksin, Hamiltonian Dynamics – Theory and Applications, Cetraro, Italy, 1999. Editor: A. Giorgilli (2005)

Vol. 1862: B. Helffer, F. Nier, Hypoelliptic Estimates and Spectral Theory for Fokker-Planck Operators and Witten Laplacians (2005)

Vol. 1863: H. Führ, Abstract Harmonic Analysis of Continuous Wavelet Transforms (2005)
Vol. 1864: K. Efstathiou, Metamorphoses of Hamiltonian Systems with Symmetries (2005)
Vol. 1865: D. Applebaum, B.V. R. Bhat, J. Kustermans, J. M. Lindsay, Quantum Independent Increment Processes I. From Classical Probability to Quantum Stochastic Calculus. Editors: M. Schürmann, U. Franz (2005)
Vol. 1866: O.E. Barndorff-Nielsen, U. Franz, R. Gohm, B. Kümmerer, S. Thorbjønsen, Quantum Independent Increment Processes II. Structure of Quantum Lévy Processes, Classical Probability, and Physics. Editors: M. Schürmann, U. Franz, (2005)
Vol. 1867: J. Sneyd (Ed.), Tutorials in Mathematical Biosciences II. Mathematical Modeling of Calcium Dynamics and Signal Transduction. (2005)
Vol. 1868: J. Jorgenson, S. Lang, $Pos_n(R)$ and Eisenstein Series. (2005)
Vol. 1869: A. Dembo, T. Funaki, Lectures on Probability Theory and Statistics. Ecole d'Eté de Probabilités de Saint-Flour XXXIII-2003. Editor: J. Picard (2005)
Vol. 1870: V.I. Gurariy, W. Lusky, Geometry of Müntz Spaces and Related Questions. (2005)
Vol. 1871: P. Constantin, G. Gallavotti, A.V. Kazhikhov, Y. Meyer, S. Ukai, Mathematical Foundation of Turbulent Viscous Flows, Martina Franca, Italy, 2003. Editors: M. Cannone, T. Miyakawa (2006)
Vol. 1872: A. Friedman (Ed.), Tutorials in Mathematical Biosciences III. Cell Cycle, Proliferation, and Cancer (2006)
Vol. 1873: R. Mansuy, M. Yor, Random Times and Enlargements of Filtrations in a Brownian Setting (2006)
Vol. 1874: M. Yor, M. Émery (Eds.), In Memoriam Paul-André Meyer - Séminaire de probabilités XXXIX (2006)
Vol. 1875: J. Pitman, Combinatorial Stochastic Processes. Ecole d'Eté de Probabilités de Saint-Flour XXXII-2002. Editor: J. Picard (2006)
Vol. 1876: H. Herrlich, Axiom of Choice (2006)
Vol. 1877: J. Steuding, Value Distributions of L-Functions (2007)
Vol. 1878: R. Cerf, The Wulff Crystal in Ising and Percolation Models, Ecole d'Eté de Probabilités de Saint-Flour XXXIV-2004. Editor: Jean Picard (2006)
Vol. 1879: G. Slade, The Lace Expansion and its Applications, Ecole d'Eté de Probabilités de Saint-Flour XXXIV-2004. Editor: Jean Picard (2006)
Vol. 1880: S. Attal, A. Joye, C.-A. Pillet, Open Quantum Systems I, The Hamiltonian Approach (2006)
Vol. 1881: S. Attal, A. Joye, C.-A. Pillet, Open Quantum Systems II, The Markovian Approach (2006)
Vol. 1882: S. Attal, A. Joye, C.-A. Pillet, Open Quantum Systems III, Recent Developments (2006)
Vol. 1883: W. Van Assche, F. Marcellàn (Eds.), Orthogonal Polynomials and Special Functions, Computation and Application (2006)
Vol. 1884: N. Hayashi, E.I. Kaikina, P.I. Naumkin, I.A. Shishmarev, Asymptotics for Dissipative Nonlinear Equations (2006)
Vol. 1885: A. Telcs, The Art of Random Walks (2006)
Vol. 1886: S. Takamura, Splitting Deformations of Degenerations of Complex Curves (2006)
Vol. 1887: K. Habermann, L. Habermann, Introduction to Symplectic Dirac Operators (2006)
Vol. 1888: J. van der Hoeven, Transseries and Real Differential Algebra (2006)
Vol. 1889: G. Osipenko, Dynamical Systems, Graphs, and Algorithms (2006)
Vol. 1890: M. Bunge, J. Funk, Singular Coverings of Toposes (2006)
Vol. 1891: J.B. Friedlander, D.R. Heath-Brown, H. Iwaniec, J. Kaczorowski, Analytic Number Theory, Cetraro, Italy, 2002. Editors: A. Perelli, C. Viola (2006)
Vol. 1892: A. Baddeley, I. Bárány, R. Schneider, W. Weil, Stochastic Geometry, Martina Franca, Italy, 2004. Editor: W. Weil (2007)
Vol. 1893: H. Hanßmann, Local and Semi-Local Bifurcations in Hamiltonian Dynamical Systems, Results and Examples (2007)
Vol. 1894: C.W. Groetsch, Stable Approximate Evaluation of Unbounded Operators (2007)
Vol. 1895: L. Molnár, Selected Preserver Problems on Algebraic Structures of Linear Operators and on Function Spaces (2007)
Vol. 1896: P. Massart, Concentration Inequalities and Model Selection, Ecole d'Eté de Probabilités de Saint-Flour XXXIII-2003. Editor: J. Picard (2007)
Vol. 1897: R. Doney, Fluctuation Theory for Lévy Processes, Ecole d'Eté de Probabilités de Saint-Flour-2005. Editor: J. Picard (2007)
Vol. 1898: H.R. Beyer, Beyond Partial Differential Equations, On Linear and Quasi-Linear Abstract Hyperbolic Evolution Equations (2007)
Vol. 1899: Séminaire de Probabilités XL. Editors: C. Donati-Martin, M. Émery, A. Rouault, C. Stricker (2007)
Vol. 1900: E. Bolthausen, A. Bovier (Eds.), Spin Glasses (2007)
Vol. 1901: O. Wittenberg, Intersections de deux quadriques et pinceaux de courbes de genre 1, Intersections of Two Quadrics and Pencils of Curves of Genus 1 (2007)
Vol. 1902: A. Isaev, Lectures on the Automorphism Groups of Kobayashi-Hyperbolic Manifolds (2007)
Vol. 1903: G. Kresin, V. Maz'ya, Sharp Real-Part Theorems (2007)

Recent Reprints and New Editions

Vol. 1618: G. Pisier, Similarity Problems and Completely Bounded Maps. 1995 – 2nd exp. edition (2001)
Vol. 1629: J.D. Moore, Lectures on Seiberg-Witten Invariants. 1997 – 2nd edition (2001)
Vol. 1638: P. Vanhaecke, Integrable Systems in the realm of Algebraic Geometry. 1996 – 2nd edition (2001)
Vol. 1702: J. Ma, J. Yong, Forward-Backward Stochastic Differential Equations and their Applications. 1999 – Corr. 3rd printing (2005)
Vol. 830: J.A. Green, Polynomial Representations of GL_n, with an Appendix on Schensted Correspondence and Littelmann Paths by K. Erdmann, J.A. Green and M. Schocker 1980 – 2nd corr. and augmented edition (2007)